Dating Quaternary Sediments

Edited by

Don J. Easterbrook
Department of Geology
Western Washington University
Bellingham, Washington 98225

SPECIAL PAPER

227

1988

Published by The Geological Society of America, Inc.
3300 Penrose Place, P.O. Box 9140, Boulder, Colorado 80301

GSA Books Science Editor Campbell Craddock

Printed in U.S.A.

Library of Congress Cataloging-in-Publication Data

Dating quaternary sediments / edited by Don J. Easterbrook.
    p.   cm. — (Special paper ; 227)
  Bibliography: p.
  ISBN 0-8137-2227-6
    1. Geology, Stratigraphic—Quaternary.   2. Radioactive dating.
I. Easterbrook, Don J., 1935–   . II. Series: Special papers
(Geological Society of America) ; 227.
QE696.D37   1988
551.7'9—dc19   88-23335
                                   CIP

Contents

Preface

Accurate determination of the age of Quaternary deposits has long been a particularly important factor in interpretation of the glacial history of the world. Radiocarbon dating has provided an exceedingly useful tool for assessing the chronology of the last 50,000 years, but development of other dating methods applicable to older sediments has proven difficult. However, in recent years, a number of techniques have been devised to address this problem.

In 1982, a symposium on Quaternary dating methods, sponsored by the Quaternary Geology and Geomorphology Division of the Geological Society of America, was held at the annual meeting in Indianapolis. Papers were presented on new progress in dating techniques of Quaternary sediments. This volume is an outgrowth of that symposium, although not all papers presented at the meeting were put into manuscript form, and a number of new innovations have been added during the time since the symposium. Attempting to update all dating methods used in Quaternary chronology was beyond the scope of this volume; it addresses new developments for selected methods.

The first paper by C. W. Naeser and N. A. Naeser includes a review of the principles of fission-track dating and points out several new, innovative techniques for using fission-tracks to estimate burial temperatures and tectonic uplift rates in the Himalayan Mountains.

G. W. Berger details the present status of thermoluminescence dating of sediments and deals candidly with some of the past problems with TL dating. He discusses how new laboratory techniques can be used to overcome technical difficulties that have led to discrepancies in some TL dates.

Amino acid dating of wood is a new technique that is presently undergoing development in several laboratories. The paper by N. W. Rutter and C. K. Vlahos addresses racemization kinetics, one of the main problems in calculating geological ages from wood.

The review of the aminostratigraphy of Quaternary mollusks from Atlantic Coastal Plain sites by J. F. Wehmiller, D. F. Belknap, B. S. Boutin, J. E. Mirecki, S. D. Rahaim, and L. L. York is a comprehensive review of the large body of amino acid data on Quaternary mollusks along the eastern coast of the United States, covering an immense amount of data accumulated over a number of years.

Paleomagnetism of Quaternary sediments, by D. J. Easterbrook, is a general overview of paleomagnetic methods and direct measurement of paleomagnetism of glacial diamictons. The paper also discusses measurement of microfabric by magnetic anisotropy.

K. L. Verosub discusses how geomagnetic secular variation preserved in lake sediments can be used for intraregional correlations. He shows how measurement of secular variation can be compared to "master" curves derived for a region and used to establish correlation and age of sediments.

In the final paper, D. J. Easterbrook, J. L. Roland, N. D. Naeser, and R. J. Carson provide examples of how application of multiple dating methods—fission-track, paleomagnetic, and tephra-chronology—can be used effectively to solve complex geologic problems.

D. J. Easterbrook

Geological Society of America
Special Paper 227
1988

Fission-track dating of Quaternary events

Charles W. Naeser and Nancy D. Naeser
U.S. Geological Survey, MS 424, Box 25046, Denver Federal Center, Denver, Colorado 80225

ABSTRACT

Fission tracks are zones of intense damage that result when fission fragments pass through a solid. ^{238}U is the only naturally occurring isotope whose decay rate results in a significant number of tracks over geologic time. Spontaneous fission of ^{238}U occurs at a known rate, and by determining the number of fission tracks and the amount of uranium present in a mineral or glass, its age may be determined. Many geologic materials contain trace amounts of uranium, but because of such factors as uranium abundance and track retention, zircon and glass are the only materials routinely dated in Quaternary samples.

Applications of fission-track dating to Quaternary studies include the dating of volcanic ash and archaeological material. The method has also been used to determine the rate of landform development in the Powder River Basin of Wyoming through the dating of clinker formed by the natural burning of coal beds. In the Himalayas of northern Pakistan, fission-track dating of zircon and apatite has shown that uplift rates during the Quaternary were as high as 1 cm/yr, which accounts for the incredible mountainous relief.

THEORY AND METHODS

A fission track is a zone of intense damage formed when a fission fragment passes through a solid. Several naturally occurring isotopes fission spontaneously: ^{232}Th, ^{235}U, and ^{238}U. Of these isotopes, only ^{238}U has a fission half-life (9.9×10^{15} yr) that is sufficiently short to produce a significant number of fission tracks over geologic time. When an atom of ^{238}U fissions, the nucleus breaks up into two subequal nuclei, one averaging about 90 atomic mass units (a.m.u.) and the other averaging about 135 a.m.u., with the liberation of about 200 MeV of energy. The two highly charged nuclei recoil in opposite directions and disrupt the electron balance of ions in the mineral lattice or glass along their path. This disruption causes the positively charged ions in the lattice to repulse each other and force themselves into the crystal lattice, forming the track or damage zone (Fleischer and others, 1975). The new track is tens of angstroms in diameter and is about 10 to 20 μm in length. The track length is longer in low-density minerals and glasses than in dense minerals like zircon.

A track in its natural state can only be observed with a transmission electron microscope, but a suitable chemical etchant can enlarge the damage zone such that it can be observed in an optical microscope at intermediate magnifications (×200 to 500) (Fig. 1). Common etchants used to develop tracks include nitric acid (apatite), hydrofluoric acid (mica and glass), concentrated basic solutions (sphene), and alkali fluxes (zircon) (Fleischer and others, 1975; Gleadow and others, 1976).

Trace amounts of ^{238}U occur in a number of common minerals and glasses. Because ^{238}U fissions spontaneously at a constant rate, fission tracks can be used to date these materials. The techniques used for dating geologic materials have been developed by physicists and geologists over the past two decades. Early development of the method has been reviewed by Fleischer and others (1975) and Naeser (1979). The age of a mineral or glass can be calculated from the amount of uranium and number of spontaneous fission tracks that it contains. Spontaneous track density is usually determined by: (1) polishing the surface of the material, (2) enlarging the fission tracks intersecting the surface by chemical etching, and (3) counting the number of tracks per unit area with an optical microscope, generally at magnifications of ×500 (glass) to ×1500 (minerals). Because the relative abundance of ^{238}U and ^{235}U is constant in nature, the easiest and most accurate way to determine the amount of uranium present in the area counted is to create a new set of fission tracks by irradiating the sample in a nuclear reactor with a known dose of thermal neutrons. This will induce fission in the ^{235}U present in the sam-

1

Naeser and Naeser

Figure 1. Etched fission tracks in apatite crystal from the radial dike at Ship Rock, New Mexico (Naeser, 1971, sample SR-6). Width of figure 0.22 mm.

ple. The resulting induced track density is a function of the amount of uranium present in the material and the neutron dose that it received in the reactor.

A fission-track age is calculated after having determined: (1) the spontaneous track density from ^{238}U (ρ_s), (2) the neutron-induced track density from ^{235}U (ρ_i), and (3) the neutron fluence (dose) (ϕ) (Price and Walker, 1963; Naeser, 1967):

$$A = \frac{1}{\lambda_D} \ln \left[1 + \frac{\rho_s \, g \, \lambda_D \, \sigma \, I \, \phi}{\rho_i \, \lambda_F} \right]$$

where,

ρ_s = spontaneous track density from ^{238}U (tracks/cm^2)

ρ_i = neutron-induced track density from ^{235}U (tracks/cm^2)

ϕ = thermal neutron fluence (neutrons/cm^2)

λ_D = total decay constant for ^{238}U (1.551×10^{-10} yr^{-1})

λ_F = decay constant for spontaneous fission of ^{238}U (6.85×10^{-17} yr^{-1}, Fleischer and Price, 1964; 7.03×10^{-17} yr^{-1}, Roberts and others, 1968; or 8.42×10^{-17} yr^{-1}, Spadavecchia and Hahn, 1967)

σ = cross-section for thermal neutron-induced fission of ^{235}U (580×10^{-24} cm^2/atom)

I = isotopic ratio ^{235}U/^{238}U (7.252×10^{-3})

g = geometry factor, and

A = age in yr.

A number of schemes have been proposed for determining neutron fluence (ϕ), involving the use of standard glass dosimeters with known uranium content or metal foil monitors (Carpenter and Reimer, 1974; Fleischer and others, 1975). Studies have shown that in a given reactor run there is a significant difference in the values of the neutron fluence determined by various glasses and foils (e.g., Hurford and Green, 1982, 1983). In practice, many laboratories have adopted the method of fluence determination that, when combined with one of the values of the fission decay constant (λ_F) listed above, consistently yields "correct" fission-track ages on standards of known age (i.e., fission-track ages that are concordant with K-Ar ages on coexisting phases in the standard) (Naeser and others, 1977). This empirical calibration effectively circumvents the problem of determining absolute values for the neutron fluence and decay constant.

Fleischer and Hart (1972) and Hurford and Green (1982, 1983) have suggested that this practice should be formalized by use of a "zeta calibration" factor in the age equation, as:

$$A = \frac{1}{\lambda_D} \ln \left[1 + \frac{\rho_s \, g \, \lambda_D \, \zeta \, \rho_d}{\rho_i} \right]$$

where,

ρ_d = fission-track density in the detector covering the glass dosimeter during irradiation (tracks/cm^2), and

ζ = the calibration factor for a given glass dosimeter, evaluated from standards of known age (A_{std}), as:

$$\zeta = \frac{[\exp(\lambda_d\ A_{std})\ -1]}{\lambda_D\ (\rho_s/\rho_i)_{std}\ g\ \rho_d}$$

(Hurford and Green, 1982).

In practice, the value of zeta adopted for a given glass dosimeter should be a mean value based on a large number of determinations on age standards. It should also be noted that each individual must determine his/her own value for zeta. The zeta method is probably the best approach to calibration for experienced observers. For those just beginning fission-track dating, this method should be used with caution because possible procedural errors will be propagated through all subsequent dating attempts.

Two factors determine if a sample can be dated by the fission-track method. First, the sample must contain a mineral or glass of appropriate uranium content. In Quaternary samples there must be enough uranium to form a significant number of tracks that can be counted in a reasonable time. Second, tracks must be completely retained once they are formed; if tracks are not retained, the calculated age will be anomalously young. Several environmental conditions, including temperature and pressure, can cause the loss, or "annealing," of spontaneous tracks once they are formed (Fleischer and others, 1975; Harrison and others, 1979). Of these environmental conditions, heating is by far the most common cause of track loss. Heating by natural processes can cause partial to complete loss or fading of the spontaneous tracks.

Data on the temperatures required for annealing have been determined by extrapolating laboratory heating experiments to geologic time (e.g., Naeser and Faul, 1969) and by measuring age-decrease with increasing depth and temperature in deep drill holes from areas where the rocks have undergone heating of known duration (Naeser, 1981). Such studies have shown that the annealing temperature depends on the mineral involved (different minerals anneal at different temperatures) and the duration of heating (the longer a mineral is heated, the lower the temperature required to anneal its tracks). Tracks are stable in most non-opaque minerals at temperatures of 50°C or less, but fission tracks in natural glasses are affected at much lower temperatures (Seward, 1979; Naeser and othes, 1980b).

Although annealing can cause problems in determining the primary age of samples, it is a powerful method for studying thermal history. Studies have been directed mainly at older rocks for determining uplift rates and thermal history of sedimentary basins and mineral deposits (e.g., Wagner and others, 1977; Naeser, 1979; Briggs and others, 1979, 1981a, 1981b; Bryant and Naeser, 1980; Naeser and others, 1980a; Gleadow and Duddy, 1981; Gleadow and others, 1983; Kohn and others, 1984; numerous papers in Fleischer and others, 1984; Naeser, 1984a, 1984b, 1986). However, several Quaternary studies, discussed later, have also made use of track annealing.

Etching studies have shown that tracks can be revealed in more than 100 different minerals and glasses (Fleischer and others, 1975), but factors such as uranium content, annealing characteristics, and abundance in the geologic environment result in use of very few minerals for fission-track dating. The only Quaternary materials dated routinely are zircon and glass.

Different procedures are required in the laboratory for the dating of glass and zircon. The reason for the different procedures is that glass from a single source tends to have a uniform uranium content, whereas zircon crystals from a single source tend to have very inhomogeneous uranium distribution.

Zircon requires the use of a technique called the external detector method (Naeser, 1979) (Fig. 2). This is because the uranium is distributed inhomogeneously both within and between zircon crystals, and therefore, the induced tracks must be counted from the same area of a crystal in which the spontaneous tracks were counted. In the external detector method, the spontaneous tracks are counted in the crystal and the induced tracks are counted in a detector that covered the crystal mount during the neutron irradiation. A low-uranium-count (<10 ppb) muscovite or a plastic can be used for a detector. Usually between 6 and 12 crystals are counted.

Glass is usually dated by a different technique—the population method (Naeser, 1979) (Fig. 3). Because all of the glass from a single source has a similar uranium concentration, the spontaneous and induced track densities may be determined from different splits of the sample. One split is mounted in epoxy, polished, and etched for the spontaneous track density determination, and a second split is irradiated, mounted, polished, and etched (standard practice is to etch both mounts at the same time). The irradiated split contains both spontaneous and induced tracks; pre-irradiation annealing to remove spontaneous tracks is not recommended because it can alter the etching characteristics and chemistry of the glass. The spontaneous track density (ρ_s) is subtracted from the total track density in the irradiated sample to arrive at the induced track density (ρ_i). The amount of glass that must be counted depends on uranium content, age, and vesicularity.

Naeser (1976) has described the laboratory procedures for the population and external detector methods.

ADVANTAGES AND LIMITATIONS OF FISSION-TRACK DATING

One advantage that fission-track dating has over most other methods is that contamination is minimized. In conventional radiocarbon and K-Ar dating, bulk samples must be analyzed. Contamination of a ^{14}C sample with older or younger carbon can result in an erroneous age, and a few older detrital grains in a K-Ar sample can have a significant effect on a K-Ar age (Naeser and others, 1981). Fission-track dating is a grain-discrete method in which individual grains are scanned and counted. In the course of dating a sample with zircons, an age is obtained on each grain that is counted. Therefore, older detrital grains often show up clearly as contamination; a grain with a Miocene age in a Pleisto-

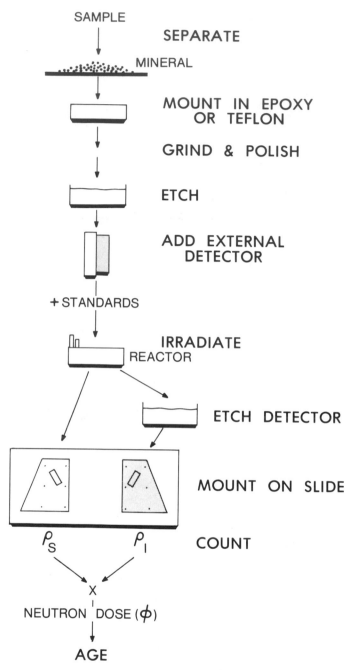

Figure 2. Steps involved in obtaining a fission-track age using the external detector method.

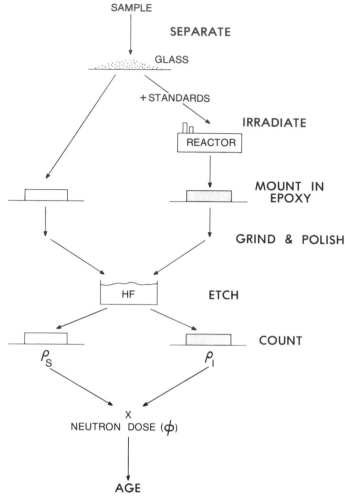

Figure 3. Steps involved in obtaining a fission-track age using the population method.

cene sample is obvious because of its older age. In addition, primary zircons in nondevitrified tephra can generally be identified by the glass adhering to them (Fig. 4). Contamination is usually not a problem with glass shards from tephra. However, N. D. Naeser has recognized older glass in a tephra sample from Saskatchewan. It composed less than 0.1 percent of the glass, but

it was obvious because of its much higher spontaneous track density; this glass was described by Westgate and Gorton (1981, Fig. 5).

A major limitation of fission-track dating in the Quaternary is that very young samples (<100,000 yr) usually have a very low spontaneous track density, which leads to long counting times and to ages with large analytical uncertainties. For example, Herd and Naeser (1974) determined the age of a population of zircons to be about 100,000 yr with a 40 percent standard deviation. They found a total of 16 tracks in 45 crystals. Naeser and others (1982) did not see any spontaneous tracks after scanning thousands of shards in one glass sample. Thus, for very young samples, the analytical uncertainty is almost always very large, but even then the result might answer a geological question.

Zircon, although preferable to glass, is not present in all

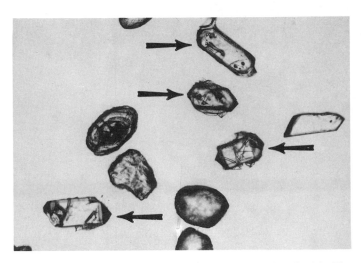

Figure 4. Zircons separated from a contaminated tephra deposit. The primary zircons (arrowed) are glass mantled; the grains without glass are detrital. Detrital contamination is found in almost all tephra deposits. Zircons are approximately 150 μm in length.

samples. In tephras, its presence depends on the chemistry of the parent magma. Silicic tephras are more likely to yield usable zircons than basic tephras. Zircons that are extremely fine grained (<75 μm) are too small to be dated by the fission-track method. This is often the case in tephra sampled far from its vent.

Natural glasses have been extensively dated because of the widespread occurrences of tephra and obsidian in Quaternary rocks, but the dating of natural glasses presents special problems. The greatest of these is the ease with which glass can lose spontaneous tracks by annealing (Fleischer and others, 1965; Storzer and Wagner, 1969; MacDougall, 1976; Seward, 1979). Hydrated glass shards, which are found in most tephra beds, are particularly susceptible to annealing (Lakatos and Miller, 1972). Work by Naeser and others (1980b) has shown that both hydrated and nonhydrated glass can lose tracks at ambient surface temperatures over periods of geologic time. In a study of 14 tephras from upper Cenozoic (<30 Ma) deposits of the western United States, only one glass had a fission-track age that was concordant with the fission-track age of coexisting zircon (Fig. 5). All other glass samples had ages that were significantly younger than the zircon ages. Seward (1979) showed that 60 percent of the glass fission-track ages of Quaternary tephras studied in New Zealand were significantly younger than the fission-track ages of coexisting zircons.

Two procedures available for checking glass for partial annealing, and then correcting the age for partial annealing, are the track diameter measurement method (Storzer and Wagner, 1969) and the plateau annealing method (Storzer and Poupeau, 1973). The plateau annealing method is much better suited for Quaternary samples than the diameter measurement method because it

can better deal with the low track densities encountered in young glasses (Naeser and others, 1980b). In the plateau method, separate splits of the irradiated and non-irradiated glass are heated together in an oven for one-hour intervals at progressively higher temperatures, and an age is determined after each heating step. If the glass has previously been partially annealed, the age will increase through the lower heating steps until it reaches a plateau, which may be equivalent to the primary age. If no annealing has occurred, progressive heating does not affect the age, even though the induced and spontaneous track densities decrease with each heating step. Figure 6 shows the results of plateau annealing determinations for 12 upper Tertiary glasses, which include both obsidians and tephra shards.

The diameter measurement and the plateau annealing methods of annealing correction are both difficult to apply to Quaternary glass because of the small number of spontaneous tracks present, which generally prevents statistically significant measurements. Furthermore, a corrected age may still be too young if any of the tracks in the glass were totally annealed (MacDougall, 1976; Naeser and others, 1980b).

Because of the ease with which glasses anneal and the uncertainty of correcting their ages, great care must be used in the interpretation of all glass fission-track ages. Fission-track ages on glass should always be considered minimum ages.

Another limitation for dating glass shards relates to their grain size and vesicularity. Large bubble-junction, platy shards are by far the easiest to date. In very fine-grained or pumiceous shards (Westgate and Briggs, 1980, Fig. 4), counting tracks and determining glass area by the conventional method of a counting grid in the ocular of the microscope is difficult. The problem of determining area can be overcome for many samples by using a point-counting technique analogous to that used in petrographic modal analysis (Seward, 1974; Briggs and Westgate, 1978; Naeser and others, 1982). Dating such glasses is very time consuming and should be undertaken only when there is no other suitable way to date the deposit.

Glass containing abundant microlites may also be difficult or impossible to date because of the close resemblance of the etched microlites to fission tracks. This problem is rarely encountered in glass shards, but it is quite common in obsidian.

In spite of these problems, glass has been used to date Quaternary samples, because in some cases it is the only datable phase present, and in some situations even minimum ages can be useful (e.g., Naeser and others, 1982).

APPLICATIONS

Tephrochronology

The major contribution of fission-track dating to Quaternary studies has been in the field of tephrochronology. Fission tracks have proved to be the most suitable method for dating tephras, particularly those older than the 40,000 to 50,000-yr upper limit of radiocarbon dating. An advantage of the fission-

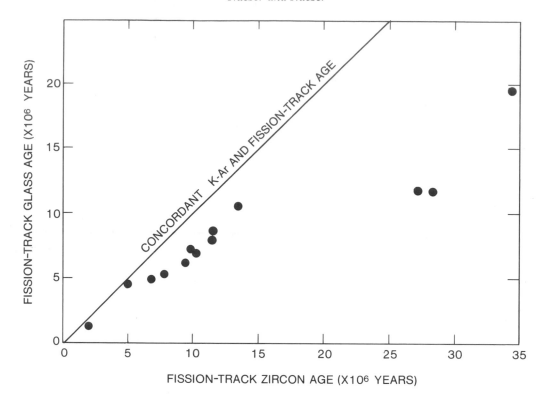

Figure 5. Plot showing the fission-track ages of coexisting glass shards and zircons from upper Cenozoic tephra deposits of the western United States (from Naeser and others, 1980b).

track method over K-Ar and conventional radiocarbon dating is that the problem of detrital contamination is greatly minimized (see discussion above). This is particularly important for dating tephras, which commonly contain detrital contaminants. Also, glass shards cannot be reliably dated by the K-Ar method because of the possible presence of excess radiogenic argon and the possibility of potassium gain and/or argon loss during or after hydration that can affect that age. Because of these problems, the K-Ar age of a glass shard separate may be too old, too young, or fortuitously correct (Naeser and others, 1981).

Since 1973, a number of examples of fission-track dating of zircon and/or glass from Quaternary tephras have been published (Naeser and others, 1973; Seward, 1974, 1975, 1979; Izett and Naeser, 1976; Aronson and others, 1977; Briggs and Westgate, 1978; Westgate and others, 1978; Gleadow, 1980; Naeser and others, 1982).

Volcanic ash beds of the Pearlette family occur in Pleistocene deposits of western North America (Izett and others, 1970, 1972). Before 1970, the Pearlette ash was considered to be a single ash bed representing a single eruption. As a result of this hypothesis, it was used as a time marker for many midcontinent Quaternary deposits. Then Izett and others (1970, 1972) recognized minor chemical differences among some of the Pearlette ash localities and found that the ashes could be correlated geochemi-

cally with deposits of the three major ash-flow eruptions that had occurred in the region of Yellowstone National Park, Wyoming. These eruptions had been dated at 2.02±0.08 Ma (Huckleberry Ridge Tuff), 1.27±0.1 Ma (Mesa Falls Tuff), and 0.616±0.008 Ma (Lava Creek Tuff) by the K-Ar method (J. D. Obradovich, written communication, 1973). Naeser and others (1973) dated zircons from two of the three types of Pearlette tephra and obtained ages of 1.9±0.1 Ma for ash correlated with the Huckleberry Ridge Tuff and 0.6±0.1 Ma for ash correlated with the Lava Creek Tuff. These ages matched the ages in the source region and confirmed the geochemical evidence of Izett and others (1970, 1972) that there are three Pearlett ashes, rather than just one. Fission-track ages obtained on glass from the Huckleberry Ridge ash bed give an excellent example of the problems caused by track annealing in glass (Table 1). Three different determinations of the ages of the glass and zircon from the Huckleberry Ridge ash bed show that the glass gives an apparent age that is about 0.6 m.y. younger than the zircon.

The Bishop ash bed, like the Pearlette family of ashes, is a widespread airfall tephra in the western United States (Izett and others, 1970; Izett and Naeser, 1976). Izett and others (1970) identified it as the airfall equivalent of the Bishop Tuff, which originated in the Long Valley Caldera about 0.74 Ma. Minerals separated from pumice in the Bishop ash bed yielded K-Ar ages

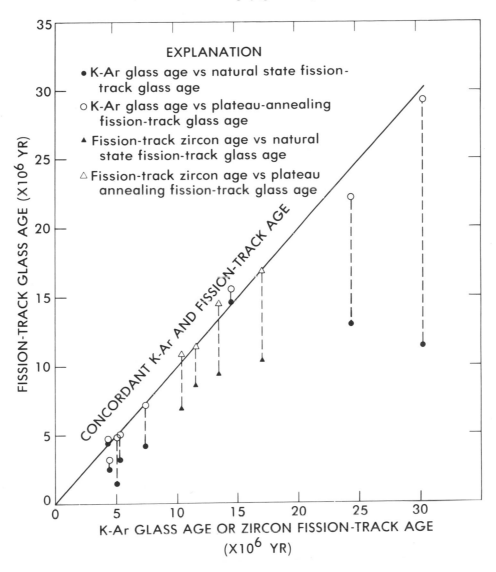

Figure 6. Diagram showing apparent and plateau-corrected ages of glasses (obsidians and tephra shards) compared to zircon fission-track or K-Ar ages (from Naeser and others, 1980b).

of 0.741±0.014 (2s) Ma (sanidine, A. J. Hurford, written communication, 1985) and 0.738±0.006 (2s) Ma (weighted average of sanidine, plagioclase, and biotite, 17 determinations, Izett and others, 1988). Both glass and zircon from the Bishop ash bed have been dated by the fission-track method (Table 2). The glass age shows the effects of annèaling and is younger than the age of the coexisting zircon.

In several cases, fission-track dating has demonstrated that tephras are considerably older than inferred from radiocarbon dating of associated material. For example, the Salmon Springs Drift at its type locality, near Sumner, Washington, consists of two drift sheets separated by about 1.5 m of peat, silt, and volcanic ash (Lake Tapps tephra) (Crandell and others, 1958; Easter-

brook and others, 1981). The peat grades downward with decreasing organic content into about one meter of silt, which in turn grades into the volcanic ash (Easterbrook, 1986). The peat was radiocarbon dated at $71,500 \,^{+1700}_{-1400}$ yr B.P. by the enrichment method (Stuiver and others, 1978), and the drift units were thus considered early Wisconsin in age. However, fission-track dating of the ash gave an age of 0.84±0.21 (1s), suggesting that the drift units are considerably older than early Wisconsin. Reversed paleomagnetism of the silts and additional fission-track dates (Table 3; Westgate and others, 1987) and geochemistry of the ash at other localities have confirmed that the radiocarbon age was incorrect (Easterbrook and others, 1981; Easterbrook, 1986; Westgate and others, 1987).

TABLE 1. FISSION-TRACK AGES OF ZIRCON AND GLASS FROM
THE HUCKLEBERRY RIDGE ASH BED, WYOMING

Investigator	Glass*	Zircon*
Naeser and others (1973, 1980b)	1.3 ± 0.17 Ma	1.9 ± 0.1 Ma
Seward (1979)	1.39 ± 0.08 Ma	1.93 ± 0.16 Ma
Naeser and others (1982); N. D. Naeser (unpub. data)	1.21 ± 0.07 Ma	1.91 ± 0.25 Ma

*Error shown is ±1 standard deviation.

TABLE 2. FISSION-TRACK AGES OF ZIRCON AND GLASS FROM
THE BISHOP ASH BED, CALIFORNIA

Investigator	Glass*	Zircon*
Izett and Naeser (1976,	—	0.74 ± 0.03 Ma
N. D. Naeser (unpub. data)	0.56 ± 0.05 Ma	—

*Error shown is ±1 standard deviation.

TABLE 3. FISSION-TRACK AGES FROM
THE LAKE TAPPS TEPHRA, WASHINGTON*

Locality	Material Dated	Age (Ma ±1s)
Sumner	zircon	0.84 ± 0.21
Auburn	zircon	0.87 ± 0.27
	glass	0.66 ± 0.04

*Data from Easterbrook and others (1981).

Landscape evolution

Fission-track dating has been used successfully to quantify certain geomorphic processes. Many of the coal basins of western North America contain large areas where the coal has burned. The heat from these natural burns has baked the overlying sandstones and shales into a brick-like rock called clinker, which is much harder and much more resistant to erosion than the original sedimentary rocks. As a result of its resistance to erosion, clinker is commonly found as a caprock on the hills and mesas in the Powder River Basin of Wyoming. In the eastern Powder River Basin, a long north-south escarpment, called the Rochelle Hills, is capped by clinker.

In the Rochelle Hills, burning of coal most likely occurred when the eastward-flowing streams cut headward through the west-dipping coal, lowering the water table and allowing the coal to dry out and become exposed to atmospheric oxygen.

Fission-track dating of zircon in the clinker has the potential for determining the time when the coal burning took place. If the detrital zircons in the clinker were heated high enough during the burning (e.g., 700°C for 1 hr; Fleischer and others, 1965), they would be totally annealed and lose all of their spontaneous fission tracks. Upon cooling, the zircons would again begin to accumulate spontaneous tracks. Thus, dating of the zircons would permit determination of the age of clinker formation.

Coates and Naeser (1984) dated zircons separated from the clinker capping the mesa north of Little Thunder Creek (Fig. 7) to see if the age of clinker formation could be determined. This study showed that the ages of zircons in the clinker become progressively younger from east to west. Zircons from eastern end of the mesas give ages of ca. 700,000 yr, while zircons at the western edge of clinker development do not contain any spontaneous fission tracks, indicating an age of <80,000 yr for that burn. The results of this study show that the burn front at Little Thunder Creek has migrated westward about 8 km, and that there has been about 200 m of downcutting into the eastern edge of the escarpment during the last 700,000 yr.

Paleomagnetic studies (Jones and others, 1982) on clinker from the same mesa show that all of the clinker formed during a period when the earth's magnetic field was normal. This result is consistent with the ages determined on the zircons separated from the clinker, which indicate that all of the clinker studied formed during the Brunhes Chron.

Fission-track dating is also useful in studying the uplift history of a mountain range. As mentioned earlier, fission tracks disappear (anneal) if a mineral is heated above a critical temperature for a period of time, and as long as the mineral is held above that temperature, any track formed will quickly disappear. Track annealing has a time and temperature relationship. Holding a mineral a long time at a low temperature will have the same effect on its tracks as a short time at a high temperature. For example, apatite will lose all of its tracks if it is held at 350°C for 1 hr (Naeser and Faul, 1969), but if it is held at about 135°C for a million years, it will also lose all of its tracks (Naeser, 1979).

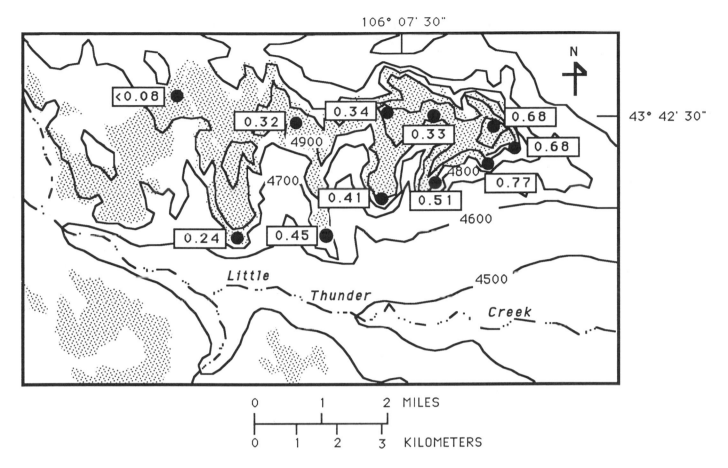

Figure 7. Map showing distribution of zircon fission-track ages (Ma) of clinker from the mesa north of Little Thunder Creek, Campbell County, Wyoming. Stippled pattern indicates area underlain by clinker.

During uplift and erosion, fission tracks will begin to accumulate in minerals as each mineral passes through its critical isotherm on its way up to the surface. Therefore, fission-track dating of these minerals will result in a series of ages that will show the uplift and erosion history of the mountain range.

In one such study, fission-track dating of apatite, zircon, and sphene separated from rocks collected in the Himalaya Mountains of northern Pakistan offers evidence of very rapid Quaternary uplift in the region surrounding Nanga Parbat (8,125 m) (Zeitler and others, 1982; Zeitler, 1985). The greatest continental relief in the world (6,930 m) is found over the 20-km distance between the Indus River (1,195 m) and the summit of Nanga Parbat (8,125 m). The Nanga Parbat–Haramosh Massif is composed of high-grade metasediments. These gneisses are uplifting and doming one of the major thrust faults related to the formation of the Himalayas, the Main Mantle Thrust. The Indus River is presently cutting a deep gorge across the northern end of the Nanga Parbat uplift. Apatite, zircon, and sphene separated from the metamorphic rocks exposed in the gorge yield fission-track ages of 0.4 Ma, 1.8 Ma, and 2.5 Ma, respectively. Zeitler and others (1982) interpreted these ages as indicating the time when these minerals passed through the following isotherms: apatite ≈150°C, zircon ≈235°C, and sphene ≈285°C. Therefore, rocks now exposed at the surface in the gorge were at a temperature of about 150°C 400,000 yr ago. This means that if the geothermal gradient is about 30°C/km, rocks now at the surface were at a depth of about 5 km as recently as 400,000 yr ago. This indicates that during the upper Pleistocene the Nanga Parbat region was being uplifted and eroded at a rate of about 1 cm/yr.

Other applications

Fission tracks are potentially useful for dating a variety of archaeological materials. They have been used to date tephras associated with archaeological remains, the material used to manufacture artifacts, and heating events in which glass or minerals in artifacts were heated to a high enough temperature to totally anneal their spontaneous tracks. A number of these studies were reviewed by Fleischer and others (1975), Wagner (1978), and Naeser and Naeser (1984).

A drawback to the use of fission-track dating in such studies is that most archaeological samples are very young and composed of material that is relatively low in uranium, so that track densities are generally extremely low. This means that counting a statistically significant number of spontaneous tracks is a tedious, time-consuming process that often involves repeated steps of polishing, etching, and counting.

CONCLUSIONS

Fission-track dating has widespread application in Quaternary studies. The major contribution has been in the field of tephrochronology, where fission-track dating has been used to date tephra horizons in Quaternary deposits and, thereby, to date Quaternary events. The method has also been used to study rates of landform development, and to determine rates of tectonic processes and how those rates relate to the overall development of the landscape. To a somewhat more limited extent, fission-track dating has been used in archaeological studies.

REFERENCES

Aronson, J. L., Schmidt, T. J., Walter, R. C., Taieb, M., Tiercelin, J. J., Johanson, D. C., Naeser, C. W., and Nairn, A.E.M., 1977, New geochronologic and paleomagnetic data for the hominid-bearing Hadir Formation of Ethiopia: Nature, v. 267, p. 323–327.

Briggs, N. D., and Westgate, J. A., 1978, A contribution to the Pleistocene geochronology of Alaska and the Yukon Territory; Fission-track age of distal tephra units, *in* Zartman, R. E., ed., Short papers of the 4th International Conference on Geochronology, Cosmochronology, and Isotope Geology: U.S. Geological Survey Open-File Report 78–701, p. 49–52.

Briggs, N. D., Naeser, C. W., and McCulloh, T. H., 1979, Thermal history of sedimentary basins by fission-track dating: Geological Society of America Abstracts with Programs, v. 11, p. 394.

——, 1981a, Thermal history of sedimentary basins by fission-track dating: Nuclear Tracks, v. 5, p. 235–237.

——, 1981b, Thermal history by fission-track dating, Tejon Oil Field area, California: American Association of Petroleum Geologists Bulletin, v. 65, p. 906.

Bryant, B., and Naeser, C. W., 1980, The significance of fission-track ages of apatite in relation to the tectonic history of the Front and Sawatch Ranges, Colorado: Geological Society of America Bulletin, v. 91, p. 156–164.

Carpenter, B. S., and Reimer, G. M., 1974, Standard reference materials; Calibrated glass standards for fission track use: National Bureau of Standards Special Publication 260–49, 16 p.

Coates, D. A., and Naeser, C. W., 1984, Map showing fission-track ages of clinker in the Rochelle Hills, southern Campbell and Weston Counties, Wyoming: U.S. Geological Survey Miscellaneous Investigation Series Map I–1462, scale 1:50,000, 1 sheet.

Crandell, D. R., Mullineaux, D. R., and Waldron, H. H., 1958, Pleistocene sequence in the southeastern part of the Puget Sound Lowland, Washington: American Journal of Science, v. 256, p. 384–398.

Easterbrook, D. J., 1986, Stratigraphy and chronology of Quaternary deposits of the Puget Lowland and Olympic Mountains of Washington and the Cascade Mountains of Washington and Oregon, *in* Quaternary science reviews: Oxford, England, Pergamon Press, p. 145–169.

Easterbrook, D. J., Briggs, N. D., Westgate, J. A., and Gorton, M. P., 1981, Age of the Salmon Springs Glaciation in Washington: Geology, v. 9, p. 87–93.

Fleischer, R. L., and Hart, H. R., Jr., 1972, Fission track dating; Techniques and problems, *in* Bishop, W. W., and Miller, J. A., eds., Calibration of hominoid evolution: Edinburgh, Scottish Academic Press, p. 135–170.

Fleischer, R. L., and Price, P. B., 1964, Decay constant for spontaneous fission of ^{238}U: Physical Review, v. 133, no. 1B, p. 63–64.

Fleischer, R. L., Price, P. B., and Walker, R. M., 1965, Effects of temperature, pressure, and ionization on the formation and stability of fission tracks in minerals and glasses: Journal of Geophysical Research, v. 70, p. 1497–1502.

——, 1975, Nuclear tracks in solids; Principles and applications: Berkeley, California, University of California Press, 605 p.

Fleischer, R. L., Harrison, T. M., and Miller, D. S., eds., 1984, Abstracts, Fourth International Fission Track Dating Workshop, Troy, New York, 1984: Troy, New York, Rensselaer Polytechnic Institute, 62 p.

Gleadow, A.J.W., 1980, Fission track age of the KBS Tuff and associated hominid remains in northern Kenya: Nature, v. 284, p. 225–230.

Gleadow, A.J.W., and Duddy, I. R., 1981, A natural long-term track annealing experiment for apatite: Nuclear Tracks, v. 5, p. 169–174.

Gleadow, A.J.W., Hurford, A. J., and Quaife, R. D., 1976, Fission track dating of zircon; Improved etching techniques: Earth and Planetary Science Letters, v. 33, p. 273–276.

Gleadow, A.J.W., Duddy, I. R., and Lovering, J. F., 1983, Fission track analysis; A new tool for the evaluation of thermal histories and hydrocarbon potential: Australian Petroleum Exploration Association Journal, v. 23, p. 93–102.

Harrison, T. M., Armstrong, R. L., Naeser, C. W., and Harakal, J. E., 1979, Geochronology and thermal history of the Coast Plutonic Complex, near Prince Rupert, British Columbia: Canadian Journal of Earth Sciences, v. 16, p. 400–410.

Herd, D. G., and Naeser, C. W., 1974, Radiometric evidence for pre-Wisconsin glaciation in the northern Andes: Geology, v. 2, p. 603–604.

Hurford, A. J., and Green, P. F., 1982, A users' guide to fission track dating calibration: Earth and Planetary Science Letters, v. 59, p. 343–354.

——, 1983, The zeta calibration of fission-track dating: Isotope Geoscience, v. 1, p. 285–317.

Izett, G. A., and Naeser, C. W., 1976, Age of the Bishop Tuff of eastern California as determined by the fission-track method: Geology, v. 2, p. 587–590.

Izett, G. A., Wilcox, R. E., Powers, H. A., and Desborough, G. A., 1970, The Bishop ash bed, a Pleistocene marker bed in the western United States: Quaternary Research, v. 1, p. 121–132.

Izett, G. A., Wilcox, R. E., and Borchardt, G. A., 1972, Correlation of a volcanic ash bed in the Pleistocene deposits near Mount Blanco, Texas, with the Guaje pumice bed of the Jemez Mountains, New Mexico: Quaternary Research, v. 2, p. 554–578.

Izett, G. A., Obradovich, J. D., and Mehnert, H. H., 1988, The Bishop ash bed (middle Pleistocene) and some older (Pliocene and Pleistocene) chemically and mineralogically similar ash beds in California, Nevada, and Utah: U.S. Geological Survey Bulletin 1675 (in press).

Jones, A. H., Geissman, J. W., Coates, D. A., Naeser, C. W., and Heffern, E. L., 1982, Paleomagnetic documentation of Brunhes and Matuyama age burning of Tertiary coals in the Powder River Basin, Wyoming and Montana: Geological Society of America Abstracts with Programs, v. 14, p. 523.

Kohn, B. P., Shagam, R., Banks, P. O., and Burkley, L. A., 1984, Mesozoic–Pleistocene fission-track ages on rocks of the Venezuelan Andes and their tectonic implications: Geological Society of America Memoir 162, p. 365–384.

Lakatos, S., and Miller, D. S., 1972, Evidence for the effect of water content of fission-track annealing in volcanic glass: Earth and Planetary Science Letters, v. 14, p. 128–130.

MacDougall, J. D., 1976, Fission track annealing and correction procedures for oceanic basalt glasses: Earth and Planetary Science Letters, v. 30, p. 19–26.

Naeser, C. W., 1967, The use of apatite and sphene for fission track age determinations: Geological Society of America Bulletin, v. 78, p. 1523–1526.

——, 1971, Geochronology of the Navajo–Hopi diatremes, Four Corners area: Journal of Geophysical Research, v. 76, p. 4978–4985.

——, 1976, Fission track dating: U.S. Geological Survey Open-File Report 76–190, 65 p.

—— , 1979, Fission-track dating and geologic annealing of fission tracks, *in* Jäger, E., and Hunziker, J. C., eds., Lectures in isotope geology: New York, Springer-Verlag, p. 154–169.

—— , 1981, The fading of fission tracks in the geologic environment; Data from deep drill holes: Nuclear Tracks, v. 5, p. 248–250.

Naeser, C. W., and Faul, H., 1969, Fission track annealing in apatite and sphene: Journal of Geophysical Research, v. 74, p. 705–710.

Naeser, C. W., Izett, G. A., and Wilcox, R. E., 1973, Zircon fission-track ages of Pearlette family ash beds in Meads County, Kansas: Geology, v. 1, p. 187–189.

Naeser, C. W., Hurford, A. J., and Gleadow, A.J.W., 1977, Fission-track dating of pumice from the KBS Tuff, East Rudolf, Kenya: Nature, v. 267, p. 649.

Naeser, C. W., Cunningham, C. G., Marvin, R. F., and Obradovich, J. D., 1980a, Pliocene intrusive rocks and mineralization near Rico, Colorado: Economic Geology, v. 75, p. 122–127.

Naeser, C. W., Izett, G. A., and Obradovich, J. D., 1980b, Fission-track and K-Ar ages of natural glasses: U.S. Geological Survey Bulletin 1489, 31 p.

Naeser, C. W., Briggs, N. D., Obradovich, J. D., and Izett, G. A., 1981, Geochronology of Quaternary tephra deposits, *in* Self, S., and Sparks, R.S.J., eds., Tephra studies: North Atlantic Treaty Organization Advanced Studies Institute Series C, Dordrecht, Netherlands, Reidel Publishing Company, p. 13–47.

Naeser, N. D., 1984a, Fission-track ages from Wagon Wheel no. 1 well, northern Green River basin, Wyoming; Evidence for recent cooling, *in* Law, B. E., ed., Geological characteristics of low-permeability Upper Cretaceous and lower Tertiary rocks in the Pinedale anticline area, Sublette County, Wyoming: U.S. Geological Survey Open-File Report 84–753, p. 66–77.

—— , 1984b, Thermal history determined by fission-track dating for three sedimentary basins in California and Wyoming: Geological Society of America Abstracts with Programs, v. 16, p. 607.

—— , 1986, Neogene thermal history of the northern Green River basin, Wyoming; Evidence from fission-track dating, *in* Gautier, D. L., ed., Roles of organic matter in sediment diagenesis: Society of Economic Paleontologists and Mineralogists Special Publication no. 38, p. 65–72.

Naeser, N. D., and Naeser, C. W., 1984, Fission-track dating, *in* Mahaney, W. C., ed., Quaternary dating methods: Amsterdam, Elsevier, p. 87–100.

Naeser, N. D., Westgate, J. A., Hughes, O. L., and Péwé, T. L., 1982, Fission-track ages of late Cenozoic distal tephra beds in the Yukon Territory and Alaska: Canadian Journal of Earth Sciences, v. 19, p. 2167–2178.

Price, P. B., and Walker, R. M., 1963, Fossil tracks of charged particles in mica and the age of minerals: Journal of Geophysical Research, v. 68, p. 4847–4862.

Roberts, J. A., Gold, R., and Armani, R. J., 1968, Spontaneous-fission decay constant of ^{238}U: Physical Review, v. 174, p. 1482–1484.

Seward, D., 1974, Age of New Zealand Pleistocene substages by fission-track dating of glass shards from tephra horizons: Earth and Planetary Science Letters, v. 24, p. 242–248.

—— , 1975, Fission-track ages of some tephras from Cape Kidnappers, Hawke's Bay, New Zealand: New Zealand Journal of Geology and Geophysics, v. 18, p. 507–510.

—— , 1979, Comparison of zircon and glass fission-track ages from tephra horizons: Geology, v. 7, p. 479–482.

Spadavecchia, A., and Hahn, B., 1967, Die Rotationskammer und einige Anwendungen: Helvetica Physica Acta, v. 40, p. 1063–1079.

Storzer, D., and Poupeau, G., 1973, Ages-plateaux de mineraux et verres par la methode des traces de fission: Paris, Académie des Sciences, Comptes Rendus, v. 276, Series D, p. 137–139.

Storzer, D., and Wagner, G. A., 1969, Correction of thermally lowered fission-track ages of tektites: Earth and Planetary Science Letters, v. 5, p. 463–468.

Stuiver, M., Heusser, C. J., and Yang, I. C., 1978, North American glacial history extended to 75,000 years ago: Science, v. 200, p. 16–21.

Wagner, G. A., 1978, Archaeological applications of fission-track dating: Nuclear Track Detection, v. 2, p. 51–64.

Wagner, G. A., Reimer, G. M., and Jäger, E., 1977, Cooling ages derived by apatite fission track, mica Rb-Sr and K-Ar dating; The uplift and cooling history of the central Alps: Memorie degli Institui di Geologia e Mineralogia dell'Universita di Padova, v. 30, p. 1–27.

Westgate, J. A., and Briggs, N. D., 1980, Dating methods of Pleistocene deposits and their problems; V, Tephrochronology and fission-track dating: Geoscience Canada, v. 7, p. 3–10.

Westgate, J. A., and Gorton, M. P., 1981, Correlation techniques in tephra studies, *in* Self, S., and Sparks, R.S.J., eds., Tephra studies: North Atlantic Treaty Organization, Advanced Studies Institute Series C, Dordrecht, Netherlands, Reidel Publishing Company, p. 73–94.

Westgate, J. A., Briggs, N. D., Stalker, A. MacS., and Churcher, C. S., 1978, Fission-track age of glass from tephra beds associated with Quaternary vertebrate assemblages in the southern Canadian Plains: Geological Society of America Abstracts with Programs, v. 10, p. 514–515.

Westgate, J. A., Easterbrook, D. J., Naeser, N. D., and Carson, R. J., 1987, Lake Tapps tephra; An early Pleistocene stratigraphic marker in the Puget Lowland, Washington: Quaternary Research, v. 28, p. 340–355.

Zeitler, P. K., 1985, Cooling history of the NW Himalaya, Pakistan: Tectonics, v. 4, p. 127–151.

Zeitler, P. K., Johnson, N. M., Naeser, C. W., and Tahirkheli, R.A.K., 1982, Fission-track evidence for Quaternary uplift of the Nanga Parbat region, Pakistan: Nature, v. 298, p. 255–257.

MANUSCRIPT ACCEPTED BY THE SOCIETY FEBRUARY 16, 1988

Geological Society of America
Special Paper 227
1988

Dating Quaternary events by luminescence

Glenn W. Berger
Department of Geology, Western Washington University, Bellingham, Washington 98225

ABSTRACT

Luminescence techniques can provide ages for deposits undatable by routine geochronometric techniques (e.g., ^{14}C, K-Ar, fission track). Two classes of events can be dated by luminescence methods: (I) growth of a mineral or its last cooling, and (II) the last exposure to sunlight. Within the past few years, significant advances in procedures, technology, and understanding of the thermoluminescence (TL) behavior of minerals have been made that place luminescence dating techniques on the verge of widespread application to Quaternary deposits.

Most progress has come from studies of known-age material deposited under known conditions. Within class I, both distal and proximal tephra deposits have been dated, using TL techniques originally developed for pottery dating. Within class II, loess, buried soils, and waterlaid silts have been successfully dated. Means have been demonstrated for isolating and controlling several major sources of error, such as the type of TL instability known as anomalous fading, as well as the effects of uncertainty about the degree of zeroing of the luminescence signal in certain depositional environments. In particular, because of different sensitivities to light of the TL of quartz and feldspars, feldspars have been shown to be the preferred component for dating most unheated sediments.

Of the competing TL methods for dating the last exposure to sunlight, the partial bleach (R-Gamma or R-Beta) technique, when properly applied, has been shown to yield the best results in general. Nevertheless, in future dating studies of unheated sediments, this laborious method may be displaced by a novel technique that uses laser light, rather than heat, to stimulate the luminescence that is a measure of the past ionizing radiation absorbed dose. This new optical (OSL) method of dating promises to be simple, sensitive, and speedy.

INTRODUCTION

The attractiveness of thermally stimulated luminescence or thermoluminescence (TL) is that it can be applied directly to deposits usually considered barren of datable material, as well as to deposits laid down within the Quaternary time window (ca. 40 to 300 ka; 1 ka = 1,000 yr) not routinely accessible by other methods (e.g., ^{14}C, K-Ar, fission-track dating). The useful upper limit of TL dating is potentially several hundred thousand years, depending on several factors.

The main purpose of this review is to discuss recent advances in our understanding of the major sources of error in TL dating methods and means to minimize their effects. In the process, applications to different types of deposits are critically summarized, and gaps in our knowledge pointed out. The Quaternary scientist interested only in a reliable, routine chronometer will find general guidelines for appropriate application of the methods.

A TL age is proportional to total ionizing radiation absorbed dose divided by the dose rate. The absorbed dose is estimated from luminescence measurements, whereas the dose rate is usually determined from radioactivity analyses. Here, particular emphasis will be placed on luminescence measurements. For sediments, relatively less error generally arises from dose-rate determination than from TL measurements. For example, accuracies of better than 5 to 10 percent in the dose-rate estimates can now be achieved routinely for many sediments. In contrast, inaccuracies of greater than 50 percent in TL ages of sediments are still relatively common, because surprisingly few rigorous tests of proper TL techniques have been made, even though such techniques were introduced several years ago (Wintle and Huntley, 1980). Consequently, an understanding of the advantages and limitations of the competing TL methods has spread only slowly. For instance, even as recently as 1987, the one TL technique (the partial bleach method) that is capable of giving accurate results for most sediments was still being under-utilized in situations where it would be useful.

Applications of TL to the dating of Quaternary deposits have been reviewed by Dreimanis and others (1978), Wintle and Huntley (1982), Mejdahl and Wintle (1984), Singhvi and Mejdahl (1985), Aitken (1985), Singhvi and Wagner (1986), Berger (1986), and Mejdahl (1986). This review has a similar orientation to that of Berger (1986), but includes greater detail in the discussion of principles and applications. Moreover, using the recommendations of Wintle and Huntley (1982) as inspiration, I have adopted a critical stance here, with the aim of sorting out the strong studies from the weak. This approach should help to minimize the sense of confusion that the nonspecialist may feel when confronted with the wide range of techniques, practitioners, and applications, and the inevitably disparate claims of success. As a consequence of this stance, TL results are not mentioned here if the authors have failed to describe their methods, even briefly, or if they have not used techniques similar to those outlined by Wintle and Huntley (1982). This deliberate omission affects most of the results from China and the Soviet Union (see Singhvi and Mejdahl [1985] for a catalog of some of this work).

Where necessary, the discussion is oriented toward two classes of event or process: (I) that which reduces the TL signal to zero at deposition time, and (II) that which reduces the signal to some non-zero level, which is unknown in advance.

The TL signal may superficially be expected to be zero for all strongly heated or shocked material, as well as biogenic and chemically precipitated material. Examples are tephra (igneous going up, sedimentary coming down), baked sediments, burnt flint, foraminifera, radiolaria, gypsum, pedogenic carbonate, and the carbonates in speleothems (cave deposits) and travertines (spring deposits). If the TL signal is zero at deposition time, then the established techniques of pottery dating can be applied. However, tests for initially zero TL intensity have not been made for all of these materials.

Within this first class, baked sediment, burnt flint, volcanic quartz, and airfall volcanic glass are most suitable for TL dating, while quartz in shocked rocks can be useful. Attempts to date quartz from hydrothermal zones have been made, and these are mentioned below. Precipitated carbonates can be useful. On the other hand, little or no work has been done on gypsum, pedogenic carbonate, or biogenic silicates and carbonates. The TL dating of speleothems, travertines, heated stones, burnt flint, and baked sediments are not discussed here because these are adequately dealt with by Aitken (1985). However, one recent application to burnt flint has geological importance, because the flints are associated with cold-climate sediments (Valladas and others, 1986), now dated indirectly by TL at 40 to 55 ka.

The second class of event or process considered is recorded only in unheated, noncarbonate, inorganic, detrital material. The last exposure to sunlight is the most important event of this type because of its universality, although, as mentioned below, shock is another (albeit rare) process that can leave a non-zero TL signal (in quartz). The central difficulty with class II processes is that the effectiveness of resetting TL mineral clocks is highly variable.

Among samples recording the last exposure to sunlight, loess and subaqueous silty clays from certain deposits are the best materials for dating. Quartz or feldspar from dune or beach deposits show much potential (unfortunately, most of the work has been done on material lacking firm independent ages). Fine-size detrital grains (largely feldspar) in buried soils and peats also offer great potential. Fluvial sediments need further study.

GENERAL PRINCIPLES OF TL DATING

A useful geologic clock must store information at a known rate, must behave as a closed system, must have a useful range, and must have a well-understood zeroing process. For TL dating, the information stored is the number of electrons that migrate to traps (particular kinds of impurities and crystal defects) as a result of absorption of the energy of α, β, γ and cosmic radiations by the crystal. The α, β, and γ radiations arise mainly from the ambient U and Th isotopes and their decay products, as well as from ^{40}K. Luminescence is emitted during laboratory heating when the electrons are liberated from traps and some recombine with opposite charges (positive holes) at luminescence centers (e.g., trace impurities). This natural TL intensity is a measure of the past radiation absorbed dose and, when combined with an estimate of the past dose rate and the sample's sensitivity to ionizing radiation, yields the time elapsed since the traps were last emptied.

The TL age, t, equals post-depositional accumulated absorbed dose (or paleodose) divided by the dose rate. The practical age equation is slightly more subtle:

$$t = \dot{D}_2 / (\dot{D}_\alpha + \dot{D}_\beta + \dot{D}_\gamma + \dot{D}_c)$$

where the equivalent dose (D_e) is the laboratory β or γ dose that produces the same TL intensity as the paleodose. $\dot{D}_\alpha$ is the effective alpha particle dose rate, and the other components correspond to the actual β, γ, and cosmic ray dose rates. The numerator and denominator are determined in separate experiments. The units are gray for D_e and gray/ka for dose rate, where 1 Gy = 1 Joule/kg = 100 rads. The cosmic ray (mostly muons below 0.5 m soil depth) component $\dot{D}_c$ is usually minor (<5 percent of the total dose rate for many sediments). This can be estimated from the data of Prescott and Stephan (1982), whereas the other components are usually calculated from measurements of the K, U, and Th concentrations.

The closed system assumption can be violated in several ways, principally by draining of electrons from traps by agents such as light (optical bleaching), heat, or even perhaps pressure and chemical changes (the effects of these last two agents have not been studied seriously). In a sense, the type of instability of TL known as anomalous fading, which occurs in some feldspars, is a violation of the closed system assumption because electrons escape from traps. Loss or gain of K, U, or Th (or their decay products) will also violate this assumption, changing the dose rate in the process, although, excluding evident translocation of fine grains, this is expected to be a problem only for U and its decay products because of the leachability of U and Ra and mobility of Rn gas.

As implied in the Introduction, the zeroing process of the TL clocks is least understood when the event being dated is the last exposure to sunlight. The problem is that the non-zero TL residual in the grains at deposition can be a highly variable fraction of the total TL intensity measured in the laboratory, depending on subtle conditions at deposition. Even for samples older than tens of thousands of years, this "inherited," "relict," or "zero-point" TL can correspond to a large fraction of the natural TL signal. In principle, the factors that determine the amount of illumination penetrating each grain at deposition are unique to each sample and cannot be determined exactly, but only by inference. Nevertheless, some study of this zeroing problem has been conducted, and attention is given to this matter below.

The analytical techniques for determining the several components of the age equation can be quite elaborate, although by modern physical research standards the cost and complexity of instrumentation is low. An excellent discussion of the relevant laboratory practices and an outline of the commonly used apparatus is given by Aitken (1985). Spanne (1984) has discussed TL readout systems in detail. Figure 1 shows a block diagram of the apparatus used in some laboratories. In principle, a computer can double as a multi-channel analyzer, and as a controller of the operation (not shown here as such).

Using unconsolidated sediment as an example, Figure 2 shows a typical general task schedule required to obtain a TL age. Problems can be encountered at any step, of course. For example radon loss during alpha counting may lead to the suspicion that the dose rate has not been constant. Also, measurements by α or γ spectrometry may indicate disequilibrium in the U, Th decay series. In these cases, burying dosimeters at the site for up to a year or elaborate isotopic analyses may provide more accurate estimates of the dose rate. However, tests have shown that such complications are not common, or that they introduce errors often much smaller than those associated with the equivalent-dose measurement. As another example of problems, the TL may be unstable and a different mineral (or mineral group) may need to be isolated (the TL from polymineralic 5 to 10-μm grains in sediments is usually dominated by feldspars). These matters are discusssed in more detail below.

DETERMINING THE DOSE RATE

The principles and practices of dose-rate evaluation have been excellently reviewed by Mejdahl and Wintle (1984) and Aitken (1985), so that only a relatively simple overview is pre-

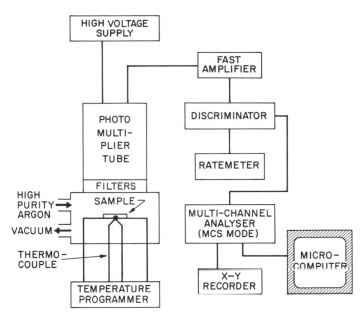

Figure 1. A block diagram of one useful arrangement of fast-photon-counting apparatus for measuring thermoluminescence. The output is a plot of light intensity (e.g., photon counts per degree C) versus temperature (a glow curve).

sented here. Some attention is given, though, to the new b-value concept, as this is discussed only briefly by Aitken (1985).

Two general approaches to dose-rate measurement are used depending largely on the accuracy sought, but also somewhat on the historical development of modern techniques. These two approaches are: (1) measurement of the concentrations of radionuclides in the sample and surroundings, and subsequent calculation of the dose rate assuming that energy absorbed (i.e., the absorbed dose) equals energy released (by the radioactive isotopes); (2) direct measurement of the β and γ dose-rate components by use of TL dosimetry, and calculation of the α dose-rate component as in approach 1. In this second approach, various TL phosphors can be placed in the deposits for up to a year or, for the γ-plus-cosmic-ray component at least, a portable γ spectrometer can be used.

The second approach can be more accurate than the first, and is preferred for dating culturally heated objects (e.g., pottery). For these, the relative uncertainty in the equivalent dose measurement can often be kept to 5 percent or less, mainly because of little or no ambiguity about the extent of TL zeroing, but also because TL growth curves are often linear for such young material (the consequences of non-linear growth curves are discussed below). Hence, in this case, uncertainties in the dose rate must be reduced as much as possible. The use of on-site dosimeters can improve the dose-rate accuracy in two ways—by directly incorporating the effects of any seasonal changes in water content (see dose-rate expressions below) and by providing a direct assess-

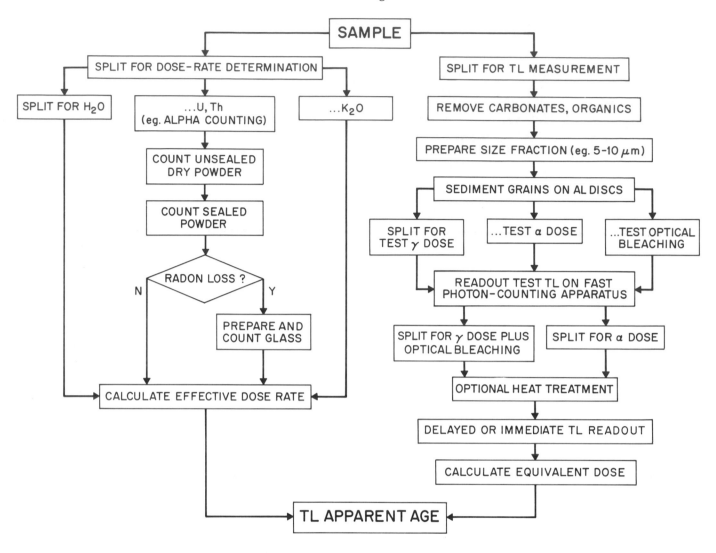

Figure 2. A typical flow chart of tasks that lead to the calculation of a TL age for unheated sediments. If the sediment has been heated (e.g., baked silts or tephra), then the optical bleaching task is eliminated. The optional heat treatment can be used primarily to remove any anomalous fading component.

ment of the dose-rate components when there is disequilibrium in the U-Th decay series (see below for additional remarks).

The first approach (the infinite matrix approach, Aitken, 1985) is often less time consuming and less costly than the second, especially if the U and Th concentrations are determined by the thick-source, alpha-particle-counting, "pairs" technique (e.g., Huntley and Wintle, 1981). This approach can yield routine accuracies of 5 to 10 percent, which may be sufficient when geologic events are to be dated. For these events, the relative uncertainty in the equivalent dose can easily exceed 20 percent, for reasons outlined below. With present techniques, reduction of the equivalent-dose error below 5 to 10 percent seems improbable for most unheated sediments.

In addition to simplicity and cost effectiveness, the alpha-counting technique has the advantage that alpha count-rates can

be used directly in the expressions for the dose-rate components. For example, with the assumption of decay series equilibrium and a uniform infinite matrix (in Gy/ka):

$$\dot{D}_\alpha = 1.34 \, (1.04 \, C_u) + C_{Th}) \, b/(1 + H_\alpha \, \Delta)$$
$$\dot{D}_\beta = (d_{\beta K} W_K + d_{\beta u} C_u + d_{\beta Th} C_{Th})/(1 + H_\beta \Delta)$$
$$\dot{D}_\gamma = (d_{\gamma K} W_K + d_{\gamma u} C_u + d_{\gamma Th} C_{Th})/(1 + H_\gamma \Delta).$$

C_u and C_{Th} are the alpha count-rates ($ks^{-1} \cdot cm^{-2}$) from the U-238, U-235, and Th-232 decay chains respectively; W_K is the weight percent of K_2O, all measured for dry sediment; and b is the b-value explained below. Δ is the ratio (water mass)/(dry sediment mass) for the sample in situ and is used to correct the dose rate for attenuation by pore water.

The values of the d and H coefficients have been given by

TABLE 1. VALUES OF THE d AND H COEFFICIENTS*

i	d_{iK}	d_{iu}	d_{iTh}	H_i
α	0	21.71[†]	19.9[†]	1.49
β	0.676	1.15	0.769	1.25
γ	0.202	0.8868	1.40	1.00

*The d values are dose rate (Gy/ka) per unit count rate (1.0 ks^{-1}·cm^{-2}) (for U and Th) or per 1% K_2O, all for dry powder. the H coefficients are the ratios of stopping powers (for α and β particles) or of absorption coefficients (γ rays), all of water to dry sediment.

[†]These values are not used in the above expressions for $\dot{D}_\alpha$ for reasons given in text.

Wintle and Huntley (1980), but their d values for U and Th must be modified to take into account the newer range-energy tables of Zeigler (1977). Their d values for K also must change slightly (Nambi and Aitken, 1986). These updated d values (in units of Gy/ka) are listed in Table 1 along with the H coefficients used by Wintle and Huntley. Note that Aitken (1985) does not use Zeigler's data, but instead employs the older values of Northcliffe and Schilling (1970), which Huntley and Wintle (1981) also employed. Huntley and others (1986) use Zeigler's data to produce new count-rate conversion values. These are combined with the dose-rate conversion values of Nambi and Aitken (1986) to yield the data in Table 1 here.

The nonspecialist or novice TL user may be forgiven for becoming confused here about how $\dot{D}_\alpha$ is calculated. For example, to measure the relative efficiency of α particles and β (or γ) rays in producing TL, historically the k-value system was used first, then the a-value, and here, a b-value. Because the b-value has been introduced only relatively recently in TL-specialist literature, an explanation of its relative significance is given here.

The a-value was introduced (see Aitken, 1985, p. 308) to supercede the k-value because the k-value is dependent on the energy spectrum of α particles. Specifically, the dimensionless a-value system was introduced because the TL per unit α particle track length is constant (to a good approximation). However, the a-value system is awkward to use (e.g., see Appendix K in Aitken, 1985), for it depends on a specific set of units. Therefore, Bowman and Huntley (1984; see also Huntley and others, 1988a) introduced the conceptually much simpler b-value system. Basically, the b-value is measured as the ratio D_e/L_e where L_e is the equivalent α-track-length density, determined as the X-axis intercept in a plot of TL intensity versus applied α irradiation. The ease of use in physical formulae of the b-value is outlined briefly below to show the origin of the constants in the above expression for $\dot{D}_\alpha$. The $d_{\alpha u}$ and $d_{\alpha Th}$ values in Table 1 would be appropriate only for the k-value system.

Essentially, these complications arise because α particles behave differently from β and γ rays. The TL for a given β or γ dose is independent of β or γ ray energy, whereas TL per unit α dose depends on α particle energy. On the other hand, TL per unit α **track length** is independent of α particle energy. The TL due to α irradiation is thus proportional to the total track length of all the α particles, rather than to their energy.

The b-value (like the a-value) exploits this fact and (unlike the a-value) can be used as a normal parameter in physical formulae and derivations. This ease of use is shown here. The b-value is defined as the β (or γ) dose per unit α particle track length per unit volume for which the two quantities yield the same TL intensity. If the b-value is in the units pGy · m^2 (= Gy · μm^2), then the b-value is related to the a-value by the expression b = 13a.

Defined as above, the b-value (in Gy · m^2) enables the dose-rate components from the U and Th decay series ($\dot{D}_{\alpha u}$ and $\dot{D}_{dTh}$) to be derived simply. Hence, for example,

$$\dot{D}_{\alpha u} = 4\eta b C_u \left[\Sigma N_i R_i / \Sigma N_i (R_i - R_o) \right]$$
$$= 4\eta b C_u \ \Psi_u^{-1} \text{ (in Gy/s)}$$

Here, unitless $\eta = 0.90$ (see Aitken, 1985, p. 311), C_u = count rate (m^{-2} · s^{-1}), N_i = number of disintegrations/g for isotope i, R_i = particle range per unit density, and R_o = residual range.

In practice, calculation of the factor Ψ_u (or Ψ_{Th}) is not necessary. As explained by Aitken (1985, p. 88), the electronics of the α counter can be adjusted to yield $\Psi_u = 0.82$ and $\Psi_{Th} = 0.85$. In other words, for a sample containing only the U series, the electronics are adjusted so that only 82 percent of the particles are counted (this is the "threshold fraction"). This practical adjustment ensures that the coefficients of b and C in the expressions for $\dot{D}_\alpha$ above do not depend on the range-energy values, and consequently the d values in Table 1 do not appear at all in these expressions.

Therefore,

$$\dot{D}_\alpha = 4\eta \ (\frac{C_u}{0.82} + \frac{C_{Th}}{0.85}) \ b \text{ (in Gy/s)}$$

Because use of b in units of pGy · m^2, C in ks^{-1} · cm^{-2}, (rather than s^{-1} · m^{-2}) and dose rate in Gy/ka is convenient, this equation then becomes:

$$\dot{D}_\alpha = 1.34 \ (1.04 \ C_u + C_{Th})b$$

as stated earlier. Methods for measuring the b-value are described in a subsequent section.

If some means other than alpha counting is used to determine the U and Th concentrations, then conversion factors can be used to calculate the dose rate per ppm of element. Dose-rate conversion values for 1 ppm U-238, 1 ppm Th-232, and 1 percent K_2O are listed in Table 2 of Mejdahl and Wintle (1984), Table 5 of Nambi and Aitken (1986) and Table 1 of Singhvi and Wagner (1986). Rb contributes a minor component, which can usually be ignored. Neutron activation techniques (NAA and

DNA) determine only the parent nuclides such that, if disequilibrium of the decay chains occurs, these techniques introduce greater errors than does alpha counting. Similarly, gamma spectrometry determination of U concentration depends on ^{214}Bi, which occurs well down the U-238 decay chain, and thus the estimation of U activity with this technique will be affected by radon loss. Gamma or alpha spectrometry (e.g., Aitken, 1985; Saro and Pikna, 1987), or alpha counting in conjunction with NAA, for example, can be used to establish the existence of disequilibrium, however.

Various aspects of disequilibrium have been discussed in detail by Mejdahl and Wintle (1984), Aitken (1985), Wintle (1978), Wintle and Huntley (1980), and Hennig and Grün (1983). Disequilibrium due only to an initial deficiency or excess of a radioisotope is dealt with by Wintle (1978) and Wintle and Huntley (1980). In this situation, the dose-rate expressions require specific modifications.

Radon loss detection by alpha counting is mentioned in Figure 2. Such radon loss may or may not indicate disequilibrium in the field sample. The practice of sample fusion (Figure 2) does not correct for the effect of in-situ disequilibrium, but instead for inaccuracies associated with alpha counting of dry powders (such as radon loss or overcounting; see Aitken, 1985; Berger and others, 1984; Huntley and others, 1986).

IMPORTANCE OF GRAIN SIZE

The average absorbed dose components to grains from external radioactivity will depend on grain size because of attenuation of the radiations. For this reason, when measuring the equivalent dose (see next section), use of grains having a narrow range of diameters is desirable. Furthermore, because the average ranges of α radiations in unconsolidated sediments is about 20 μm, much less than the averages of 2 mm and 30 cm for β and γ radiations, respectively (Aitken, 1985), common practice is to use either fine (2–10 μm) or coarse (about 100–300 μm) grains for equivalent-dose measurements.

For example, for a uniform spread of grains between 5 and 10 μm, the average reduction in the external α dose will be about 10 percent compared to that for negligibly small grains. However, because α particles are less effective than β or γ rays in producing TL, this 10 percent reduction may produce only a 1 to 2 percent effect in the total dose and can often be ignored. For coarse grains the attenuation of both the α and β dose components becomes significant. In this case, for quartz and feldspar, we usually remove the effects of external α irradiation by HF acid etching. However, the reduction in the β dose component must be calculated. To illustrate, the attenuation will be 5 to 25 percent for 100 to 300 μm radioactivity-free grains (e.g., most quartz), depending on the source of the β rays (K, U, or Th). Calculation of corrections for β attenuation can be difficult for much larger grains (Mejdahl, 1983).

We generally assume, for expediency, that the HF acid etching removes a surface layer uniformly. However, in fact, natural minerals are etched along defect channels and imperfections, producing etch channels that penetrate the grain (e.g., Berner and Holden, 1977). Such anisotropic etching introduces no significant error into calculations of the dose rate in most cases. However, Goedicke (1984) has shown that the residual alpha doses remaining after etching of quartz can amount to 10 percent of the total dose if standard etching practice is followed. He suggests that, in some cases, etching be avoided and that the alpha dose component to quartz be calculated. This problem is less important for large feldspar grains (>100 μm) because they can have a substantial internal β dose rate (from potassium), thus reducing the relative significance of the α dose-rate component.

In summary, the use of fine grains (poly- or monomineralic) for equivalent dose measurement (next section) permits relatively straightforward application of the above dose-rate expressions to the age equation. The age equation can be further simplified if only large grains of quartz are used for measuring the equivalent dose, because then (after acid etching) the α component of the dose rate may be zero. The difficulty with using quartz, however, is that for dating the last exposure to sunlight, large grains of quartz can have several crippling disadvantages (outlined below) compared to feldspars. On the other hand, large grains of feldspar have the encumbrance of requiring calculation of an attenuated β dose-rate component. Furthermore, large grains of either quartz or feldspar in waterlaid sediments are associated only with deposits that have been transported short distances or at depth. This association is a fatal weakness in the use of large grains for dating the last exposure to sunlight of most subaqueous deposits.

METHODS FOR MEASURING THE EQUIVALENT DOSE

Different techniques are required for measuring the equivalent dose for events I or II.

Class I events

The additive dose procedure developed for pottery dating (Atiken, 1985) is appropriate for this type of zeroing event. Known γ (or β) doses are added, as illustrated in Figure 3, to different subsamples of the unirradiated (natural or N) sample. This TL buildup curve is extrapolated to zero TL intensity at each temperature point on the resulting glow curves (plots of TL intensity against laboratory heating temperatures), yielding an equivalent dose value (if no supralinearity correction is made). A plateau in a plot of these intercept values against temperature is thought to indicate the region of the glow curve corresponding to electron traps that are thermally stable over the time span of interest. This plateau value is the equivalent dose in the age equation.

For young samples (less than a few hundred or few thousand years), the additive dose curve could be expected to be supralinear (upward curving; for discussions see Aitken, 1985) over the initial part of its range. This has, in fact, been observed

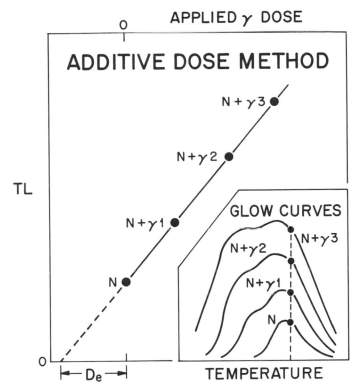

Figure 3. The preferred method for measuring the equivalent dose (D_e) for heated, biogenic, authigenic, or precipitated material.

for precipitated carbonate (Debenham and others, 1982), heated limestone (Berger and Marshall, 1984), one sample of airfall tephra (Berger, 1985a), and one sample of loess (Berger, 1987), although in these examples, the causes of the supralinearity may be different from the theoretical explanations reviewed by Aitken (1985). Whether or not supralinearity is observed in the additive dose curve, neglect of its effect can lead to underestimation of the equivalent-dose value for young samples (e.g., <5,000 yr).

The common procedure used to correct for this effect is to construct a second-glow growth curve, i.e., an additive dose curve for subsamples whose TL has been drained by laboratory heating. In addition to estimating any supralinearity correction, such a second-glow or regeneration growth curve could also be used to estimate the equivalent dose values by matching (interpolating) the TL intensity from an unirradiated subsample to this curve. However, this regeneration curve often has a different slope from the additive dose growth curve for pottery (Aitken, 1985) and tephra (Berger and Huntley, 1983), thus indicating a change in sensitivity to γ or β radiations brought about by heating. For this reason, only the additive dose procedure is recommended for dating class I events.

How to make an accurate extrapolation when the additive dose curve becomes sublinear, as it does for old samples, is of greater concern than supralinearity effects, if geological samples are being dated. The D_e values can have very large uncertainties if the extrapolations are large, unless the exact form of the build-up curve is known. This problem of extrapolation from non-linear growth curves will be given more attention below. Here, the simplest physical model for such sublinearity is outlined.

If the grains contain a single type of electron trap (N_s being the maximum number available), if n is the number of electrons in traps (the number of filled traps) at a given time, and if δ is the (constant) probability per unit β or γ dose D of filling an empty trap, then the rate of trap filling is $dn/dD = \delta (N_s-n)$, where N_s-n is the number of unfilled traps. With the condition that n = 0 at D = 0, the solution of the equation is $n = N_s (1-exp[-\delta D])$. For simplicity, suppose that the TL intensity I_γ is proportional to n at a given glow-curve temperature (first-order kinetics). Then $I_\gamma = I_s (1-exp[-CD])$ where I_s is the saturation intensity of TL and C is a constant. For a sample having a natural TL intensity I_n, to which laboratory β or γ doses (D_a) are applied, the equation becomes $I_\gamma = I_s (1-exp [-C(D_e + D_a)])$ where D_e is the equivalent dose (the extrapolated intercept on the applied dose axis).

For low doses, this saturating exponential growth curve can be successively approximated by a straight line, a quadratic, or by higher-order polynomials, depending on the ratio $(I_s-I_n)/I_s$. Clearly, supralinearity is ignored here, but its effect can be negligible, depending on the uncertainty in the extrapolation. On the other hand, supralinearity effects can be approximated by the use of growth curves derived from non-first-order-kinetics (e.g., second-order) expressions. Such curves have not yet been used in dating studies (e.g., Berger, 1987).

Huntley and others (1988a) have shown that a saturating exponential can also describe the TL growth curve resulting from alpha irradiations. In this case, if L is the length of α tracks per unit volume and A is the cross-sectional area of each track, then $I_\alpha = I_s (1-exp [-A(L_e+L_a)])$. They also showed that the b-value simply equals the ratio D_e/L_e, where D_e and L_e are the intercept values for the growth curves on the β dose and α-track-length axes respectively. This simple relation for b holds whether the TL growth curves are linear or obey a saturating exponential model. The suggestion (Aitken, 1985, p. 139; Singhvi and others, 1987, p. 51) that the nature of the TL growth curves for alpha and beta (or gamma) irradiations is different, and that this difference introduces an error into TL age estimates, is not confirmed in practice or in theory. For sublinear growth curves, no correction is required to the simple relation $b = D_e/L_e$ if a saturating exponential model is used.

To summarize, the rates of curvature of the TL buildup curves are not usually known independently, but a simple model can be used to justify extrapolations of additive dose growth curves to obtain equivalent dose and b-values. Clearly, uncertainties in extrapolated D_e and L_e values will be largest for large extrapolations of data showing poor reproducibility. Sometimes young samples are used to test this saturating exponential model (Berger, 1987; Lamothe and Huntley, 1988; see also below).

Class II events

Exposure to light readily reduces the TL signal initially, but, even after long light exposures, some electrons remain in traps at deposition. Wintle and Huntley (1980) introduced the partial bleach or R-γ method (R-β if a β source is used) and the regeneration technique to correct for this non-zero component. Singhvi and others (1982) employed a total bleach or residual technique, which is a special case of the partial bleach procedure. These three methods are illustrated in Figure 4.

With the partial bleach method, a short illumination is given to some of the subsamples (both irradiated and unirradiated). Tests by Wintle and Huntley (1980) have shown that a given illumination reduces the TL from the different irradiated subsamples by the same fraction. The resulting two growth curves (Fig. 4a) are extrapolated to their intersection and, as before, these intercept values (D_e) are plotted against temperature. In essence, this technique determines the equivalent dose at which the observed reduction R in TL is zero. The main advantage of the partial bleach technique is that arbitrarily short light exposures can be used, thus avoiding draining electron traps unaffected at deposition.

Alternately, as a special case of the partial bleach method, one can use the total bleach technique, wherein some low intensity of TL (I_r in Fig. 4a) is extrapolated to the additive dose curve to also yield an equivalent dose value. Finally, all the subsamples can be given a long light exposure and a TL growth curve regenerated to match the intensity of TL in the unirradiated (N) sample (Fig. 4b). The applied dose that produces a matching intensity of TL is again the equivalent dose value, which is plotted against temperature as before. The principal advantage of the regeneration and total bleach techniques is their ease of use. All three methods should yield the same result for samples that have experienced long light exposures at deposition, in the absence of uncorrected sensitivity changes.

Both the total bleach and this regeneration method share a major weakness. The required long light exposures can reduce the intensity of the natural TL below that attained at deposition and thereby produce equivalent dose values that are too large. Furthermore, this regeneration technique can suffer from sensitivity changes (Wintle and Huntley, 1980; Rendell and others, 1983) manifested by different growth rates in the unbleached and regeneration curves, which can be difficult to detect for non-linear growth curves. Which portions of the curves are to be compared can be unclear when the curves are non-linear. Moreover, imprecision of TL reproducibility introduces large uncertainties into such a comparison.

As with the additive dose technique, both the partial and total bleach methods depend on accurate extrapolation of the TL growth curves. For geologically young samples (<10 to 20 ka), or non-quartz samples that have received relatively low natural doses (e.g., <200 Gy), the growth curves are generally linear so that linear extrapolations are justified. On the other hand, for older samples (or the quartz component of some deposits as

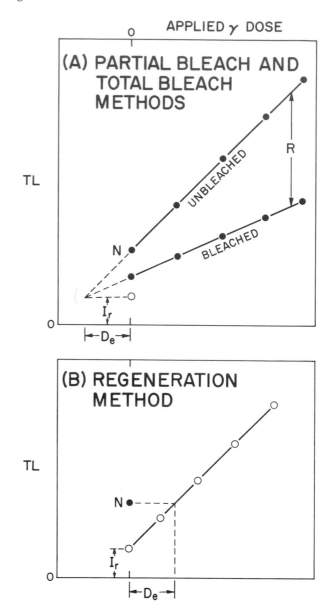

Figure 4. Three methods suitable for measuring the equivalent dose (D_e) in samples whose TL clocks have been zeroed by sunlight. The partial bleach method (plot of R versus applied dose) is preferred for most such sediments. The total bleach regeneration method shown here (B) is a special case of the original method of Wintle and Huntley (1980), and this simpler version has been widely (and uncritically) applied.

young as 5 to 10 ka), the TL growth curves often become sublinear, primarily because of saturation of the electron traps. Under these circumstances, the form of the extrapolated buildup curves may not be immediately obvious, although as discussed above, a saturating exponential model may be appropriate. Three approaches to extrapolation have been followed for sublinear buildup curves. In one, advocated by Prescott (1983), additive dose growth curves are constructed and combined with growth curves built up from thermally or optically drained samples. This combined extended curve is fitted by a polynominal (or, for examples of extreme sublinearity, an exponential). The disadvantage of this approach is that sensitivity changes upon heating or intense optical bleaching can change the shape of the buildup curve from what may have occurred in nature. How such sensitivity changes can be measured accurately for old samples is difficult to see, because (as stated above) one must compare rates of change of curvature having ambiguous relationships.

The second (and preferred) approach is to obtain natural samples that are similar to (but much younger than) those to be dated, and then to generate a TL additive dose growth curve from the younger sample, which ought to indicate the form of the extrapolation required for the older samples. For example, a choice between a polynomial and an exponential could be made. This approach has been advocated by Huntley and others (1983) and Smith (1983), and demonstrated for a marine mud by Berger and others (1984), for a lacustrine silt by Lamothe and Huntley (1988), and for a loess by Berger (1987). The difficulty with this approach for terrestrial deposits is that the younger equivalent of the older sample of interest is very often impossible to obtain. Testing of this approach is needed, using samples of known lithology. However, the expected general applicability of the saturating exponential model makes this routine use of such young samples seem superfluous.

A third approach to extrapolation of TL growth curves may well be the most expedient because measurements are made only on the sample to be dated, but without the worry of the kind of sensitivity changes inherent in the first approach. In this case, the form of the required extrapolation is assumed to be the same as the statistically determined, high additive dose growth curves (Berger and others, 1988, provide a rigorous treatment of sublinear extrapolation and error analysis). As discussed above (class I events), if the non-linearity is not extreme, a quadratic function might be fitted to the data to yield a good approximation of the true (natural) form of the extrapolation. This approach has yielded satisfactory results for waterlaid silts younger than ca. 70 ka (Huntley and others, 1983; Berger and others, 1984, 1987; Berger, 1984, 1985b, 1985c; Lamothe, 1984; Lamothe and Huntley, 1988). For older samples exhibiting extreme non-linearity, or where large laboratory doses are applied, higher order polynomials might be used, or preferably, saturating exponentials. Such exponentials and polynomials have given satisfactory results for volcanic ash (Berger, 1985a, 1987) and for waterlaid silty clay (Berger and others, 1987, 1988), but not necessarily for waterlaid quartz (Belperio and others, 1984; Berger, 1984, 1985b), presumably because of other problems (such as insufficient resetting of the TL of quartz at deposition, discussed below).

The useful upper age limit of these methods for measurement of equivalent dose will probably be determined by two principal factors: the onset of saturation and the mean thermal lifetimes of the relevant electron traps. Neither of these factors has yet been investigated thoroughly for most of the mineral varieties likely to be encountered in practice, but a practicable age limit of ca. 500 ka is possible (see Mejdahl, 1986).

With all of these qualifications, the remainder of this review deals primarily with efforts to apply the methods of Figures 3 and 4. Two important problems will be considered in detail: uncertainty in the extent of "zeroing" of the TL clocks in class II events, and anomalous fading. The former has been perhaps the major stumbling block to progress in the TL dating of unheated sediments. Anomalous fading refers to an unstable TL component, associated primarily with feldspars, that can lead to gross underestimation of the true age of an event unless a correction is made. Recent research presents hope that routine correction for, or elimination of, anomalous fading is now possible.

EFFECTIVENESS OF ZEROING

Two approaches to testing the effectiveness by which any mechanism reduces the TL, or zeros the mineral clock, are: (1) to measure the TL equivalent dose for zero-age samples, the most direct approach for most materials; and (2) to simulate a natural zeroing mechanism in laboratory experiments. The second approach has been little used, although it is attractive.

Class I events

That crystallization leaves zero latent TL (McDougall, 1968) has generally been assumed. However, this assumption has not yet been tested on biogenic carbonates and silicates. A systematic study of a young (4 ka) stalagmitic calcite (Debenham and others, 1982) illustrates a problem that could occur with other forms of precipitated minerals. Debenham and others (1982) observed an equivalent dose ten times greater than expected. Examination of the spatial TL emission showed that most of the signal arose from small bright specks (e.g., Aitken, 1985, p. 206). Debenham and others suggested two possible origins for this relict TL: incorporation of old limestone dust during formation, or incorporation of old sediment dust. The second explanation is very like operative for most environments; airborne sediment grains smaller than 10 μm are ubiquitous. The possibility that such relict grains always contaminate young precipitated carbonates, for example, should be considered.

Fortunately, the above contamination was not an intractable problem, at least for stalagmitic calcite. Debenham and others were able to selectively block the TL attributed to the contamination, and to obtain a correct young TL age by using a blue-pass filter on the photomultiplier tube.

The problem of non-zero TL remaining in heated minerals has been discussed by Aitken (1985). For dating tephra, glass grains have been selected (Berger and Huntley, 1983) because glass would not be expected to retain relict TL. Indeed, the measurement of an accurate TL age for the A.D. 1480 Mount St. Helens W_n tephra (Berger, 1985a) confirms the assumption of adequate zeroing of this airfall glass at eruption time.

Sutton (1985a) recently observed that quartz from shock-metamorphosed rocks can have a negligible apparent initial TL level if the shock heating is severe enough. That is, for a suitable shock class, the additive dose method can be applied, while samples below this shock class are rejected. In his attempt to date the Meteor Crater impact event, Sutton extracted 44 to 100-μm-size grains of quartz from shock-metamorphosed sandstone and dolomite, and observed that the equivalent dose values decreased until the petrographic shock class 2 was attained, corresponding to an equilibrated state at ca. 10 GPa and 700 °C. Fragments of shock class 3 or greater were suitable for TL dating, the threshold for total resetting of the quartz TL by shock heating having been exceeded.

Class II events

Soviet geologists (see Wintle and Huntley, 1982) first suggested that mechanical abrasion might be an agent for resetting TL mineral clocks. If such an agent is effective, then this would suggest that, for example, glacial abrasion in lodgement till components could be dated directly. This matter has received little attention for dating applications, perhaps because grinding can both reduce TL and augment it, depending on the material. The existence of a class of materials (molecular crystals, salts, glasses) that emits TL when ground (triboluminescence) (e.g., Beese and Zink, 1984) complicates the process. A preliminary investigation of the effects of pulverizing LiF crystals and bricks (Nakajima, 1984) illustrates further complications that may be relevant to geological samples. The abrasion-related processes that can reduce the TL are not known. Very likely, the effect on TL arises from localized heating or other processes, rather than pressure itself.

An attempt to test the effectiveness of mechanical abrasion for resetting TL mineral clocks has been made by Berger (unpublished data, 1983). Irradiated and unirradiated polymineralic sediments were ground in an agate mortar under acetone, and the results suggested that although cool grinding can produce significant reduction in TL, the process is far less effective than is a brief exposure to sunlight. Thus, in view of the common intermixing of waterborne insolated grains with subglacial sediments, useful application of this grinding effect seems unlikely.

The effectiveness of sunlight in reducing TL in sediments (Wintle and Huntley, 1980) is now well known, and forms the basis of most of the dating applications discussed in this review. Although direct insolation is very effective in resetting TL mineral clocks, many sediments are deposited under conditions that apparently screen most or all of the individual grains from significant illumination. Attempts have therefore been made to identify the sediment types and facies that most likely have been well insolated, and that are thus datable by the methods in Figure 4. The approach has mainly been to measure equivalent dose values in zero-age samples or in young known-age samples. Only recently has a laboratory simulation experiment been attempted.

Eolian material. Loess and dune sands seem at first glance to be more likely to be well insolated than any other sediments.

We would therefore expect the TL to be reduced to very low levels at deposition. Oddly, studies of zero-age loess have just begun; however, several workers have measured equivalent dose values for the ca. 100-μm quartz from the top few millimeters of dune sands, with surprising results.

As expected, a near-zero equivalent dose (1.5 ±1.0 Gy) was measured for one dune by Singhvi and others (1983), using the total bleach method and a Hg-based lamp (which has several UV spectral lines, and therefore does not approximate insolation). On the other hand, Prescott (1983) and Readhead (1984) observed a TL signal from coarse quartz from the surfaces of dunes that, unexpectedly, could easily be reduced further by additional light exposures, natural or artificial. Huntley (1985a) deduced from optical bleaching tests on 100-μm quartz grains from beach and dune sands that these grains had received *effectively* only short sunlight exposures at deposition, despite intensive insolation. Overestimates of the equivalent dose were obtained when long light exposures were used.

Similarly, in an investigation of a zero-age proximal loess, Berger (1987) showed that a large, easily bleachable TL signal remained in polymineral fine grains at deposition and that overestimates of the equivalent dose were obtained even with relatively feeble optical bleaching. Figure 5 shows that only when the briefest illumination (2 min) was used were equivalent dose values obtained that were indistinguishable from zero. These results for zero-age eolian silt and dune sand require some comment.

This "overbleaching" effect can best be understood from Figure 6, which represents quartz. In this example, any light exposure longer than a few minutes would reduce the TL in the collected grains below the intensity at deposition (simulated by the dashed curve). The obvious implication is that, of the methods in Figure 4, only the partial bleach technique can yield the correct equivalent dose for any quartz sample exposed briefly to light. Thus far, the only direct support for this implication for quartz has been provided by Huntley (1985a); other workers inexplicably have avoided the partial bleach technique for such samples.

The fine grain samples of Malan Loess studied by Lu Yanchou and others (1987b) were probably transported tens to hundreds of kilometers, implying an insolation time of perhaps 12 to 24 hr. To check the TL zeroing assumption, these authors compared the relict TL intensity in fine grains (thought to be mainly quartz) removed by adhesive tape from surface scarps, with the TL remaining in fine grains extracted from adjacent buried samples and given an artificial exposure to sunlight of 15 hr, which removed almost all of the light-sensitive TL. The similarity in these relict intensities suggested that for these samples an artificial optical bleaching (using sunlight) of 15 hr could be used to estimate the natural relict TL in the buried samples without the risk of overbleaching.

In subsequent TL dating studies of coarse-grain quartz from dune sands and loess, Australian and Chinese workers (Lu Yanchou and others, 1987a; Gardner and others, 1987) have apparently sidestepped the over-bleaching problem by using the TL

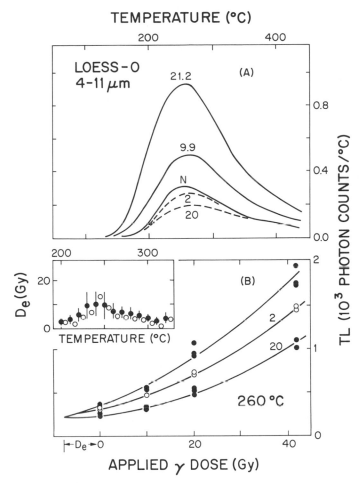

Figure 5. Glow curves (A), partial bleach dose-response curves (B), and equivalent dose plots (B insert) for zero-age proximal (1 to 3 km) loess from Alaska. Only for the shortest laboratory illumination (2 min, >550-nm wavelengths) were D_e values indistinguishable from zero obtained (open circles in insert, large error bars not shown). The finite D_e values for longer illumination times correspond to an apparent age of ca. 1 ka for this ca. 50 to 100-year-old sediment. Additive dose (9.9 Gy and 21.2 Gy), unirradiated (N), and optically bleached (2 and 20 min) N glow curves in A (after Berger, 1987).

intensity in modern surface samples to estimate the relict TL signal. They then employed a modified regeneration technique (Readhead, 1984). However, there is an unexplained curiosity in the above two sets of results obtained by Lu Yanchou and colleagues. Specifically, they observed that almost all of the light-sensitive TL had been removed in the fine grains (mostly quartz) of scarp surface samples, yet the coarse-grain quartz from the same scarp surface samples retained a large fraction of their light-sensitive TL. This curiosity requires further investigation. Did these coarse grains have only a local origin, unlike the fine grains? If so, what are the implications for the accurate TL dating of such deposits?

None of the authors who studied quartz from dunes suggested a cause for the observed ineffectiveness in the resetting of the TL for such intensively insolated sands. However, Readhead (1984) described some tests on quartz extracted from the upper few millimeters of dunes that suggest that the amount of reduction in TL may depend on the in situ close-packing geometry of 100-μm grains relative to larger and smaller grains. Perhaps the explanation for this ineffective zeroing of quartz TL, in spite of intensive insolation, can be found in the observation of Berger (1985b) that the TL of quartz can be insensitive to wavelengths of light longer than ca. 450 nm (indigo) (Fig. 7).

The implication of Figure 7 for the dune quartz of Readhead is that some factor may have attenuated the shorter wavelengths before they penetrated the bulk of each 100-μm dune-sand quartz grain. One factor may speculatively be related to scattering and absorption of sunlight among a closely packed mixture of different grain sizes and different minerals. Another factor may be the presence of surface coatings on the grains (e.g., desert polish, iron staining), which are routinely removed by HF acid before TL studies. The sample treatment used by Berger (1985b) (0.1N HCl followed by hydrofluosilicic acid) may not have removed much of any pre-existing iron stainings (although they were not observed to be significant on large grains). Thus, the fine-grained quartz of Figure 7 is not **unambiguously clean**.

Amorphous iron oxide coatings such as goethite and limonite do not pass much visible light above about 450 nm in wavelength (Hunt and others, 1971; Townsend, 1987). This fact suggests that TL experiments on both HF acid etched (or CBD treated, Berger, 1984) and unetched dune quartz, using different grain sizes and different wavelength windows, may elucidate this phenomenon of ineffective zeroing in dune sands. Indeed, a preliminary experiment by Singhvi and others (1986a) using both HF etched and unetched 100 to 150-μm quartz grains (and a "naked" Hg lamp) produced a dramatic difference in bleaching. However, in this example there is good reason to think the grain coatings are post-depositional.

Berger (1987) recently suggested that fine-grain clumping during transport, rather than pre-depositional iron staining, is the likely reason that a feldspar-dominated, "zero-age, eolian proximal silt retained a large, bleachable relict TL signal (Fig. 5).

To summarize, sand-sized dune quartz grains and silt-sized proximal eolian feldspar grains can be screened from light before final burial, but the screening mechanisms probably differ and are poorly understood. On the other hand, silt-sized feldspar grains in some soils may not be screened. For example, Huntley (1985a) measured an equivalent dose of <1 Gy for a surface soil when he used the partial bleach technique and wavelengths of light as short as 360 nm (UV).

Waterlaid material. Studies of the TL in waterborne sediments have suggested that the factors controlling the effectiveness of zeroing during water transport can be quite different from those outlined above for eolian grains. The evidence is, however, indirect. For example, with samples ranging from suspended river silts (Berger and Huntley, 1982; Huntley and others, 1983; Hunt-

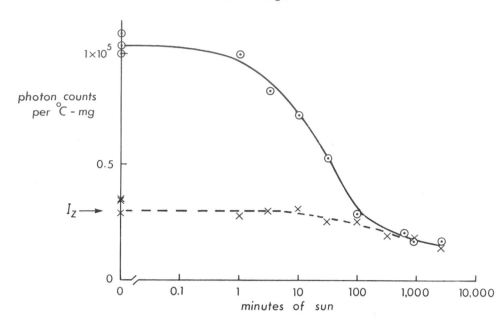

Figure 6. The reduction in TL at 400°C versus sunlight exposure time for 100-μm quartz from a beach sand. Solid line and circles represent a sample first given a laboratory dose of 38 grays. The dashed line is the same after shifting to the left by 120 min. The crosses represent the sample without the initial applied dose. Thus, a zero-age depositional event has been simulated by ca. 120 min of light exposure. Any subsequent additional light exposure less than ca. 10 min (total <130 min) would have little effect on the TL intensity (dashed curve), while additional light exposures longer than ca. 100 min would have a measurable effect (after Huntley, 1985a).

ley, 1985a) to marine mud (Berger and others, 1984), correct equivalent dose values could be obtained only when the partial bleach technique and long wavelengths of bleaching light (above 550 nm) were used. Here again, as for the eolian material above, Figure 7 provides the key to an understanding, when combined with knowledge of the behavior of light in water.

In cloudy or turbid water the shorter wavelengths of visible light are relatively attenuated through absorption (primarily) and scattering by solid particles (e.g., Jerlov, 1976). Consequently, the ambient solar spectrum appears shifted toward the red, the effect being greater for more turbid water (Fig. 8) or water having a high organic content, such as estuarine waters. Clear water, in comparison, has only a small effect on the relative attenuation of different wavelengths in the region ca. 325 to 625 nm. Therefore, why the TL of waterborne quartz grains will be reduced ineffectively, if at all, at deposition can be easily understood. Furthermore, because the turbidity of the water cannot be known in advance, the relative attenuation of the shorter wavelengths will be unknown.

Hence, employing the longer wavelengths (>about 550 nm) and using only the partial bleach technique is prudent for waterborne feldspars. For example, using laboratory bleaching wavelengths below about 550 nm would be improper for grains transported mainly in water as turbid as that represented in Figure 8. This explains why, for example, Belperio and others (1984) measured very large equivalent doses for zero-age waterlaid

quartz—they used too-short wavelengths and too-long light exposures. Similarly, the use of bleaching wavelengths as short as 350 nm for feldspar-dominant suspended river silt probably accounts for the non-zero TL ages (4 ± 1 ka) measured by Lamothe and Huntley (1988). However, this zero-point error of about 5 ka would not be too important for similar sediments older than ca. 70 ka. Nevertheless, this zero-point error should not be assumed to be small for all "old" sediments because, in general, a great many subaqueous depositional modes can be represented in any particular stratigraphic sequence.

Finally, another approach to investigation of the effectiveness of zeroing in waterlaid sediments is to simulate sediment burden and flow rates in laboratory experiments, while exposing the sediment to controlled lighting. Gemmell (1985) has attempted to do this with polymineralic fine silts, but with no consideration of the spectrum of the light source. Not surprisingly, in view of the information in Figure 7, his results are uninformative—he concluded only that the rate of transport has a marked effect on the rate of reduction of TL. Selective blocking of different portions of the spectrum of his light source might have produced more useful information.

ANOMALOUS FADING
The problem

The TL produced in volcanic feldspars by laboratory irradiation is unstable and decays rapidly (Wintle, 1973, 1977a). Be-

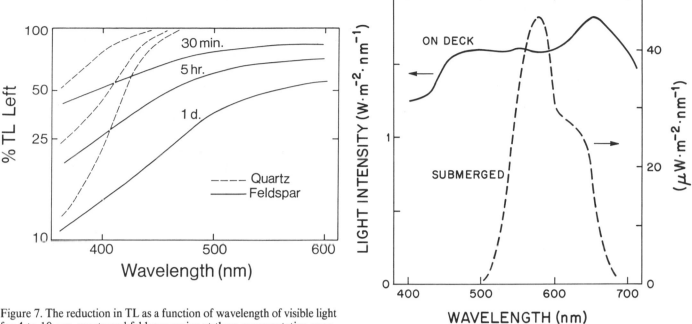

Figure 7. The reduction in TL as a function of wavelength of visible light for 4 to 10-μm quartz and feldspar grains at three representative exposure times to a laboratory light source. These curves are generalizations from data of Berger (1985b; unpublished data). The results of Berger (1985c) extend the feldspar lines to above 660 nm (deep red).

Figure 8. Plots of downwelling spectral irradiance at the sea surface (late afternoon) and at a depth of 4 m in a turbid river (after Berger and Luternauer, 1987).

cause this decay affects all regions of the glow curve and does not behave as a thermal instability governed by kinetic theory, it has been called anomalous fading—first described by Hoogenstraaten (1958) for zinc sulphide. Besides the volcanic feldspars studied by Wintle, some zircons extracted from ceramics and zircon-rich sand have exhibited this type of fading, as have apatite and calcite (Aitken, 1985). Quartz has generally possessed only stable TL, although Readhead (1984) has recently claimed that some sedimentary quartz also shows this effect, and Mejdahl (1983) has observed such fading in culturally heated quartz and feldspars.

Anomalous fading presents a serious problem for TL dating because the short-lived unstable component is present only in the artificial (laboratory-induced) TL, having mostly disappeared from the natural thermoluminescence. Thus, when growth curves such as those in Figures 3 and 4 are constructed, the slope of the curves will change as the unstable component decays. Clearly, measurements made on such samples immediately after irradiation will underestimate equivalent dose values.

Little is known about which types of nonvolcanic feldspars are free of fading, although feldspars are present in most sediments and their TL dominates that of the other major mineral components (Fig. 9). Wintle (1981, 1982) and Berger (1985b) observed little or no fading in their samples, while Wintle and others (1984) observed it only in a sample containing some vol-

canic ash. On the other hand, loess from Alaska (Berger, 1987), and waterlaid sediments from eastern Canada (Lamothe, 1984; Lamothe and Huntley, 1988; Berger, 1984), the Gulf of Mexico (Berger and others, 1984), and other sites in British Columbia (Berger and others, 1987) have all exhibited 5 to 20 percent fading over 2 to 6 weeks. Lundqvist and Mejdahl (1987) observed up to 22 percent fading in 100 to 300-μm grains of unspecified potassium feldspar stored up to 11 months. Wintle and Westgate (1986) reported no significant anomalous fading for Alaskan loess samples equivalent to those examined by Berger (1987) (who has attributed this discrepancy to differences in laboratory technique; see below).

Hitherto, not much effort has been made to identify the varieties of feldspars in TL samples (but see Lundqvist and Mejdahl, 1987), largely because most samples have been fine grained (<10 μm), or the occurrence or absence of fading in sediments has not commonly been reported. Consequently, for fine-grained, polymineralic, feldspar-dominant samples, it is not known whether the detected fading represents a few grains that fade strongly, or whether all the grains have a short-term fading component. A variety of feldspar types would be expected in such sediments, mostly the low- and intermediate-temperature structural forms typical of plutonic and metamorphic rocks. Knowledge of which nonvolcanic feldspar types in polymineralic

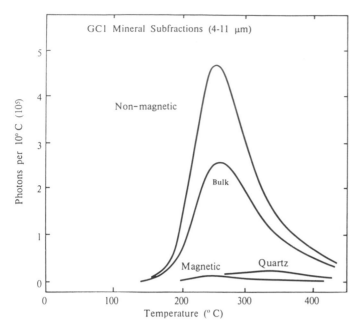

Figure 9. The relative TL intensity (shown as glow curves) for three mineral fractions (nonmagnetic, magnetic, and quartz) from the bulk, polymineralic, 4 to 10-μm grains of a waterlaid silt. The TL is dominated by the feldspars (unspecified plagioclases in this example; after Berger and Huntley, 1982).

samples are most prone to fading and what their TL idiosyncrasies are is essential to the development of objective procedures for circumventing the problem of anomalous fading.

Overcoming the problem

Two general approaches to this perennial problem are available: utilization of nonfading minerals, and exploitation of the TL behavior of the fading minerals. Recent successful TL dating of tephra (discussed below) has exploited the nonfading characteristics of clear airfall glass. In unheated deposits, quartz has long been touted as the best alternative when feldspars exhibit fading; however, the TL of most sedimentary quartz saturates at relatively low doses (Wintle, 1982; Berger, 1984) and is often ineffectively zeroed at deposition. Berger (1984) has detected little or no anomalous fading in the fine-grain magnetic fraction of silts, suggesting that this hard-to-isolate component (see also Fig. 9) might function as another alternative to feldspars in sediments. Generally, limitations presently exist to the use of nonfading minerals in most deposits. Until recently, therefore, samples showing fading were either rejected or only an uninformative "minimum-estimate" TL age could be produced.

Within the last few years, however, techniques have been demonstrated for overcoming the problem of anomalous fading without mineral separations (for some sediments at least). Much more progress in this area is expected in the near future, and recent progress is outlined here.

Lamothe (1984), Lamothe and Huntley (1988), and Berger (1984) have shown that acceptable TL ages can be obtained if TL readout is delayed several weeks or months after laboratory irradiations. This delayed-glow approach can be successful only if most of the fading component has a reasonably short lifetime. Such a delayed-glow approach apparently was not successful for several bulk tephra samples containing feldspars (Berger, 1985a).

Room-temperature storage tests of only a few days duration may fail to reveal the presence of a significant anomalous fading component. The use of different storage durations appears to explain why, in tests on nearly identical loess samples, Wintle and Westgate (1986) did not detect any significant fading component, whereas Berger (1987) found a 25 percent component. Comparisons of this sort are rarely possible, however, because few authors have provided quantitative results of their tests, few have stated how long their storage intervals were, and some have not even mentioned whether or not such fading tests were carried out, despite the clear admonition of Wintle and Huntley (1982).

Evidently, some means of accelerating the decay of the fading component is needed to reduce sample turn-around time. Templer (1985) in a study of zircons extracted from a zircon-rich sand, convincingly demonstrated that this decay can be accelerated by storage for only a few days at suitably chosen elevated temperatures.

Berger (1987) and Forman and others (1987) demonstrated that a similar acceleration of the fading can be produced in polymineralic, feldspar-dominant fine silts. Whereas Berger chose storage at 75°C for 8 days because it left the natural TL signal unaffected, Forman and others chose storage at 150°C for 16 hr, which reduced both artificial and natural TL signals. The efficacy of this latter choice remains to be demonstrated. That is, no evidence exists that the procedure of Forman and others has in fact separated the unstable and stable TL components. An example of the results from Berger (1987) is shown in Figure 10. Here, the use of an elevated temperature has eliminated an unstable component (22 percent of initial TL in irradiated samples) in only 4 to 8 days, compared to several months required with room-temperature storage.

Some independent observations and speculations are relevant here. Akber and Prescott (1985) attempted to relate fading characteristics to feldspar compositions. Specifically, they attempted to classify the TL emission spectra of plagioclase feldspars having different proportions of Na and Ca. A 3-D TL spectrum of annealed Amelia albite (Fig. 11) shows two distinct emission bands (around 375 nm [UV] and 530 nm [yellow]). They stated that the yellow emission progressively decreased as the Ca concentration increased (in a comparison of known feldspars). They also suggested that the more Ca-rich members exhibited stronger fading. This trend implies that if the most intense fading in a mixture of feldspar grains is associated with only the Ca-rich grains, then their TL can be selectively blocked with a suitable yellow filter in conventional TL dating.

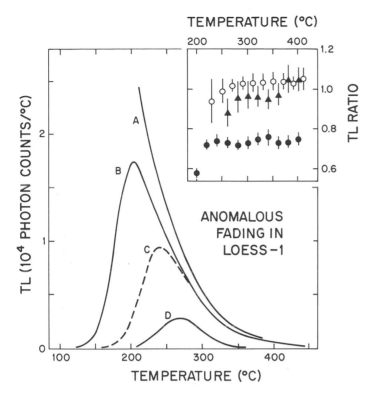

Figure 10. Glow curve plots and ratios thereof (inset), demonstrating the removal of an anomalous fading component (ca. 22 percent of total TL) by two methods: sample storage at room temperature for ca. 6 months (B), and storage at 75°C for 8 days (C). Glow curves measured 1 day after irradiation (2 kGy) (A), and without any irradiation (D), are also shown. D has been subtracted from both the numerator and denominator in the plotted ratios. The ratios shown are: B/A (●), C/B (▲), and heated D/unheated D (○) (after Berger, 1987).

Although attractive and promising, the results and suggestions of Akber and Prescott should not be taken at face value without confirmation, for two reasons: (1) they annealed their samples prior to measurement, and (2) the feldspar-fading connection may be through structural state rather than through composition.

Hasan and others (1986) hypothesize that the high-temperature (i.e., disordered) structural forms of the feldspar family would not exhibit fading. If the degree of fading is related to the structural state, then an association between emission spectra and degree of fading would not be expected (emission spectra reflect elemental impurity differences), unless an indirect association of impurity type and concentration with structural state exists.

Huntley and others (1988b) recently demonstrated the use of a sensitive TL spectrometer for classification of emission spectra from 12 feldspars of known structural and compositional state, as well as spectra from quartz and polymineralic sediments.

Because studying the anomalous fading problem was not one of their objectives, their data are not presently relevant to this problem. Clearly, however, such sensitive spectrometers should be used to seek potential associations between degree of fading, emission spectra, and feldspar character.

To summarize, the hypothesis of Hasan and others (1986) implies that heat treatments are able to remove unstable TL components in polymineralic samples only when a favorable mixture of high- and low-temperature structural forms is present. Whether these structural forms can occur as separate grains is unknown, but because the grains in Berger's (1987) study were small (4 to 11 μm), these forms likely were separate. Such grain distinctions could possibly be identified by advanced petrographic or micro-analytical techniques.

We presently stand only on the edge of a relatively unmapped territory of feldspar TL behavior, not to mention the other potentially useful minerals within sediments. This is not surprising, given that TL is a solid state phenomenon reflecting characteristic defect structures and trace impurity contents, which for minerals, depend on complex thermomechanical histories that are unknown in detail. The TL of even well-studied quartz and silica is but poorly comprehended (McKeever, 1984, 1985). Thus, significant new discoveries related to dating in the TL behavior of complex feldspars seem likely in the next few years.

Mechanisms of fading

The mechanisms of anomalous fading are poorly understood. Two classes of models exist: (1) those invoking leakage of trapped electrons by a quantum-mechanical tunneling to adjacent recombination centers, and (2) the localized transition model. As ordinarily pictured, the probability of an electron recombining with an opposite charge (positive hole) by tunneling through the energy barrier separating the two is independent of temperature. Such athermal fading has been observed for volcanic feldspars (Wintle, 1977a) and for calcite (Visocekas and others, 1976; Visocekas, 1979), but only at temperatures below ca. −20°C. At higher temperatures, temperature-dependent anomalous fading has been reported for labradorite (Wintle, 1977a), zircon (Templer, 1985), and calcite (Visocekas and others, 1976). Clearly, if tunneling is involved, it has a temperature dependence at ordinary temperatures.

The localized transition model (Chen and Kirsch, 1981) has been proposed (Templer, 1986) to account for this temperature dependence. In this case, electrons hop over the energy barrier (Fig. 12), but instead of entering the conduction band (as in path a, Fig. 12) or recombining directly with a nearby luminescence center (as in path b), they recombine with such a center via a shared excited energy state. Such an excited state is not shown in Figure 12 but would occur just above the top of the energy barrier, and distinctly below the conduction band. If the electron traps are not local to the recombination centers, but rather are spatially remote (e.g., more than a few hundred angstroms) from the nearest luminescence centers, then recombination can occur

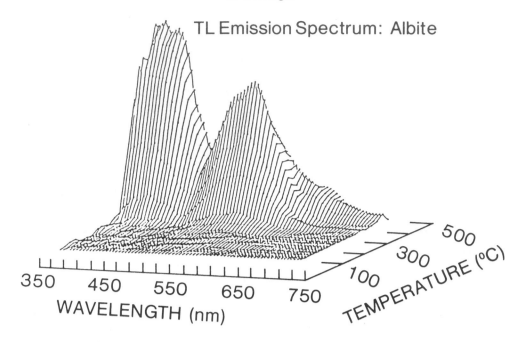

Figure 11. A 3-D TL spectrum for 100-μm grains of albite from Amelia County, Virginia, annealed at 500°C prior to a β dose of 0.2 grays. This spectrum, obtained by a Fourier-transform spectrometer, shows two distinct emission bands at ca. 375 nm (UV) and at ca. 530 nm (yellow). The heating rate was 5°C/sec (after Akber and Prescott, 1985).

only via the conduction band (path a). That is, for such nonlocalized pairs the electron traps are "stable" (i.e., subject to only the usual kinetics rules). Templer (1986) has investigated this localized transition model and demonstrated that it accounts for the characteristics of anomalous fading in a zircon sample. The tunneling model of Visocekas and others (1976, see below) did not satisfy the observed data as well.

In the two variations of the tunneling model to be outlined here, the probability of tunneling by the electron is inversely proportional to the area of the energy barrier above the energy level of the electron. Therefore, the higher the energy of the electron, or the lower and narrower the energy barrier, then the greater is the probability of the electron tunneling through the barrier (Fig. 12).

The model of Visocekas and others (1976) envisages a thermally assisted tunneling that competes with athermal transitions at low temperature shortly after irradiation, but which dominates the athermal process at longer times and at higher temperatures. In this model, the energy barrier separating the opposite charges is fixed in height and shape, and the electron can be nudged by thermal agitation to a higher energy (but less than the barrier height) sufficient to increase the probability of tunneling. One difficulty with Visocekas' 1976 model is that the predicted logarithmic dependence of fading with time leads to infinitely large fading at very large storage times, although a plausible corrective argument is given by Visocekas (1985).

An alternative picture of the process of tunneling has been suggested by Huntley (1985b), borrowing from work in chemical kinetics and solid state physics (McKinnon and Hurd, 1983; Hurd, 1985). He pointed out that although the mean width of the energy barrier itself is not temperature dependent, the width of the barrier vibrates due to thermal oscillations of the atomic sites. These vibrations of the barrier give rise to a change in the probability of tunneling, and because the amplitude of the vibrations is temperature dependent, this vibrating barrier model admits a temperature dependence for the escape of susceptible electrons from traps.

The observed acceleration of fading in certain materials with light perhaps complicates the phenomenology of anomalous fading. Hess and others (1984) observed optically assisted fading in a gallium-doped silicate. They drained elections from shallow traps (associated with thermally assisted fading below –228°C) using infra-red (900 nm) light. Templer (1985) demonstrated that >550 nm light removed a fading component from zircons in a few days. However, such long-wavelength light can separate the unstable from the stable TL in zircons only because the stable (natural) TL of zircons is apparently unaffected by light >550 nm. This is not the case with feldspars (Fig. 7).

All of these observations provide a rich panorama of intriguing physical problems related to anomalous fading. The very complexity of luminescence phenomena enables a wide variety of tests to be applied to the problem of modeling anomalous fading. In one approach, Templer (1986) measured the phosphorescence due to anomlous fading in a zircon at a storage tempera-

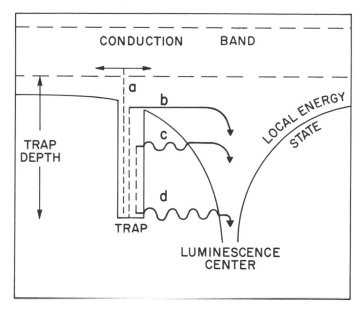

Figure 12. Possible escape paths of an electron from a trap: thermal escape (a, b), thermally assisted tunnelling (c), athermal tunnelling (d) (modified from Aitken, 1985). "Trap depth" simply means thermal activation energy.

ture of 100°C, and showed that the localized transition model satisfies the data well. Experiments could also be performed on well-characterized fading zircons using light of various wavelengths with the sample held at various elevated temperatures. Clearly, tests of diverse types are necessary to distinguish among the different models. Because of its predictive power, an acceptable model would benefit TL dating methodology.

APPLICATIONS TO CLASS I EVENTS

Biogenic and chemically precipitated minerals

Wintle and Huntley (1980) reported that biogenic silica (mainly radiolaria) appears to emit a broad TL signal at about 370°C (3°C/s) of the glow curve, but that this signal was much smaller and occurred at a higher temperature than the signal for the bulk samples of their marine sediments. They inferred that the bulk signal was largely from detrital minerals, and subsequent work (e.g., Fig. 9) has shown that feldspars can be expected to dominate the TL from detrital minerals. No further research on the TL from biogenic silica has been reported, although this could seemingly be a useful TL clock in certain circumstances. Unpublished data of Berger show that freshwater diatoms have negligible natural TL and negligible sensitivity to beta irradiations.

Biogenic and precipitated carbonates, on the other hand, have received much more attention. Little work on biogenic carbonates has been reported since that of Bothner and Johnson

(1969), but studies of the TL properties of precipitated carbonates, such as speleothems and travertines, by Wintle (1978) and Debenham (1983) suggest that the lifetimes of electrons in some traps should be as long as 1 m.y. at ambient temperatures, and that anomalous fading is not common. In this context, the effort of De and others (1984) to measure TL ages for pre-Pleistocene oceanic carbonates would appear to have been futile, although it does not detract from the potential usefulness of the foram (and carbonate nannofossil) clock in sediments younger than 1 Ma.

Several significant factors that have to be considered in any application of TL to marine biogenic carbonate have been outlined by Wintle and Huntley (1983): principally, (1) correct evaluation of the dose rate to include contributions from excess Th-230, (2) proper laboratory handling to avoid exposure to sunlight, and (3) consideration of the thermal lifetimes of relevant electron traps.

Therefore, forams and coccoliths seem to have the potential to provide new TL clocks for marine sediments younger than 1 Ma. The most useful role for such an absolute clock will likely be in the dating of deep-sea cores having intermittent, rather than continuous, deposits of such carbonates. For cores having a continuous record of carbonate, the stratigraphy can often be constructed from oxygen isotope measurements. On the other hand, oxygen isotope measurements may be of little use for cores with intermittent deposits of forams, such as are recovered from the Arctic Ocean and adjacent waters.

Unlike most marine sediments, Quaternary deposits on land are inherently discontinuous, and thus often difficult to correlate from region to region without a direct quantitative chronometer. Hence, the following discussion concentrates almost exclusively on applications of TL to terrestrial material.

Some buried soils contain grains representing both class I and class II events, namely, precipitated carbonates and detrital silicates respectively. Because the latter are ubiquitous and because most TL studies have concentrated on application of class II methods to the detrital grains, attempts to date buried soils are discussed in more detail in the next section. In this section, an attempt by May and Machette (1984) to use pedogenic carbonate as a TL clock is discussed briefly. This example provides a good illustration of a violation of the closed-system assumption (dose-rate only), and of a system for which a single zeroing point is not defined (sequential generations of TL clocks are intermixed).

Such carbonate occurs in soils developed in arid to semiarid climates, and most of it accumulates slowly (present surface soils have been developing for 20,000 to 200,000 yr). Accumulation is principally by dissolution of soluble calcium carbonate at or near the soil surface and by downward transport by water to lower levels where it is precipitated. From the TL dating perspective, this carbonate has four attractive qualities: (1) growth of TL from zero at the time of crystallization, (2) absence of anomalous fading in calcium carbonate in general, (3) high sensitivity to ionizing radiation, and (4) an expected mean lifetime of electron traps of ≈1 m.y. The serious disadvantage of such soil carbonate is its quasi-open-system behavior, with solution and redeposition occurring over a long time and leading to both a nebulous zero point and a time-varying dose rate (U is soluble) within a given horizon.

Aware of these advantages and limitations, May and Machette analyzed several samples from New Mexico having mean ages of 6 to 500 ka. They concluded that the TL intensity (normalized to a rough estimate of dose rate) increased with age and suggested that relative age estimates could be obtained with their technique. That is perhaps optimistic. The principal limitation to their data was their TL measurement technique. Their methods have three shortcomings: (1) a mixture of carbonate, feldspar, and quartz was analyzed, (2) to compare samples of different ages they calculated the ratio of TL intensity from an unirradiated sample to that from the same sample after heating and re-irradiating, and (3) for this ratio, the TL at 300°C from the unirradiated sample was compared to that at 150°C in the regenerated signal (heating rate 20°C/s). These procedures are inadequate because: (1) the TL intensity at these temperatures very likely represents contributions from more than one mineral (feldspars and carbonate can have more than one glow-curve peak, and are brighter than quartz), and (2) changes in opacity and TL sensitivity in the reheated "carbonates" were not considered. For pedogenic carbonate, the only appropriate TL method is the additive dose technique (Fig. 3) applied to a mono-mineralic sample.

Tephra

A more interesting application of TL within class I is tephra, because ash layers often provide pronounced time-stratigraphic marker horizons. The time window of ca. 40,000 to 500,000 yr, between the normal ranges of radiocarbon, and fission-track and K-Ar methods (e.g., Westgate and Briggs, 1980; Naeser and others, 1981) needs a precise, accurate chronometer of tephra deposition.

This need motivated the study of Berger and Huntley (1983) and Berger (1985a) of several well-characterized tephra deposits from North America. Figure 13 illustrates the use of the additive dose technique to obtain a plateau in equivalent dose values, yielding a TL age of 7,800 yr for the well-known Mazama ash in accord with associated (corrected) radiocarbon ages of 7,400 to 8,000 yr. Although the TL growth curves for most younger tephra are nearly linear, growth curves for older samples indicate (as expected) the onset of saturation of electron traps, one of the main factors (besides thermal instability) determining a practical upper age limit for this technique.

Berger (1985a) demonstrated that the main obstacle presently limiting the usefulness of this new method of tephrochronology is the common occurrence of significant anomalous fading in bulk fine-grained samples. Because this fading was not removed by storage tests several months long, he suggested that isolation of fine-grained glass from feldspars (present mostly as detrital contamination, although some primary volcanic feldspars might also be expected) is necessary to obtain reasonable age accuracy. Fol-

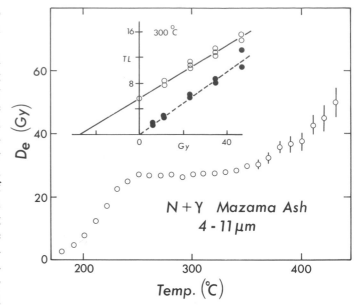

Figure 13. The use of the additive dose technique (open circles, inset) to obtain equivalent dose values for 4 to 11-μm grains from the ca. 7,500-yr-old Mazama ash. In the inset the solid circles represent a TL buildup curve constructed after unirradiated subsamples were heated to ca. 450°C. The different slopes indicate a change in TL sensitivity caused by this laboratory heating, so that a regeneration technique cannot be used to measure equivalent dose values for airfall glass (this unseparated sample is ca. 100 percent glass) (after Berger and Huntley, 1983).

lowing up this suggestion, Berger (1987) demonstrated that separation of glass apparently has eliminated the anomalous fading component from a sample of Old Crow tephra in Alaska. Growth curves for the glass-rich fraction were quite sublinear (Fig. 14), reflecting the older age (ca. 110 ka) of this sample compared to the ash in Figure 13. Support for this TL age is provided by a concordant result of 108 ka from contiguous loess, using the partial bleach technique (Fig. 15).

The principal conclusion from these tephra studies is that clear 4 to 10-μm-sized airfall glass provides a new and well-behaved TL clock, but that the presence of feldspars in the bulk material will frequently lead to underestimation of the deposition age because of associated anomalous fading. Separation of clear glass from crystals by heavy liquid procedures can eliminate such anomalous fading. An additional and important benefit from glass separation is the removal of any relict TL that would be borne by detrital crystals. The difficulties encountered by Berger (1985a) and Wintle (1977a) in removing anomalous fading by long-term storage or heat treatment suggest that the only effective treatment for tephra is physical separation of glass.

Shocked material

In a novel application of TL dating techniques, Sutton (1985b) obtained a TL age of 49±3 ka for the shock-

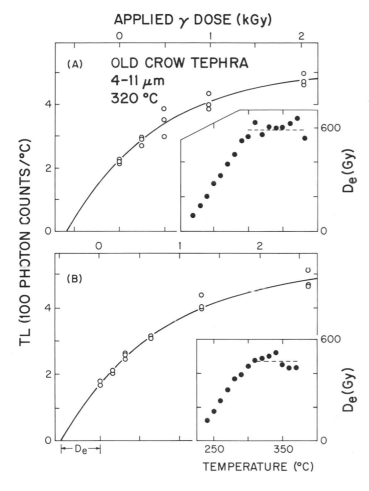

Figure 14. Two separate measurements of equivalent dose values (D_e) for a glass-rich sample of the Old Crow tephra. Saturating exponential regressions have been used. The dashed lines (insets) show the inferred plateaus used to calculate an average TL age of 110 ±12 ka (after Berger, 1987).

metamorphosed quartz from sandstones and dolomites associated with Meteor Crater, Arizona. This is much older than the 25 ka estimate previously inferred for this impact event. The equivalent dose values used to calculate the TL age were measured for grains corresponding to petrographic shock classes >3, for which the associated heating was sufficient to reset the quartz TL clock.

Heated quartz

Aitken (1985) has adequately reviewed TL applications up to 1985. Two additional applications of TL dating to geological quartz are summarized here. The first example illustrates the hazard of applying TL dating to deposits for which the dose-rate closed-system assumption has likely been violated, even though the zeroing assumption is probably valid. The second example is a novel application of TL to a complex geological system, hydrogeothermal areas.

In attempting to date archeological sites in New Guinea, Groube and others (1986) recovered 2 to 10-μm (reviewer's inference) quartz from heavily altered and reworked (clay-like) tephra deposits containing artifacts. The TL behavior of these quartz extracts was ideal (no fading, linear dose-response curves, broad plateaus), but the derived TL ages are uncertain by at least about 25 to 30 percent (ca. 45 ± 15 ka). Unlike most examples discussed in this review, here the dose rate dominates the uncertainty. The main problem was that much of the K had apparently been leached from the tephra beds. These authors reported only TL ages calculated with the assumption of zero water content. Hence, the true ages of these tephra layers are probably greater than 40 to 50 ka, unless the layers contain significant translocated, U-bearing, clay-size particles, a possibility not addressed by Groube and co-workers.

Based on development work by Takashima (1979), Takashima (1985) and Takashima and Honda (1986) measured TL ages for quartz phenocrysts from volcanic rocks that were altered during hydro-geothermal activity. By plotting TL intensities (actually, the area under the glow curve peak) against associated fission-track ages, Takashima (1979) showed that hydrothermal activity can reset the quartz TL clock. In a later study, Takashima (1985) described his procedures in detail, and discussed results from quartz phenocrysts and secondary quartz from altered volcanic rocks, as well as from mixtures of alteration minerals. He concluded that residual quartz phenocrysts are preferred for dating alteration events. From their examination of quartz phenocrysts from altered rhyodacitic rocks, Takashima and Honda (1986) reported TL ages ranging from 57 ka to 1.8 Ma, with an uncertainty of about 30 percent.

Three weaknesses are apparent in these studies: (1) most samples were heated in air for TL readout, (2) an artificial TL signal was induced in some samples by grinding the rocks to extract quartz (that a correction was made for this induced signal is not clear), and (3) to estimate the sensitivity of the samples to laboratory ionizing radiations, only two points on a TL growth curve were used. This procedure is valid, however, only if the growth curve is known to be linear. For these reasons, the results must be considered preliminary.

In summary, over the time range up to perhaps 1 Ma, several useful TL clocks exist for dating class I events. Airfall volcanic glass from both proximal and distal tephra deposits is among the most promising of these clocks. Further investigation of the effects of hydro-geothermal alteration on the TL of quartz is warranted.

APPLICATIONS TO CLASS II EVENTS

The growth in applications to this category of events has accelerated dramatically since the review of Wintle and Huntley (1982), although the early apparent success with strongly insolated loess (Wintle, 1981) was not duplicated for waterlaid

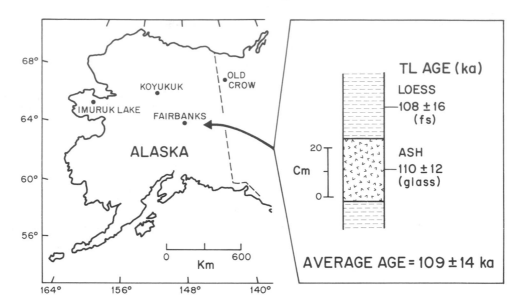

Figure 15. The results of both an indirect (loess sample) and direct (glass-rich sample) TL dating of the Old Crow tephra from the Fairbanks, Alaska, area. The source of this widespread tephra probably lies in the eastern Aleutian Islands (e.g., see Miller and Smith, 1987).

deposits until recently. The main reasons for this are summarized in the section above on zeroing.

For convenience of discussion, applications are grouped under the headings eolian sediments, waterlaid sediments, and buried soils and peats. The TL in the noncarbonate minerals within the first two groups is reset by sunlight during transport, whereas for soils and peats, exposure to light may occur during transport (by wind or water) as well as during bioturbation (probably mainly so for soils). The possibility that the TL of noncarbonate minerals within some soils is reset by the cumulative effects of natural fires has not been investigated, although Jungner (1985) reported that small pieces of charcoal were observed during collection of one Finnish dune deposit. More attention should be paid to the possible effects of natural fires.

Eolian sediments

For loess older than ca. 40 ka, controversy within the TL dating community presently exists over the appropriate choice of methods (Fig. 4) to be used for measuring equivalent doses, even in the "simple", fine-grained, loessic feldspars. The root problem is that few comparisons have been made between TL age estimates and sound, independent, absolute deposition ages for loess. Most TL workers have applied only the regeneration method of Figure 4B to older loess and have been content merely to report that their TL age estimates are stratigraphically self-consistent (e.g., Bronger and others, 1987; Singhvi and others, 1986a, 1987). Although this approach provides numerical age estimates where none existed before, it does not contribute to the develop-

ment of TL dating methods. The accuracy of such results is difficult to assess until further checks on the methods are made, using known-age samples older than 100 ka.

Several methodological factors need to be isolated to attain a clear understanding of this controversy. Most of these factors have been emphasized by Wintle and Huntley (1982) as prerequisites to a reliable TL date, yet many are still ignored, or assessed with inadequate rigor. (1) Anomalous fading must be demonstrated to be insignificant or to have been removed. Relevant tests are often not described. (2) Adequate regression analysis and extrapolation techniques must be used. A misconception exists that the partial bleach method is not valid when TL growth curves are sublinear (Aitken, 1985, p. 139; Wintle and others, 1984; Lu Yanchou and others, 1987b). (3) Adequate tests for sensitivity changes due to optical bleaching must be demonstrated for the regeneration technique. (4) If the simple regeneration technique of Figure 4 is used, its result must be compared to that from careful partial bleach experiments to assess any possible overbleaching effect in the former, as well as any sensitivity effect. (5) All other requirements for a closed system must be met.

Before discussing specific TL dating applications, a quick look at Figure 16 is helpful. Published TL ages (as of early 1987) for both coarse grains and fine grains are plotted against associated ^{14}C ages. Results obtained by the three methods in Figure 4 are distinguished, and some results from particular studies are emphasized (circled dots, triangles, and circled x's). All of the results indicated by circled dots and triangles represent large grains of quartz or feldspar from dune deposits, as do some of those represented by open circles, and are discussed below. Al-

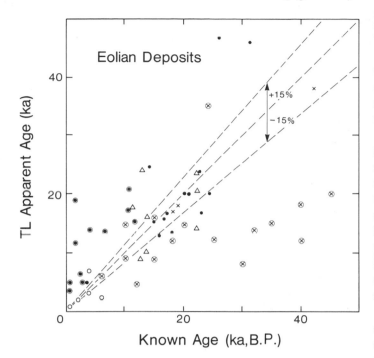

Figure 16. Published TL ages versus known ages (based almost entirely on uncorrected C-14 dates, shown only if the uncertainty in the latter is <10 percent) for loess and dune deposits. Analyses for both fine (e.g., 5 to 10 μm) and coarse (ca. 100 μm) grains are represented. Four methods of measuring the equivalent dose are indicated: (·) partial bleach, (O) total bleach, (x) regeneration, and (Δ) partial bleach (Mejdahl version). The results of Bluszcz and Pazdur (1985) (⊙) and of Debenham (1985) (⊗) are distinguished from those of other workers. Clearly, differences in TL methods and practitioners can greatly affect the results.

most all of the results denoted by dots, x's, and circled x's represent fine grains of loessic feldspar.

Fine grains. In Figure 16, reliable results for loess younger than 50 ka were obtained most often with the partial bleach technique. If the results of Debenham are excluded, the regeneration technique has also yielded accurate ages. What then distinguishes Debenham's results? The answer is related to the controversy of appropriate choice of technique mentioned above.

The basic controversy was created when Debenham (1985) applied the regeneration technique to European loess believed to span the age range from about 5 ka to ca. 700 ka. Both Debenham (1985) and Wintle (1985a) (with a smaller number of samples) were unable to obtain TL ages greater than about 120 ka, and invoked a physical phenomenon (instability of luminescence centers) to explain the results. The effect of this putative phenomenon would be that all TL ages from loess (regardless of technique) would be progressively too low, up to a usual limit of 100 to 120 ka. Ages under 50 ka, however, should be accurate within 5 to 10 percent. Oddly, Debenham (1985) did not comment on the inconsistency with this argument that was presented by his own data for young loess (Fig. 16). Since then, other

equally plausible explanations have been provided for this "effect."

Berger (1986) and Johnson and others (1985) suggested that this underestimation of deposition ages is largely a consequence of failure to correct for the well-known, common occurrence of sensitivity changes observed with the regeneration technique (Wintle and Huntley, 1980; Rendell and others, 1983; Berger, 1984; Wintle, 1985b). Indeed, Wintle (1985b) has shown that such corrections would be difficult to apply accurately when the relevant dose-response curves are sublinear (as they usually are) for samples older than 50 to 100 ka. In addition, the reliability of the quoted geologic ages of some of Debenham's older samples is questionable (Mejdahl, 1986). Furthermore, several workers have subsequently obtained TL regeneration ages for other samples up to 300 ka (Singhvi and others, 1986b, 1987) (without, however, correcting for anomalous fading or for sensitivity changes). Additionally, anomalous fading may have affected some of Debenham's results.

The approach used by various workers to detect sensitivity changes in the regeneration method is worthy of comment. Debenham (1985), Singhvi and others (1987), and Canfield (1985) compared the slope of a tangent to the sublinear regeneration curve with the tangent to an (incomplete) additive-dose growth curve. Such a procedure is highly subjective and therefore likely to be uninformative. For example, these workers observed variations of 20 to 30 percent in slopes, but stated that such variations are not significant. This is strange because for most samples older than about 100 ka, a slope variation of as little as 10 percent in the interpolated TL can introduce a dramatic shift in the equivalent dose estimates if no correction is applied.

To summarize, Occam's Razor has not been applied to this controversy. That is, the simplest interpretation of Debenham's (1985) results is that ambiguities with the methodology, not the physics, are apparently the problem. Clearly, a comparison of the partial bleach and regeneration techniques using loess having soundly determined, independent deposition ages older than 100 to 200 ka (as suggested by Berger, 1986) is needed. Although rare, such samples exist. Until this is accomplished, any TL ages obtained for loess older than 50 to 80 ka, using the regeneration technique of Figure 4B, should be considered only relative ages (e.g., Wintle and others, 1984; Singhvi and others, 1986b, 1987; Bronger and others, 1987).

Lu Yanchou and others (1987b) have compared equivalent-dose values from the total bleach, partial bleach, and simple regeneration methods for quartz-rich fine grains from two Malan Loess samples (Z-03 and Z-06), although there was no reliable independent age estimate. For the younger Z-06 sample (probably younger than <35 ka) the three techniques gave concordant D_e values, whereas for the older Z-03 sample (probably 50–70 ka) the results from the total and partial bleach methods were 50 percent greater than the D_e values from the regeneration method. These authors therefore deduced that only the regeneration method would yield accurate results for loess older than about 40 to 50 ka. This deduction must be treated with a healthy skepticism because only linear extrapolations were used for these total and partial bleach data (not shown by the authors). As discussed above, linear extrapolations will overestimate the correct D_e values for most samples older than 30 to 40 ka, so the observed discordancy for sample Z-03 is hardly surprising.

Because of this use of the regeneration method for samples putatively older than 50 ka, these authors' particular TL age estimates (up to 85 ka) are subject to the same doubts as those of other workers mentioned above (no tests for sensitivity changes were mentioned). A subsequent dating of the same loess samples using this regeneration method applied to the 90 to 125-μm quartz grains (Lu Yanchou and others, 1987a) yielded concordant results and will be discussed below under the heading "Large grains." This concordancy between results obtained from fine and coarse grains is not unexpected because both size fractions were dominantly quartz.

Rather more detailed comparison of results from the total bleach, partial bleach, and regeneration methods have already been made for waterlaid silts (Berger and others, 1984; Berger, 1984, 1985b) (see below). Only the partial bleach technique proved to be generally accurate, although, occasionally the others also produced expected equivalent dose values. The main limitation of the total bleach method was its concomitant overbleaching, while the regeneration technique suffered both from this effect and variable sensitivity changes.

In the context of the above remarks, one example of extremely questionable results is that of Buraczynski and Butrym (1984), who obtained TL age estimates from 28 to 280 ka using the ca. 50-μm-sized grains from loess at Achenheim, France. Problems with these results are: (1) no sound independent ages are cited, (2) they did not correct for any relict or residual TL component, (3) their dose-rate calculations are questionable, and (4) few details of their methods are provided.

On the other hand, little reason exists to question several results less than 40 ka obtained for unknown-age samples with the partial bleach technique (Wintle, 1981, 1982; Wintle and Brunnacker, 1982; Wintle and others, 1984; Wintle and Catt, 1985a). The general reliability of results for loess younger than ca. 40 ka has been demonstrated by Norton and Bradford (1985), who applied the partial bleach method to loess having associated ^{14}C dates (these are plotted in Fig. 16). Their few "incorrect" results may be the product of failure to correct for anomalous fading (tests were not described), use of large linear extrapolations, or overbleaching effects.

TL dating of loess from Fairbanks, Alaska, has recently provided an illustration of how different workers can obtain discrepant results, even when applying the same, or similar, techniques to identical samples. Wintle and Westgate (1986) applied the total bleach method (Fig. 4a) to four samples of loess straddling the Old Crow tephra. They also applied the partial bleach method to one of these samples (71H) and obtained the same equivalent dose, thus demonstrating no intra-laboratory difference. However, Berger (1987) applied the partial bleach technique to a similar sample and obtained a TL age of about 110 ka

(Fig. 15), compared to about 85 ka by Wintle and Westgate (1986). The difference, according to Berger (1987), can be attributed to the counteracting effects in the work of Wintle and Westgate of the use of linear extrapolations and of uncorrected anomalous fading (effects pointed out by Berger, 1984). Additional comparisons of this type are needed.

In summary, an unambiguous demonstration of TL dating capability and accuracy for loess older than about 100 ka is needed. The kinetic studies of Stickertsson (1985) on a potassium feldspar sample predict a mean lifetime of 10 m.y. for electron traps corresponding to the 280°C glow curve peak, implying that other nonvolcanic feldspars will have similar thermally stable traps. However, preliminary experiments by Mejdahl (1986) suggest that the thermal stability of electron traps is lower than this and that the effective upper age limit for TL dating of loess may be only about 500 ka. This important inference needs confirmation. Certainly, the absence of any significant anomalous fading effect in the samples of Mejdahl must first be clearly demonstrated before this suggestion can be taken at face value.

Another approach to dating fine-grained eolian sediments is to utilize the quartz component, its main attractions being the near absence of anomalous fading and the long mean lifetime of electron traps (ca. 10 m.y. for the 325°C peak; Wintle, 1977b). Fine-grained quartz, however, has generally been avoided because of the onset of nonlinearity in TL growth curves at relatively low applied doses (e.g., <200 Gy) compared to feldspars (e.g., Wintle, 1982; Berger, 1984, 1985b), as well as a greater resistance to optical bleaching than is shown by feldspars (Fig. 7).

The easiest way to avoid any quartz TL in polymineralic samples is to select samples having a large fraction of feldspars. Because feldspars as a group emit about 20 to 50 times more TL per gram than quartz (e.g., Fig. 9; Berger, 1984), this approach may be adequate. However, since the glow curve peak of quartz occurs at higher glow curve temperatures than for some feldspars, and because the most useful feldspar TL for dating also often occurs at high glow curve temperatures (e.g., >300°C), then in some samples the unwanted quartz TL may represent a significant component.

To cope with such samples, Debenham and Walton (1983) proposed the routine use of a UV-pass optical filter (e.g., Schott UG11) during TL readout. This suggestion was based on the spectrometric observation of a reduced TL emission at <400 nm for **one** quartz sample compared to some feldspars. Thus, such a filter would hopefully suppress any quartz TL relative to that of the feldspars. However, in an examination of quartz extracts from three sediments from Canada, Berger (1985c) found that the UG11 filter did not reduce the quartz TL. Presently, little is known about the variety of quartz emission spectra likely to be encountered in sediments (Huntley and others, 1988b). Indeed, a mixture of different populations of quartz TL emitters (e.g., red and blue emitters) can be expected in some samples (Hashimoto and others, 1986). Insofar as some quartz is known to emit significant TL at wavelengths below 400 nm (McKeever, 1984, 1985), the use of a filter such as UG11 should not be assumed to be effective in rejecting all quartz TL emissions. Caution must be used in interpreting glow curves from fine-grain sediments containing more than a little quartz.

Large grains. For some deposits, such as beach and dune sands, quartz offers several advantages. It is usually much more abundant than feldspars (and other minerals), primarily because of its greater resistance to weathering over geologic time. An additional advantage of large-grained quartz is the simplified dosimetry—quartz often has negligible internal radioactivity. Also, ca. 100-μm-sized grains of quartz can be separated and visibly inspected with relative ease. Furthermore, the expected lower dose rate in such deposits compared to other sediments should ensure that such quartz can be used for dating well beyond the 20 to 30-ka limit placed on quartz extracted from some fine-grained feldspar- and clay-rich sediments (Berger, 1984, 1985b). At present, the practicable upper age limit for beach and dune quartz (limited by saturation effects) has not been determined. Applications to beach sands will be mentioned in the section on waterlaid sediments (below).

Prescott (1983) and Readhead (1984) attempted to date dune deposits with ca. 100-μm quartz, and concluded that estimating the residual intensity at deposition from equivalent zero-age samples was necessary. However, as discussed by Huntley (1985a), an alternative and often preferable procedure is to use the partial bleach technique with short laboratory light exposures.

In an attempt to date a series of beach dunes from ca. 3 to 11 ka, Jungner (1985) applied both the partial bleach technique and the total bleach method to both quartz and feldspar 100 to 300-μm grains. Although he did not block the UV component from his laboratory light source, or discuss anomalous fading, his results do support the conclusions of Berger (1984) and Huntley and others (1985a) (see the above discussion on zeroing). For both quartz and feldspar, the total bleach method generally produced larger apparent ages than the partial bleach technique, indicating overbleaching with the former. With the partial bleach technique, 70 percent of the feldspars produced agreement with the expected ages, whereas only 20 percent of the quartz samples produced such agreement—80 percent consistently overestimated the ages (Fig. 16). This suggests that overbleaching of the quartz has occurred by use of inappropriate wavelengths of light (see Fig. 7).

Bluszcz and Pazdur (1985) have applied a variation (only two dose points were used) of the partial bleach technique to the 100 to 150-μm quartz extracted from eolian dune deposits in Poland. They calculated TL ages for 11 samples that can be compared with ^{14}C dates, using an exposure time of 24 hr with an unfiltered Hg-based lamp. All but one or two of the TL ages were much greater than the known ages (Fig. 16). Again, these results suggest that inappropriate illumination times and wavelengths were chosen.

Mejdahl (1985) has proposed a new partial bleach method for 100 to 300-μm quartz and feldspar grains deposited less than 25,000 yr ago. This quartz-feldspar method attempts to exploit the different rate of TL reduction in quartz and feldspar grains exposed to sunlight, and the different ionizing radiation dose rates for each mineral because of their different radioelement concentrations. In this method, unirradiated samples of quartz and feldspar are exposed to sunlight in air for different times. An equivalent dose value for each of these minerals and times is estimated by subtracting different baselines of TL intensity from the TL buildup curve, in a fashion analogous to that shown in Figure 4a for the total bleach method. The correct sun exposure is assumed to be that which produces a D_e (feldspar)/D_e (quartz) ratio equal to the dose-rate ratio. This dose-rate ratio, measured separately by radioactivity analysis, is greater than unity because of the self dose to feldspar from its high potassium concentration.

Mejdahl (1985) presented preliminary results for nine samples from eolian dune sands. The scatter of these results about the concordance line in Figure 16 suggests that this method has unresolved problems. In a later novel application of this method to sandy frost-wedge casts from western Denmark, Kolstrup and Mejdahl (1986) obtained TL ages for the 100 to 300-μm quartz and feldspar grains of 39 ± 5, 24 ± 3, and 17 ± 3 ka, compared to expected ages of 20 ± 5 ka. Although promising, this technique has several shortcomings (see below) in addition to its practicable complexity.

Although Mejdahl (1985) acknowledged the present applicability of his method only to eolian deposits, in view of the above discussion on zeroing and the rather different wavelength dependency of the TL reduction in quartz and feldspar (Fig. 7), this quartz-feldspar method cannot be expected to yield correct results, even for most eolian material, if the interiors of grains are screened from a portion of the daylight spectrum, and if the experimenter does not allow for such screening. With our presently limited understanding of these matters, only the partial bleach method of Figure 4a seems likely to be generally reliable; it is certainly easier to use than this quartz-feldspar method.

Singhvi and others (1986a) have recently reported the first TL dating results for coastal dune sands in Sri Lanka. They applied the total bleach method (Fig. 4a) (and to one sample, the regeneration method) to ca. 100-μm quartz and obtained TL ages ranging from 23 to 74 ka. However, these results must be considered preliminary in view of the above discussion of applications to dune sands. Firm, independent age checks are needed, extrapolation of growth curves should be kept small (and should be shown), and a constant dose rate must be probable. The partial bleach method should also be used. In this particular application, radioactivity measurements on both reddened and cleaned sands would have been informative. For example, Southgate (1985) observed a significant reduction in alpha particle count rates after the removal of iron staining from one dune-sand sample. Nevertheless, the study of Singhvi and others is important because it demonstrates the feasibility of dating red sand beds by TL, and it points out some of the problems to be considered.

A TL dating study of dune sands from central Sweden has been reported by Lundqvist and Mejdahl (1987). They applied the total bleach method (with linear extrapolations, although no data were shown) to 100 to 300-μm grains of potassium feldspar extracted from both dune sands and associated eolian silt. The TL age estimates of about 9,000 yr for six of the eight samples were somewhat younger than the exepcted ages (based on corrected radiocarbon and clay-varve dates). The authors suggest that uncorrected anomalous fading (tests were made) was responsible for the too-young ages. One of the two remaining samples gave a TL age very close to the expected 500 yr, while the other TL result of 11 ka was significantly greater than the expected 9,300 yr. These results are quite encouraging, and several improvements in techniques could produce more reliable TL ages. For example, the authors recognize that anomalous fading effects could perhaps be eliminated by use of elevated storage temperatures (e.g., Berger, 1987). The use of the partial bleach technique with sublinear extrapolations also ought to be attempted. The single too-old result (11 ka) may be attributable to an overbleaching effect with the total bleach method, or to use of linear extrapolation.

TL age estimates for dune sands from Australia have been obtained by Gardner and others (1987) using a different TL method. They applied the modified regeneration method of Readhead (1984) to 90 to 125μm grains of quartz. With this approach, the relict TL intensity is assumed to equal the TL in modern surface samples; in this way they hoped to get around the overbleaching effect common with the regeneration method in Figure 4b. This modified regeneration method needs comparative testing with careful partial bleach experiments, and also with the original regeneration technique of Wintle and Huntley (1980), which is potentially capable of avoiding the effects of sensitivity changes. In spite of these caveats and the lack of strong, independent age estimates (discussed by Gardner and colleagues), these results are encouraging. TL age estimates up to about 250 ka suggest the existence of an old dune-building phase in this region (Strezelecki Dunefield). An important methodological aspect of this study is the demonstration (by use of gamma spectrometry) of the absence of any decay-series disequilibrium in these samples.

The modified regeneration method of Readhead has also been applied to the 90 to 125-μm quartz grains from loess near Beijing, China (Lu Yanchou and others, 1987a). Unlike the complementary study of fine grains from the same samples (Lu Yanchou and others, 1987b), here checks for sensitivity changes were made, and one sample (Z-05) was consequently rejected. The resulting TL age estimates of 20 to 85 ka agree with the fine-grain results (not surprisingly, because the fine grains were also mainly quartz). Notwithstanding the need for confirmation by other TL techniques and other dating methods, these results are encouraging.

Lu Yanchou and others (1987a) draw two significant inferences from this study. First, the apparent saturation of the TL of this quartz at a beta dose of 500 to 600 Gy suggests that (with a natural dose rate of 3 to 4 Gy/ka) this regeneration method is applicable to such quartz as old as 150 to 200 ka. Second, the 85 ka result for the top of paleosol S_1 suggests that this is the approximate ending time of the last interglaciation in north China. Although this is a reasonable deduction, the authors' citation of a

similar (85 ka) TL age estimate from Normandy (Wintle and others, 1984) as support for their deduction is curious because Wintle (1985b) has considered her own results to be too young.

In summary, several independent studies on different continents suggest that dune sands are good candidates for TL dating. Nevertheless, some improvements in TL procedures (suggested above) are still needed, as are additional comparisons with independent dating techniques. Coarse-grain quartz from loess (if available) also appears to be a good TL sediment clock, provided either that the partial bleach method is used or the modified regeneration method of Readhead can be justified. Certainly the original regeneration technique of Wintle and Huntley (1980; not discussed explicitly in this review, but see also Berger, 1984) also needs to be applied to eolian sediments because of its potential capability for sidestepping the effects of sensitivity changes that plague the simpler regeneration methods (e.g., Fig. 4b). Some thought should also be given to the likelihood that any coarse quartz from loess has a different transport history than the fine grains (whether quartz or polymineral).

Waterlaid sediments

Since the pioneering work of Wintle and Huntley (1980) on 4 to 10-μm-sized noncarbonate grains from deep sea sediments, several attempts have been made to utilize both large (>100 μm) and small (2 to 10 μm) grains of quartz and feldspar from waterlaid sediments. The principal problem has been uncertainty about the extent of resetting of the TL at deposition, which was recognized in a few reconnaissance studies of different types of waterlaid sediments (Berger and Huntley, 1982; Kronborg, 1983; Huntley and others, 1983; Mejdahl and others, 1984). A secondary problem has been anomalous fading in the polymineralic fine grains. Kronborg (1983) and Mejdahl and others (1984) used 100 to 300-μm feldspars but did not comment on anomalous fading.

Most of the published TL ages for known-age waterlaid and beach deposits (included here because water is an agent in deposition) are plotted in Fig. 17). Several unsuccessful partial bleach and total bleach results are shown. The circled dots represent careful studies of fine grains from different facies of the same age unit, facies that were deliberately selected to check the zeroing assumption. The immediate message of this figure is that overestimation of deposition ages is the greatest risk for such sediments, a direct consequence of the highly variable effectiveness of zeroing of the TL mineral clocks when transported in water.

Fine grains. Berger (1984, 1985b, 1985c) and Berger and others (1984, 1987) compared results from the different methods in Figure 4 for a series of studies of glaciolacustrine, lacustrine, and marine sediments and showed that only the partial bleach technique is generally reliable and that waterlaid quartz is not a suitable TL clock. The waterlaid quartz studied by Lamothe (1984) appears to give a correct age, but because the extent and nature of extrapolation of the TL growth curves were not specified, this result is difficult to assess. The studies of Berger also

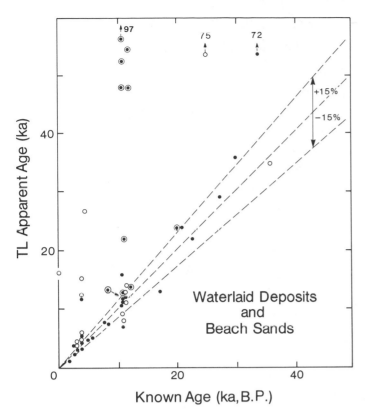

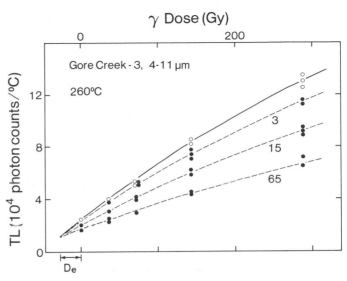

Figure 18. The use of the partial bleach technique for a mudflow silt deposited at the average rate of ca. 1 mm/yr. Both long wavelengths (>550 nm) and relatively brief light exposures (3 to 65 min) were employed to determine the equivalent dose plateau value of 27 grays (TL age = 7,800 yr) for this known-age (≤7,500-yr-old) feldspar-dominated silt (after Berger, 1985b).

Figure 17. Published TL age versus known age (as in Fig. 16) for waterlaid and beach deposits. To reduce the scatter, the results from only two TL methods (partial bleach (·, ⊙) and total bleach (o) are shown. Most of the total bleach results represent ca. 100-μm feldspars, the others represent 2 to 10-μm feldspar-dominated bulk samples. Total bleach and regeneration results for some of these samples from Berger and others (1984) and Berger (1984, 1985b) would plot well above the concordancy. Only one result for a zero-age sample (Belperio and others, 1984) is plotted here—others are discussed in the text. Some partial bleach results (⊙) are discussed explicitly in the text.

clearly demonstrated that for general reliability of the partial bleach method, only light having wavelengths above about 550 nm (see also Fig. 8) should be used in the optical bleaching.

The preferred facies in the glaciolacustrine sediments studied by Berger was the thin (e.g., <1 to 5 cm) clayey layers. Samples containing sand, or even an abundance of coarse or medium silt, produced gross overestimates of the deposition age. These results of Berger are shown by the circled dots in Figure 17, where only the clayey samples (with the exception of a single thin silt layer) yielded the expected ages. Some specific examples of results from these studies are discussed below, to illustrate several effects.

One conclusion reached by Berger (1985b) and Berger and others (1987) is that use of the partial bleach method with only one set of bleached subsamples (e.g., Fig. 4a) can often lead to subjective interpretation of the TL apparent age, i.e., with only a single illumination time, estimation of the correct equivalent dose value may be impossible. Consequently, two or more substantially different illumination times should be employed, at least for waterlaid sediments (see above discussion of zeroing). This procedure, although labor intensive, is expected to produce constant equivalent dose values for the different illumination times if the sediment has been well exposed to light. Decreasing equivalent dose values for decreasing illumination times are expected if the sample has been only weakly illuminated at deposition. In this case, the smallest equivalent dose value would represent an upper limit to the correct value. Such an "internal" consistency test was first attempted by Berger (1985b) using a polymineral (feldspar-dominant) fine-grained mudflow deposit (Fig. 18). This type of test is difficult to apply fully, but is necessary for most waterlaid deposits if overestimation of the true deposition age is to be recognized without reference to independent dating methods. At the very least, only two illumination times could be used if one of these is the shortest practicable exposure consistent with avoiding unacceptably large errors in the equivalent dose values (e.g., the 3-min exposure in Fig. 18 was too brief).

A particular illustration of the effect of sediment facies on TL results is shown in Figure 19. Here, for the parent varves, only the clayey layer produced the correct age (ca. 11,000 yr B.P.), even though short illumination times and long-wavelength light were used for the silt-rich sample. The TL data that produced the correct age for the mudflow silt (ca. 7.5 ka) are shown in Figure 18 (see also Fig. 13).

TL DATING OF WATERLAID SILTS AND TEPHRA
FROM GORE CREEK, BRITISH COLUMBIA

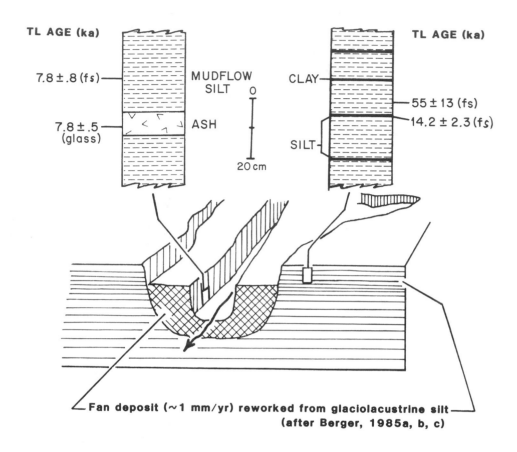

Figure 19. The results from several TL studies of known-age silts, clays, and ashes at one site. The TL data for ash are in Figure 13, those for the mudflow silt are in Figure 18, and those for the clay are in Figure 22. The expected age of the ash is 7,400 to 8,000 yr, that for the varved sediments is ca. 11,000 yr B.P.

To follow up this site-specific study, Berger and others (1987) applied the partial bleach method to feldspar-dominated fine grains (2 to 4 μm and 4 to 11 μm) from various known-age facies of glaciolacustrine sediments from another area of central British Columbia (Fig. 20). Both rhythmites and massive deposits were sampled, having textures ranging from silty clay through to sandy silt. The results shown in Figure 17 clearly infer that for such sediments, only grains deposited by slow rain-out from suspension (silty-clay layers) will be datable with present methods.

Even the most favorable of these small-basin samples were difficult to date, even by careful use of present TL methods. For example, the data in Figure 21 represent a silty clay likely deposited by slow rain-out; yet, as indicated by the different equivalent dose values for the two different illumination times, resetting of

the light-sensitive TL feldspar clocks was not entirely effective. These data also clearly illustrate the onset of saturation and the appropriateness of the saturating exponential model.

Sediments deposited in larger lakes or in near-shore marine environments might be expected to be more easily datable by present TL methods because such environments offer the opportunity for more effective zeroing of the TL feldspar clocks. The studies of Berger (1984), Lamothe (1984), and Lamothe and Huntley (1988) on fine grains from lacustrine sediments and shallow marine clay seem to confirm this expectation, as does that of Berger and others (1984) on continental slope muds. The results of Lamothe and Huntley (1988) deserve special comment because they showed that a shallow marine clay, a lacustrine clayey silt, and a glaciolacustrine clayey silt yielded expected

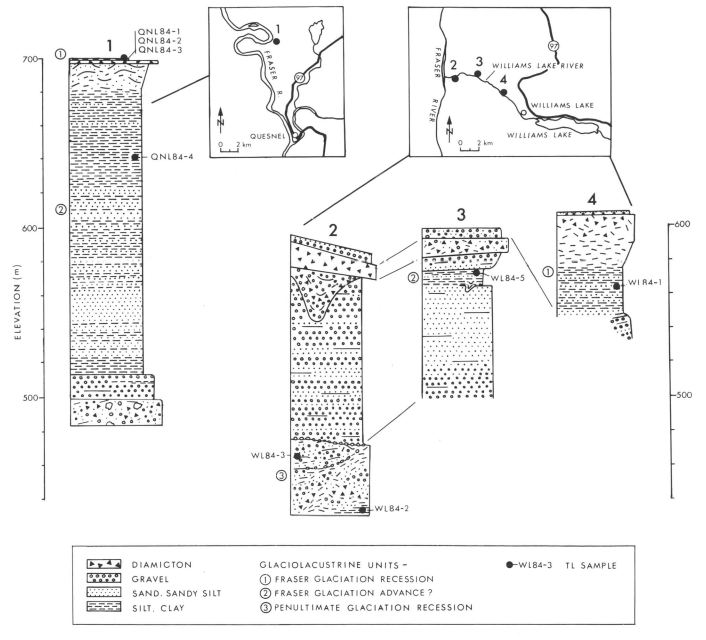

Figure 20. Stratigraphic sections of known-age glaciolacustrine sediments from central British Columbia, showing the positions of samples collected for TL dating. Samples QNL84-1,2,3 and WL84-1 were deposited at 11 ±1 ka B.P. Samples QNL84-4 and WL84-5 were laid down at 20 ±2 ka B.P. whereas samples WL84-3 and WL84-2 are >60 ka in age (after Berger and others, 1987). Only a few of the TL samples gave the expected deposition ages.

ages, whereas a pebbly, deep, marine clay produced an overestimate of the deposition age. The TL age for this last sample would probably have been even greater had sublinear regressions been used for both TL growth curves. The experiment should also be repeated using only >550 nm light for optical bleaching.

Recently, Forman and others (1987) investigated the feasibility of dating near-shore marine sediments (now above sea level) from the Hudson Bay Lowland. Their comparison of different methods for one known-age (ca. 6 ka) sample confirms the conclusion reached by the studies of Berger cited above—that only the partial bleach method with long-wavelength bleaching light is reliable. They applied this method to an older sample and obtained a TL apparent age of about 70 ka. However, because they used large linear extrapolations from data representing only a limited applied dose range, and because their only independent

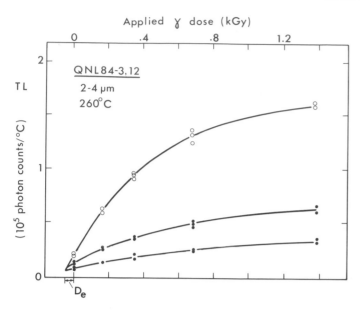

Figure 21. Partial bleach growth curves for subsample 12 (a clayey silt layer) of sample QNL84-3 (Fig. 20). The bottom curve represents a laboratory illumination of 21 hr, the middle curve represents 1 hr, both using >550-nm light. Saturating exponential regressions are shown (after Berger and others, 1987).

age check was an amino-acid-racemization-based age estimate, this promising result must be considered only preliminary.

Bryant and others (1983) produced overestimates of the deposition age for two fine-grain feldspar-dominant samples. This effect probably arose from the use of too-long illumination times and too-short wavelengths of light.

A peculiar problem encountered by Berger and Huntley (1982), Berger (1985b, c), and Berger and others (1987) was the failure to obtain a plateau in equivalent dose plots. This was associated with apparent shifts in the peaks of additive dose glow curves and interpreted as a manifestation of non-first-order kinetics behavior. Evidence for the occurrence of non-first-order kinetics in feldspars has been given by Chen and others (1983) and by Ahmed and Garta (1985). The adopted solution to the above problem was to align the glow curves prior to calculation of the equivalent dose values, a procedure with some theoretical justification (Chen and others, 1983). An example of such results is shown in Figure 22 for the 5-mm-thick clay-rich layer from Figure 19. Other workers have not reported this effect, and indeed, samples from outside of British Columbia measured by the reviewer have not shown it (Berger and others, 1987).

Results using partial bleach methods on fine-grained material from three sandy samples from Utah (Colman and others, 1986) were inaccurate. One sample yielded an excessive equivalent dose, whereas two produced gross underestimates. The authors suggest that the zeroing assumption may have been violated for one sample, while either unrecognized anomalous fading

or thermal instability may have affected the other two. Whatever the likely cause of these inaccuracies, the previous discussions suggest that most sand-bearing deposits are poor candidates for TL dating (unless they are unambiguously dune or beach sand deposits).

An exception to the above proscription of sandy samples may be braided channel deposits. Nanson and Young (1987) report TL age estimates for polymineralic fine grains from fluvial gravels and sand-clay sediments that are, to a first order, in agreement with radiocarbon dates on associated or correlated wood and charcoal. The implication of this rough agreement is that TL sediments deposited in braided channels may be sufficiently zeroed at deposition to be accurately datable by present TL methods. Notwithstanding this promise, the limitations of the study of Nanson and Young (linear extrapolations of TL dose-response curves, no mention of anomalous fading) suggest that additional, detailed, TL dating studies of fluvial sediments are needed to demonstrate this potential. Another class of sandy waterlaid deposit that has potential for TL dating is marine beach sediment. Data (Berger, unpublished data) show that the TL of polymineral fine grains from the sandy surf zone of one modern beach has been reasonably well zeroed when >550 nm light is used for optical bleaching.

One problem, encountered with waterlaid silts but rarely with loess, is loss of radon during thick-source alpha-particle counting of dry powders to determine the U and Th concentrations—a loss that precludes accurate calculation of the dose rate from these measurements. Berger and others (1984), Berger (1984), and Huntley and others (1986) have shown how glassing of a sample prior to alpha counting can overcome this problem using a modification of the recipe of Jensen and Prescott (1983). Correction for the effects of polonium loss by sublimation during fusion (Akber and others, 1985; Huntley and others, 1986) is necessary for increased accuracy. The effect of this Po loss on the calculated total dose rate will probably be less than 10 percent because such loss affects only the U decay chain.

A possible explanation for the occurrence of laboratory radon loss from some fine-grain waterlaid sediments and organic-rich soils (Berger, unpublished data), compared to eolian silts, is the greater proportion of clay-sized or smaller grains in the former. If much of the uranium is concentrated preferentially on such very small grains, then the probability of radon loss related to recoil effects (e.g., Fleischer, 1983) is enhanced. Interestingly, an association exists between uranium concentration and fine-sized organic-matter content (e.g., Cochran and others, 1986). Significant amounts of organic matter occur even in glaciolacustrine silts. The origin of much of this organic matter may be micro-organisms such as bacteria and algae, which have been shown to fix or take up uranium (Beveridge and Fyfe, 1985; Mann and Fyfe, 1985). Such uranium could then coat the smallest mineral grains (having the largest surface-to-volume ratios), or be readily absorbed into clay minerals.

Large grains. Several workers have attempted to date waterlaid sediments by using large (e.g., 100 to 300 μm) grains of

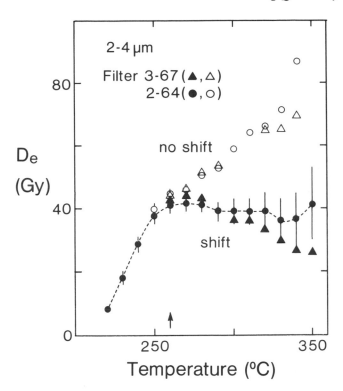

Figure 22. A plot of partial bleach, equivalent dose values versus temperature for the bulk, feldspar-dominated, 2 to 4-μm grains from a 5-mm-thick, clay-rich, annual layer from a ca. 11,000-yr-old glaciolacustrine silt (Fig. 19), illustrating the effects of realignment of glow curves. Two experiments are represented: one used wavelengths only above 660 nm (red filter 2-64); the other used wavelengths only above 550 nm (orange filter 3-67). The plateau values correspond to a TL age of 14 ±2 ka (after Berger, 1985c).

feldspar and quartz, presumably because these are easy to isolate and identify. Although these workers have interpreted many of their results as deposition ages, in view of the near absence of fundamental studies of such large grains from zero-age deposits (unlike for silt-sized grains) and considering the understanding gained from fundamental studies of known-age silts (discussed above), alternative interpretations are preferred. In particular, large grains require high-energy deposition modes (such as fluviatile deposition) to ensure long transport, and by implication, lengthy exposures to light. Turbidity flows in large lakes can also keep such grains in suspension for lengthy times (e.g., Hsü and Kelts, 1985), but will preclude significant exposure to light. In small steep-walled lakes, grain flows will be common and also preclude much illumination. Thus, overestimation of the true deposition age is expected to be common for large grains, especially when the total bleach and regeneration techniques are used.

Such overestimation by overbleaching explains the 40-ka age observed by Belperio and others (1984) for the 100-μm quartz from a zero-age coastal marine sediment. Indeed, with the same total bleach method for the same sample, they obtained a >16-ka age for the feldspar-dominated fine grains (Fig. 17). For other large-grain quartz samples of this study, some yielded excessive apparent ages while some gave reasonable results. Unfortunately, these samples have had rather complicated depositional histories so that interpretation of reliability is difficult.

In other studies of 100 to 300-μm feldspars, Kronborg (1983) and Mejdahl and others (1984) used the total bleach method, but with linear extrapolations of TL buildup curves. Consequently, the apparent agreements of their TL results with the often poorly known true deposition ages are uninformative. Jungner (1983) explained his partial bleach results for 100 to 300-μm feldspars from sands in terms of a mixture of unbleached and bleached grains. However, these results can also be explained by limitations in the experimental technique. Specifically, arbitrarily lengthy laboratory illuminations with UV-bearing light were used, whereas work on waterlaid silts (discussed above) suggests that short illuminations with long-wavelength light are prudent. Furthermore, linear extrapolations of TL growth curves were used. Finally, the selection of equivalent dose values to support his interpretation was somewhat arbitrary (his sample 4).

Huntley and others (1985b) report preliminary tests of several assumptions of TL dating related to beach deposits. The ca. 100-μm quartz grains from several known-age, Australian raised beach deposits are being studied, and initial results are encouraging.

Miscellaneous anomalies. In view of the foregoing discussion of the dependence of the TL signal at deposition upon mineral type and ambient light spectrum, that apparent stratigraphic age reversals have sometimes been measured (e.g., Hütt and others, 1983; Berger and others, 1984), or that large fluctuations in total TL signal have been recorded along young marine cores (Cini Castagnoli and others, 1982, 1984) is not surprising. In this latter example, the authors have invoked extraterrestrial causes (solar and/or supernova bursts of ionizing radiation affecting atmospheric dust, which then dominates the TL contributors in marine mud) for the observed fluctuations. However, some hint of a later moderation in this rather fanciful interpretation is given (Cini Castagnoli and Bonino, 1984).

Summary. Apparently, the best candidates for TL dating of waterlaid sediments are clay-rich deposits, such as found in varves, distal lake beds, or coastal marine muds. Most sand-bearing or silt-rich deposits should be avoided (e.g., glaciofluvial sand, grain flow sand, turbidity flow sediment). Lodgement till is also not datable (Berger 1984; Wintle and Catt, 1985a) because grains are not well insolated before final burial. The fine-grained component of some alluvial deposits—such as braided channel sediments— has promise. Detailed TL-dating studies of such sediments are needed.

Buried soils and peat

The prospect of directly dating the burial age of buried soils and peat beyond the radiocarbon range is of great interest because

G. W. Berger

these deposits can serve as chrono-stratigraphic markers as well as paleoclimatic indicators. In this section the use of only the detrital silicates—feldspar and quartz—is discussed because these will record class II events. An alternative TL clock, pedogenic carbonate, has been discussed above, under class I events.

The first TL dating studies of paleosols were carried out at Simon Fraser University (Berger and Huntley, 1982; Divigalpitiya, 1982; Huntley and others, 1983; Huntley, 1985a), using the partial bleach technique of Figure 4 applied to fine grains. These studies demonstrated the feasibility of dating such deposits. The principal conclusions are as follows: (1) fine silt-sized feldspars in A horizons of soils can be dated, (2) correction to dose-rate calculations for the effects of the high organic content must be made (Divigalpitiya, 1982), (3) only the partial bleach technique with low-energy photons is routinely reliable, and (4) unknown changes in past water content (compaction of peat) or in U concentration (enrichment by complexing on organics) may invalidate present-day estimates of the past dose rate. This last problem remains unresolved but may be severe only for peat.

Other workers have also reported reasonable TL ages for paleosols. Kronborg (1983) extracted 100 to 300-μm feldspar grains and used the total bleach technique with an unfiltered Hg-based lamp (containing significant UV components). However, because he used linear extrapolation to TL growth curves, a total bleach technique with a UV component, and did not discuss anomalous fading, his ca. 70-ka TL ages may be only fortuitously in accord with the approximately known deposition ages. Wintle and others (1984) reported reasonable ages for two paleosols developed in loess, using the fine-grain regeneration technique with UV components. These results are difficult to assess because the deposition ages are not well known independently, but the measurement of a 0.7 ± 0.1 ka apparent age for a near-surface buried soil in loess at another site (Wintle and Catt, 1985b) supports the zeroing assumption for such A horizons and lends credibility to the results of Wintle and others.

Most of the above studies have thus demonstrated that the burial time of A horizons in paleosols developed in dune, fluviatile, and loess deposits may be dated with acceptable accuracy by TL methods, provided experimental rigor is maintained. Apparently, some A horizons developed in alluvium and colluvium are not suitable for TL dating (Berger and Huntley, 1986). A large fraction of the fine-grain TL clocks in such deposits apparently are not effectively zeroed before final burial.

A horizons are not preserved in some buried soils. Can B horizon material be used to place an acceptable limit on the burial time? Alternatively, can the TL dating methods be used to provide direct information on rates of paleosol development through application to B subhorizons? Presently, almost nothing is known of the effects of processes associated with soil formation on the TL mineral clocks (in particular, feldspars). Exposure to sunlight during bioturbation is the most favored hypothesis for A horizons (Huntley and others, 1983), although many grains are likely to be added by wind.

The potential of TL dating methods to provide direct infor-

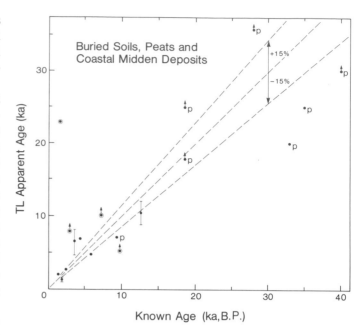

Figure 23. TL age versus uncorrected C-14 age for organic-rich deposits: A horizons of paleosols (·), coastal middens (⊙), and peats (p). Samples with arrows exhibited uncorrected anomalous fading. Only the fine-grain, partial bleach method is represented. Data are from Huntley and others (1983), Wintle and Catt (1985b), and Berger (with error bars, unpublished).

mation on rates of paleosol development has not yet been exploited. Some evidence (Wintle and Catt, 1985b) suggests that some process, probably associated with soil development, has partially reset the TL feldspar clock in B subhorizons of a soil developed in loess. Whether this process involves exposure of some of the grains to daylight or some other action is not known. However, the full capability of the partial bleach TL technique has not yet been utilized in the study of paleosols. For example, the stratigraphic age reversals in the profile examined by Wintle and Catt (1985b) could be experimental artifacts caused by over-bleaching effects. This possibility must be considered for each sample in such partially reworked material as B subhorizons. Recently Singhvi and others (1987) reported regeneration TL ages for paleosols (both A and B horizons) developed in loess. As none of their samples had sound, independently determined burial ages, these TL results cannot be judged. That the results do fall in stratigraphic sequence is at least somewhat reassuring.

All known published partial bleach TL ages for buried soils that have reasonable independent age determinations are plotted in Figure 23, along with partial bleach results for coastal midden and peat deposits. Three unpublished results (with error bars) for Ab horizons from temperature paleosols are also included (Berger, unpublished data). A horizons yield good results. Sometimes the comparison age from ^{14}C dating does not refer to the same event, so that some divergence is expected in a plot such as this. Coastal midden material seems unsuitable, and peat gives variable results. Further testing of peat is required.

One interesting result not shown in Figure 23 is the 480 ± 90 ka minimum TL age (the sample had uncorrected anomalous fading) obtained by Divigalpitiya (1982) and Huntley and others (1983) for the Salmon Springs peat. Stuiver and others (1978) calculated an enriched ^{14}C age of 72 ka for this peat, but it grades into silt and tephra fission-track dated at 1.06 ± .11 Ma (Westgate and others, 1987).

RESULTS HAVING GEOLOGICAL SIGNIFICANCE

A minimum TL age of 670 ka for the previously undated Coutlee tephra (Berger, 1985a) extends the chronology of Pleistocene deposits in southern British Columbia.

Proszynska (1983) obtained a regeneration TL age estimate of 115 ± 20 ka for loess from Komarow Gorny (Poland), which is associated with the beginning of the Blake Event (e.g., Jacobs, 1984). Berger (1987) dated the widespread Old Crow tephra from Fairbanks, Alaska, at 109 ± 14 ka. This tephra is found above a paleomagnetic excursion at three widely separated sites, which has been suggested to represent the Blake Event (e.g., Wintle and Westgate, 1986).

Wintle and others (1984) demonstrated, by TL dating, that deposition of loess in Normandy was a rapid process, and inferred that intervals of loess accumulation probably occupied less than 10 percent of late Pleistocene time at their sites.

Singhvi and others (1987) produced regeneration TL age estimates of ca. 110 ka for a Bt paleosol horizon from Kashmir, thus relating this paleosol to the last interglacial.

Lamothe and Huntley (1988) determined an early Wisconsin age (60 to 85 ka) for glaciolacustrine sediments underlying the St. Pierre Interstadial sediments of the St. Lawrence Lowland, Quebec. Berger (1984) obtained a partial bleach TL age estimate of 66 ± 7 ka for the regionally extensive Sunnybrook diamicton in southern Ontario (Eyles and Westgate, 1987).

Sutton (1985b) dated the Meteor Crater event in Arizona at ca. 50 ka, based on measurements of TL from shock-heated quartz grains.

SEDIMENTS AND LASERS

A new and exciting technique (which exploits a decades-old principle) for measuring equivalent doses in unheated sediments has been demonstrated by Huntley and others (1985a). Rather than heat, laser light is used to detrap electrons with far greater sensitivity than is possible with the conventional procedures of Figure 4. The signal of the optically stimulated luminescence (OSL) is illustrated in Figure 24 for a sequence of raised beach sands of known age. Grains of quartz were exposed to a 514.5-nm, ca. 50-mW/cm^2, argon-ion laser beam. At the instant of switching on the laser, an intense luminescence signal was observed in the spectral region of ca. 380 to 440 nm (suitable filters

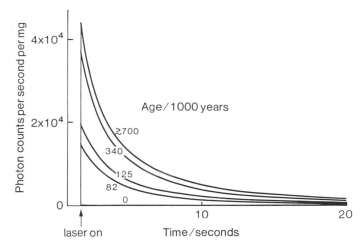

Figure 24. Luminescence intensity versus time after switching on an argon-ion laser, for 100-μm quartz grains extracted from a sequence of stranded beach dunes, showing an increasing intensity with known age. (Reprinted by permission from Nature, v. 313, p. 105–107, Copyright 1985, Macmillan Journals Limited.)

were used in front of the photomultiplier tube and in the incident laser beam). This signal was distinct for different ages, and for each sample decayed rapidly within a few seconds.

Apart from this new luminescence readout technique, almost all other steps in the OSL dating process are similar to those used in conventional TL dating of unheated sediments (e.g., Fig. 2). However, as emphasized below, great care is needed during sample handling to minimize exposure of the sample to laboratory lighting, even red light. Furthermore, because this technique empties electrons from both thermally stable and thermally unstable traps at the same time, irradiated samples must be heated moderately before being illuminated by laser light. For example, Smith and others (1986) pre-heated all samples for 5 min at 220°C.

The remarkable characteristic of this OSL phenomenon is that the initial decay represents extremely light-sensitive traps. For the samples of Figure 24, ca. 50 percent of these traps were emptied in about 2 sec (ca. 100 mJ/cm^2 of input energy). Roughly 90 percent of these traps would be emptied in 7 sec of the laser exposure, with the assumption that decay is exponential (only the initial decay is close to being exponential). In comparison, only about 10 sec of unattenuated sunlight (33 mW/cm^2, 0 to 515 nm) would be required to empty the same 90 percent fraction. The implication is that this technique can measure equivalent doses for minerals that have received very brief light exposures (e.g., < 1 min) at deposition, something the partial bleach technique cannot do well (e.g., Fig. 18), or at all (e.g., the 12-cm-thick silt in Fig. 19). Luminescence corresponding to the most light sensitive traps is measured directly with the optical technique, rather than as the difference between two large signals (e.g., Fig. 18).

A practical demonstration of the accuracy of this OSL dating technique was given for a ca. 60-ka-old pond silt, an Ah paleosol, and a present-day beach sand (Huntley and others, 1985a). An age of 62 ± 8 ka was measured for the feldspar-dominated pond silt, in agreement with the partial bleach result of 62 ± 12 ka and comparable to an associated (uncorrected) ^{14}C age of 58.8 +2.9/−2.1 ka (QL-195, Clague, 1980) for wood just below. The two other comparison tests were not so successful. The optical technique gave an equivalent dose for quartz from an Ah paleosol that, as for the partial bleach method, was double the expected value. Although quartz from a present-day bleach gave equivalent dose values (8 ± 2 Gy) several times lower than did the partial bleach technique, these were not zero as expected. The authors offered no explanation for the Ah paleosol result, but suggested that the beach quartz was not well exposed to light and that this could be recognized through the absence of constant equivalent dose values as the laser illumination time increased.

Another interpretation of their quartz results is possible, considering the information in Figure 7, above, and the success of the laser technique with the feldspar-dominant fine grains of the pond silt. The interiors of these quartz grains may not have been exposed significantly to 515-nm light at deposition, but perhaps only to longer-wavelength light that failed to detrap electrons. Such long-wavelength light could readily have detrapped electrons in feldspars. Application of this OSL dating technique to feldspars (if available) from the problematic Ah paleosol and beach sand would be instructive. Alternatively, some longer wavelength of laser light (such as from a He-Ne laser) applied to these quartz grains may be more appropriate. Certainly the effects of many parameters must be investigated before this new technique becomes routinely applicable, as Huntley and others (1985a) recognized.

A group at Oxford, England, has also begun testing the OSL dating technique (Smith and others, 1986), using samples of quartz and zircon and the 515-nm Ar laser line. Several detailed effects associated with OSL are discussed by Smith and others. In one application, the expected equivalent dose plateau (ca. 20 Gy) was obtained with illuminations of up to 200 mJ/cm^2 using 100-μm quartz from an archaeological site dated by TL at 14 ± 3 ka (Aitken and others, 1986).

This optical dating technique is revolutionary, if only because it has the potential to provide rapid and automated readout of luminescence signals. However, this highly sensitive method will likely contribute much more, perhaps opening a new door to the understanding of some natural luminescence behaviors. Two aspects of such behavior worth investigating are the non-exponential decay in Figure 24, and the wavelength-dependent sensitivity of different minerals in Figure 7.

Early work on infrared stimulated luminescence in phosphors (e.g., Garlick, 1949) produced either exponential decays or inverse-power-law decays, or combinations of these. The former are characteristic of conditions where no electrons are retrapped after being liberated by photons. The power-law decays are expected whenever electron retrapping occurs. For some models producing power-law decays, the thickness of the phosphors and the ratio of absorption coefficients for the emitted and stimulating light are important parameters. The discussion of Garlick suggests that some help in distinguishing these two hypotheses (no retrapping versus retrapping) for minerals could be obtained by using light of varied intensities and wavelengths, and mineral grains of varied size ranges. As an alternative hypothesis, the power-law decays could result from the combined effect of a distribution of trap types, the decay for each type being itself exponential (e.g., Garlick, 1949, p. 169 and 33). Indeed, Smith and others (1986) have suggested that a distribution of traps of different bleachability may account for this non-exponential decay of OSL.

The wavelength-dependent sensitivity illustrated in Figure 7 deserves study with narrow-band or monochromatic optical stimulation, using a range of wavelengths directed at different minerals. However, because the OSL technique is so sensitive, one must be wary of the possibility of effects other than stimulation when different wavelengths are used. One possible effect, which may be subtle in minerals, is photoquenching. This is known to occur in phosphors in a wavelength region that may overlap with that associated with photostimulation (Leverenz, 1968, p. 299).

Notwithstanding the above potentially complicating effects, in any application of OSL to dating, one may be tempted to seek an optimum stimulating wavelength for each mineral, using as a guide the simple relation between thermal and optical activation energies propounded by Mott and Gurney (1948, following the original suggestion by D. Curie). Such a simple relation, however, has been questioned by Garlick (1949), and so a relatively wide range of stimulating wavelengths may be "optimum" for a given mineral. From tests on one zircon sample, Smith and others (1986) suggested that the use of stimulating wavelengths below about 440 nm would be the most efficient in producing OSL for that mineral. However, to justify the use of any such particular range of wavelengths in OSL dating, one would have to infer what the ambient light spectrum was during deposition. As an example, for some subaqueous sediments, the data in Figure 8 suggest that use of wavelengths outside the range of about 550 to 650 nm would be inappropriate.

SOME LIMITATIONS TO ACCURATE DOSE-RATE DETERMINATION

Although this review is concerned almost exclusively with progress in the accurate measurement of equivalent dose values, two particular sources of error in TL apparent ages—variable dose rate due to migration of elements, and poor knowledge of the past pore-water content—must be mentioned. These two factors can be expected to receive increasing attention now that the major limitations to accurate measurement of equivalent doses (by the methods of Fig. 4) seem to be surmountable. Keep in mind, however, that the pore-water content need not be known if minerals such as apatite or zircon can be used for dating unheated sediments (although this has yet to be demonstrated). The relatively high internal radioactivity of these minerals precludes dependence of the paleodose upon the surrounding dose rate.

Decay series disequilibrium or varying dose rates can arise in several ways. For example, progressive U enrichment can occur in association with iron staining on grain surfaces. U, Th, and Ra can also migrate through sediments with percolating ground waters. Mobility of such elements is expected to be less for fine-grain than for coarse-grain deposits because of the lower permeability of the former. However, the presence of organic material (see above discussion of fine-grain waterlaid sediments) and of hydraulic gradients also effects post-depositional elemental mobility. Causse and Hillaire-Marcel (1986) examined the geochemical behavior of U and Th in unconsolidated organic-bearing sediments, and concluded that the isotopes of these elements are least mobile in cohesive clayey sediments that experience only weak hydraulic gradients. Fortunately for TL dating, clayey sediments are preferred for other reasons, too—among waterlaid deposits they are the most likely to satisfy the zeroing assumption.

The dose rate can also change if decay series disequilibrium is established through radon loss (especially ^{222}Rn, half life 3.82 days). Radon loss can be enhanced by fluctuations in the water content of the deposit. The diffusion coefficient of radon in water is about four or five orders of magnitude smaller than in air (Tanner, 1964); thus, the mere presence of intergranular water is unlikely to enhance radon loss. Rather, as the recent experimental and theoretical work of Fleischer has shown (e.g., Fleischer, 1983), it is the cyclical flushing of the nanopores, micropores, and intergranular spaces that enhances radon loss. For this reason, such radon loss is likely to be most significant for terrestrial waterlaid deposits exposed to fluctuating water tables or susceptible to percolation effects.

The uncertainty in estimating a realistic average value of the past water content also contributes greatly to the uncertainty in the dose rate (see dose-rate equations above). Inaccuracy in estimates of past pore-water content will be least (5 to 10 percent) for marine muds and greatest (30 to 50 percent) for terrestrial waterlaid sediments now above the water table. The effective past water content is usually <5 percent for coarse-grained eolian and beach deposits.

With terrestrial fine-grained deposits, two approaches to estimation of past water content have been tried: (1) sampling away from desiccated section faces in combination with measurements of seasonal fluctuation in water content (e.g., Wintle and others, 1984), and (2) development of reliable present indicators such as laboratory compaction tests together with field measurements of bulk density and detailed observations of sedimentology and microstructure (Rendell, 1983). Both approaches have been applied only to loess, and the second one, albeit promising, has not been fully developed. Only careful field sampling coupled with some knowledge of the history of water table fluctuations seems likely to provide useful estimates of past water content for waterlaid deposits (including beach sands). In this case, present practice is

often to choose an estimate between the present in situ water content and the saturation value (such as the mean of the two).

SOME GUIDELINES TO USEFUL APPLICATION OF TL METHODS

Suitable samples

Because TL is a solid-state physical phenomenon, it occurs in many materials. The two main classes of zeroing agents, heat (or crystallization) and light, should be kept in mind when searching for suitable samples in a given stratigraphic section. Only a few types of Quaternary deposits will presently provide reliable TL dates, and only if TL methods (Figs. 3 and 4) are applied prudently.

Specifically, heated quartz in volcanic deposits (or burned flint in anthropogenic contexts), some precipitated calcites, and distal tephra can be dated reliably using the additive dose method. Sample selection is less straightforward in unheated sediments. Suitable samples are: fine (distal) loess; thin clayey layers in lacustrine deposits; clayey muds in off-shore or near-shore marine sediments; and organic-rich (A horizon) paleosols developed in loess, lacustrine, fluviatile, or dune deposits. Dune sands are very promising, but unshakable reliability has yet to be demonstrated. For all such unheated sediments, the partial bleach method (Fig. 4) is the preferred and prudent technique to use. Finally, consideration during sample selection of how the different closed-system assumptions might have been violated is most important.

Sample collection

Aitken (1985, Appendix D) discussed sample collection at length. However, his remarks about exposure of samples to sunlight during collection may not be stringent enough. Less than 3 to 5-sec exposure is desired, unless the sample can be carved out as a cohesive block. In this case the outer 3 to 5 mm can be removed later in the TL laboratory.

Suitable sample containers are opaque, 100 to 300-ml, plastic or tin cans. Under near darkness, such as at night or under a black shroud, one-quarter-liter paint cans, preferably lined with double-walled, sealed, plastic bags to prevent moisture loss, can also be used to collect noncohesive sediment.

Sample analyses

TL dating of pottery shards is routine, but this is not yet true for geologic samples. A TL specialist must be involved in most stages of the work, including sample collection if possible. Analytical criteria that must be met to ensure reliability of the TL result are similar to those outlined by Wintle and Huntley (1982), and are discussed at length in this review.

Specifically, checks must be made for: (1) proper zeroing of the TL clocks, (2) the presence of anomalous fading, (3) sublinear dose-response curves, (4) laboratory-induced TL sensitivity changes, (5) radon loss, and most importantly, (6) the presence of a stable region of the glow curves (the plateau test). Any anomalies should be reported so that the development of TL methodology can proceed.

SUMMARY AND CONCLUDING REMARKS

New applications of TL dating (tephra, buried soils, peats) have been demonstrated in recent publications, and notable progress has been made in understanding the limitations to accuracy, as well as in the development of new techniques. For example, sidestepping the ill effects of anomalous fading in polymineralic samples without necessarily isolating a nonfading mineral is now sometimes possible by delaying TL readout until any short-term (weeks) fading component has decayed. Alternatively, this decay can sometimes be accelerated by holding samples at elevated temperatures. Additionally, ongoing high-resolution investigations of the TL emission spectra from several minerals may provide new laboratory procedures for discriminating against the malign fading minerals, thereby obviating the time-consuming approaches now employed (e.g., delayed glow, accelerated fading, mineral separation).

Another major limitation to accurate measurement of the age of last exposure to sunlight, especially for waterlaid material, has been the uncertainty in the extent of resetting of the TL at deposition. Frequently there has been a misconception that this uncertainty is of serious concern only for Holocene samples, perhaps for samples even somewhat older. Some workers have assumed that as deposits get older, the relative contribution of the relict, or zero-point, TL remaining at deposition becomes smaller, and therefore correction for this relict TL is unnecessary in sediments older than, for example, 30 to 50 ka. This idea is misleading. Very likely, the relative contribution of this unbleached TL component for such older material will sometimes be more than 50 percent, depending on several factors. Recent work has shown that these factors include duration of exposure to light (obviously), mineral type, and ambient light spectrum—factors strongly dependent on mode of deposition.

A corollary to the observations of Berger (1985b) (e.g., Fig. 18 above) and Huntley (1985a) (e.g., Fig. 6) is that, for waterlaid silts, at least two quite different illumination times should be used with the partial bleach technique to check for an overbleaching effect; if the equivalent dose decreases with decreasing exposure time, then the correct value is probably less than the smallest measured value. An alternative objective criterion for acceptability of equivalent dose values would be observation of the same equivalent dose from a short-exposure partial bleach experiment as from a total bleach experiment.

The emphasis of this review has been on ways to improve the accuracy of measurement of the equivalent dose term in the TL age equation. The conclusion is that means have been demonstrated for overcoming the largest sources of inaccuracy in this term. Consequently, the major remaining sources of inaccuracy and imprecision in a TL age will often be in the dose-rate term, particularly the possible migration of elements such as uranium, radon, and radium, and the uncertainty in estimation of the past pore-water content. Variation in dose rate due to mobility of elements is likely to be most significant for coarse-grained (therefore permeable) terrestrial deposits, and least for marine mud. Ignorance of the history of pore-water content is greatest for terrestrial waterlaid material, and some more accurate means for estimation of this parameter are needed.

Perhaps outshining all of these new applications and understandings of the methodologies proposed in 1980 by Wintle and Huntley is the potentially revolutionary demonstration of luminescence dating by optical stimulation, using lasers for example. Although this OSL dating technique may be limited by many of the same problems afflicting conventional TL methods (e.g., anomalous fading, varied spectral sensitivity of different minerals, unknown ambient light spectrum at deposition), it potentially represents a quantum leap in sensitivity, simplicity, and speed.

In closing, the "hard sledding" in the last few years by several laboratories seems to have cumulatively generated sufficient understanding of the TL behavior of minerals and of methodologies that the time is auspicious for a far greater involvement of geologists with these luminescence-dating tools.

ACKNOWLEDGMENTS

The many insightful contributions of D. J. Huntley to the amber art of TL dating of geological material, through his published efforts and his personal interactions, are recognized here. I also thank him for helpful comments made on a very early version of this manuscript as well as for several fruitful discussions. Very constructive reviews were provided by S. R. Sutton and R. M. Farquhar.

REFERENCES CITED*

Ahmed, A. B., and Garta, R. K., 1985, Trapping levels in $KAlSi_3O_8$: Physica Status Solidi (a), v. 94, p. 645–651.

Aitken, M. J., 1985, Thermoluminescence dating: New York, Academic Press, 351 p.

Aitken, M. J., Huxtable, J., and Debenham, N. C., 1986, Thermoluminescence dating in the Palaeolithic; Burned flint, stalagmitic calcite, and sediment: Bulletin de l'Association Francaise pour l'Etude du Quaternaire, v. 26, p. 7–14.

Akber, R. A., and Prescott, J. R., 1985, Thermoluminescence in some feldspars; Early results from studies of spectra: Nuclear Tracks, v. 10, p. 575–580.

Akber, R. A., Hutton, J. T., and Prescott, J. R., 1985, Thick source alpha counting using fused glass discs; Corrections for loss of radon and polonium: Nuclear Instruments and Methods in Physics Research, v. A234, p. 394–397.

Beese, W., and Zink, J. I., 1984, The intensity of triboluminescence: Journal of Luminescence, v. 29, p. 119–122.

Belperio, A. P., Smith, B. W., Polach, H. A., Nittrouer, C. A., DeMaster, D. J., Prescott, J. R., Hails, J. R., and Gostin, V. A., 1984, Chronological studies of the Quaternary marine sediments of northern Spencer Gulf, South Australia: Marine Geology, v. 61, p. 265–296.

Berger, G. W., 1984, Thermoluminescence dating studies of glacial silts from Ontario: Canadian Journal of Earth Sciences, v. 21, p. 1393–1399.

——, 1985a, Thermoluminescence dating of volcanic ash: Journal of Volcanology and Geothermal Research, v. 25, p. 333–347.

——, 1985b, Thermoluminescence dating studies of rapidly deposited silts from south-central British Columbia: Canadian Journal of Earth Sciences, v. 22, p. 704–710.

——, 1985c, Thermoluminescence dating applied to a thin winter varve of the Late Glacial South Thompson silt, south-central British Columbia: Canadian Journal of Earth Sciences, v. 22, p. 1736–1739.

——, 1986, Dating Quaternary deposits by luminescence; Recent advances: Geoscience Canada, v. 13, p. 15–21.

——, 1987, Thermoluminescence dating of the Pleistocene Old Crow tephra and adjacent loess, near Fairbanks, Alaska: Canadian Journal of Earth Sciences, v. 24, p. 1975–1984.

Berger, G. W., and Huntley, D. J., 1982, Thermoluminescence dating of terrigenous sediments: PACT Journal, v. 6, p. 495–504.

——, 1983, Dating volcanic ash by thermoluminescence: PACT Journal, v. 9, p. 581–592.

——, 1986, San Diego research and chronology revisited: Current Research in the Pleistocene, v. 3, p. 39–40.

Berger, G. W., and Luternauer, J. J., 1987, Preliminary field work for thermoluminescence dating studies at the Fraser River delta, British Columbia: Geological Survey of Canada Paper 87–1A, p. 901–904.

Berger, G. W., and Marshall, H., 1984, The thermoluminescence of some limestone rocks: Ancient TL, v. 2, p. 1–6.

Berger, G. W., Huntley, D. J., and Stipp, J. J., 1984, Thermoluminescence studies on a C-14-dated marine core: Canadian Journal of Earth Sciences, v. 21, p. 1145–1150.

Berger, G. W., Clague, J. J. and Huntley, D. J., 1987, Thermoluminescence dating applied to glaciolacustrine sediments from central British Columbia: Canadian Journal of Earth Sciences, v. 24, p. 425–434.

Berger, G. W., Lockhart, R. A., and Kuo, J., 1988, Regression and error analysis applied to the dose-response curves in thermoluminescence dating: Nuclear Tracks and Radiation Measurements, (in press).

Berner, R. A., and Holden, G. R., Jr., 1977, Mechanism of feldspar weathering; Some observational evidence: Geology, v. 5, p. 369–372.

Beveridge, T. J., and Fyfe, W. S., 1985, Metal fixation by bacterial cell walls: Canadian Journal of Earth Science, v. 22, p. 1893–1898.

Bluszcz, A., and Pazdur, M. F., 1985, Comparison of TL and C-14 dates for young eolian sediments; A check of the zeroing assumption validity: Nuclear Tracks, v. 10, p. 703–710.

Bothner, M. H., and Johnson, N. M., 1969, Natural thermoluminescent dosimetry in late Pleistocene pelagic sediments: Journal of Geophysical Research, v. 74, p. 5331–5338.

Bowman, S.G.E. and Huntley, D. J., 1984, A new proposal for the expression of alpha efficiency in TL dating: Ancient TL, v. 2, p. 6–11.

Bronger, A., Pant, R. K. and Singhvi, A. K., 1987, Pleistocene climatic changes and landscape evolution in the Kashmir Basin, India; Paleopedologic and chronostratigraphic studies: Quaternary Research, v. 27, p. 167–181.

Bryant, I. D., Gibbard, P. L., Holyoak, D. T., Switsur, V. R., and Wintle, A. G., 1983, Stratigraphy and palaeontology of Pleistocene cold-stage deposits at Alton Road quarry, Farnham, Surrey, England: Geological Magazine, v. 120, p. 587–606.

Buraczynski, J., and Butrym, J., 1984, La datation des loess du profil d'Achenheim (Alsace) à l'aide de la méthode de thermoluminescence: Bulletin de l'Association Francaise pour l'Etude du Quaternaire, v. 20, p. 201–209.

Canfield, H., 1985, Thermoluminescence dating and the chronology of loess deposition in the central United States [M.Sc. thesis]: Madison, University of Wisconsin, 159 p.

Causse, C., and Hillaire-Marcel, C., 1986, Géochemie des familles U et Th dans la matière organique fossile des dépôts interglaciaires et interstadiaires de l'est et du nord du Canada; Potential radiochronologique: Geological Survey of Canada Paper 86–1B, p. 11–18.

Chen, R., and Kirsch, Y., 1981, Analysis of thermally stimulated processes: London/New York, Pergamon Press, 335 p.

Chen, R. Huntley, D. J., and Berger, G. W., 1983, Analysis of thermoluminescence data dominated by second-order kinetics: Physica Status Solidi (a), v. 79, p. 251–261.

Cini Castagnoli, G., and Bonino, G., 1984, Thermoluminescence in recent sediments, *in* Proceedings, National Symposium on Thermally Stimulated Luminescence and Related Phenomena: Ahmedabad, India, Physical Research Laboratory, p. P63–P65.

Cini Castagnoli, G., Bonino, G., and Miono, S., 1982, Thermoluminescence in sediments and historical supernovae explosions: Il Nuovo Cimento, v. 5C, p. 488–494.

Cini Castagnoli, G., Attolini, M. R., Golli, M., and Beer, J., 1984, Solar cycles in the last centuries in Be-10 and O-18 in polar ice and in thermoluminescence signals of a sea sediment: Il Nuovo Cimento, v. 7C, p. 235–244.

Clague, J. J., 1980, Late Quaternary geology and geochronology of British Columbia; Part I, Radiocarbon dates: Geological Survey of Canada Paper 80–13, 28 p.

Cochran, J. K., Carey, A. E., Sholkovitz, E. R., and Surprenant, L. D., 1986, The geochemistry of uranium and thorium in coastal and marine sediments and sediment pore waters: Geochimica et Cosmochimica Acta, v. 50, p. 663–680.

Colman, S. M., Choquette, A. F., Rosholt, J. N., Miller, G. H., and Huntley, D. J., 1986, Dating the upper Cenozoic sediments in Fisher Valley, southeastern Utah: Geological Society of America Bulletin, v. 97, p. 1422–1431.

De, R., Rao, C. N., Kaul, I. K., Sunta, C. M., Nambi, K.S.V., and Bapat, V. N., 1984, Thermoluminescence ages of oceanic carbonate sediments; Oozes and chalks, from DSDP sites 216, 217, 219, and 220 in the northern Indian Ocean: Modern Geology, v. 8, p. 207–215.

Debenham, N. C., 1983, Relativity of thermoluminescence dating of stalagnitic calcite: Nature, v. 304, p. 154–156.

——, 1985, Use of uv emissions in TL dating of sediments: Nuclear Tracks, v. 10, p. 717–724.

Debenham, N. C., and Walton, A. J., 1983, TL properties of some wind-blown sediments: PACT Journal, v. 9, p. 531–538.

---

*Several new applications and developments have been reported at the 1987 TL and ESR Specialists' Seminar, Cambridge, U.K., after this review was completed. These reports will be published in *Quaternary Science Reviews* and in *Nuclear Tracks and Radiation Measurements*.

Debenham, N. C., Driver, H.S.T., and Walton, A. J., 1982, Anomalies in the TL of young calcites: PACT Journal, v. 6, p. 555–562.

Divigalpitiya, W.M.R., 1982, Thermoluminescence dating of sediments [M.Sc. thesis]: Burnaby, British Columbia, Simon Fraser University, 93 p.

Dreimanis, A., Hütt, G., Raukas, A., Whippey, P. W., 1978, Dating methods of Pleistocene deposits and their problems; 1. Thermoluminescence dating: Geoscience Canada, v. 5, p. 55–60.

Eyles, N., and Westgate, J. A., 1987, Restricted regional extent of the Laurentide Ice Sheet in the Great Lakes basins during early Wisconsin glaciation: Geology, v. 15, p. 537–540.

Fleischer, R. L., 1983, Theory of alpha recoil effects on radon release and isotopic disequilibrium: Geochimica et Cosmochimica Acta, v. 47, p. 779–784.

Forman, S. L., Wintle, A. G., Thorleifson, L. H., and Wyatt, P. H., 1987, Thermoluminescence properties and age estimates of Quaternary raised-marine sediments, Hudson Bay Lowland, Canada: Canadian Journal of Earth Sciences, v. 24, p. 2405–2411.

Gardner, G. J., Mortlock, A. J., Price, D. M., Readhead, M. L., and Wasson, R. J., 1987, Thermoluminescence and radiocarbon dating of Australian desert dunes: Australian Journal of Earth Sciences, v. 34, p. 343–357.

Garlick, G.F.J., 1949, Luminescent materials: Oxford, Clarendon Press, 254 p.

Gemmell, A.M.D., 1985, Zeroing of the TL signal of sediment undergoing fluvial transportation; A laboratory experiment: Nuclear Tracks, v. 10, p. 695–702.

Goedicke, C., 1984, Microscopic investigations of the quartz etching technique for TL dating: Nuclear Tracks, v. 9, p. 87–93.

Groube, L., Chappel, J., Muke, J., and Price, D., 1986, A 40,000-year-old human occupation site at Huon Peninsula, Papua, New Guinea: Nature, v. 324, p. 453–455.

Hasan, F. A., Keck, B. D., Hartmetz, C., and Sears, D.W.G., 1986, Anomalous fading of thermoluminescence in meteorites: Journal of Luminescence, v. 34, p. 327–335.

Hashimoto, T., Koyanagi, A., Yokosaka, K., Hayashi, Y., and Sotobayashi, T., 1986, Thermoluminescence color images from quartzs (sic) of beach sands: Geochemical Journal, v. 20, p. 111–118.

Hennig, G. J., and Grün, R., 1983, ESR dating in Quaternary geology: Quaternary Science Reviews, v. 2, p. 157–238.

Hess, H., Kahlert, H., and Krautz, E., 1984, Tunnelling luminescence in zinc silicate doped with gallium: Journal of Luminescence, v. 31, p. 311–313.

Hoogenstraaten, W., 1958, Electron traps in zinc-sulphide phosphors: Philips Research Reports, v. 13, p. 515–693.

Hsü, K. J., and Kelts, K., 1985, Swiss lakes as a geological laboratory; Part I, Turbidity currents: Naturwissenschaften, v. 72, p. 315–321.

Hunt, G. R., Salisbury, J. W., and Lenhoff, C. J., 1971, Visible and near-infrared spectra of minerals and rocks; III, Oxides and hydroxides: Modern Geology, v. 2, p. 195–205.

Huntley, D. J., 1985a, On the zeroing of the thermoluminescence of sediments: Physics and Chemistry of Minerals, v. 12, p. 122–127.

—— , 1985b, A note on the temperature dependence of anomalous fading: Ancient TL, v. 3, p. 20–21.

Huntley, D. J., and Wintle, A. G., 1981, The use of alpha scintillation counting for measuring Th-230 and Pa-231 contents of ocean sediments: Canadian Journal of Earth Sciences, v. 18, p. 419–432.

Huntley, D. J., Berger, G. W., Divigalpitiya, W.M.R., and Brown, T. A., 1983, Thermoluminescence dating of sediments: PACT Journal, v. 9, p. 607–618.

Huntley, D. J., Godfrey-Smith, D. I., and Thewalt, M.L.W., 1985a, Optical dating of sediments: Nature, v. 313, p. 105–107.

Huntley, D. J., Hutton, J. T., and Prescott, J. R., 1985b, South Australian sand dunes; A TL sediment test sequence; Preliminary results: Nuclear Tracks, v. 10, p. 757–758.

Huntley, D. J., Nissen, M. K., Thomson, J., and Calvert, S. E., 1986, An improved alpha scintilation counting method for determination of Th, U, Ra-226, Th-230 excess and Pa-231 excess in marine sediments: Canadian Journal of Earth Sciences, v. 23, p. 959–969.

Huntley, D. J., Berger, G. W., and Bowman, S.G.E., 1988a, Thermoluminescence responses to alpha and beta irradiations, and age determinations when the high dose response is nonlinear: Radiation Effects (in press).

Huntley, D. J., Godrey-Smith, D. I., Thewalt, M.L.W., and Berger, G. W., 1988b, Thermoluminescence spectra of some mineral samples relevant to thermoluminescence dating: Journal of Luminescence, v. 39, p. 123–136.

Hurd, C. M., 1985, Quantum tunnelling and the temperature dependent DC conduction in low-conductivity semiconductors: Journal of Physics, C, Solid State Physics, v. 18, p. 6487–6499.

Hütt, G., Punning, J-M., and Mangerud, J., 1983, Thermoluminescence dating of the Eemian–Early Weichselian sequence at Fjosanger, western Norway: Boreas, v. 12, p. 227–231.

Jacobs, J. A., 1984, Reversals of the Earth's magnetic field: Bristol, Adam Hilger Ltd., 230 p.

Jensen, H. E., and Prescott, J. R., 1983, The thick-source alpha particle counting technique; Comparison with other techniques and solutions to the problem of overcounting: PACT Journal, v. 9, p. 25–35.

Jerlov, N. G., 1976, Marine optics (second edition): New York, Elsevier Scientific Publishing Company, 231 p.

Johnson, R. J., Stipp, J. J., Wintle, A. G. and Tamers, M. A., 1985, Thermoluminescence dating of sediments; A re-extension of age range of loess: Geological Society of America Abstracts with Programs, v. 17, p. 620.

Jungner, H., 1983, Preliminary investigations on TL dating of geological sediments from Finland: PACT Journal, v. 9, p. 565–572.

—— , 1985, Some experiences from an attempt to date post-glacial dunes from Finland by thermoluminescence: Nuclear Tracks, v. 10, p. 749–756.

Kolstrop, E., and Mejdahl, V., 1986, Three frost wedge casts from Jutland (Denmark) and TL dating of their infill: Boreas, v. 15, p. 311–321.

Kronborg, C., 1983, Preliminary results of age determination by TL of interglacial and interstadial sediments: PACT Journal, v. 9, p. 595–605.

Lamothe, M., 1984, Apparent thermoluminescence ages of St.-Pierre sediments at Pierreville, Quebec, and the problem of anomalous fading: Canadian Journal of Earth Sciences, v. 21, p. 1406–1409.

Lamothe, M., and Huntley, D. J., 1988, Thermoluminescence dating of late Pleistocene sediments, St. Lawrence Lowland, eastern Canada: Géographie Physique et Quaternaire (in press).

Leverenz, H. W., 1968, An introduction to luminescence of solids: New York, Dover Publications Inc., 569 p.

Lundqvist, J. and Mejdahl, V., 1987, Thermoluminescence dating of eolian sediments in central Sweden: Geologiska Föreningens i Stockholm Förhandlingar, v. 109, p. 147–158.

Lu Yanchou, Mortlock, A. J., Price, D. M., and Readhead, M. L., 1987a, Thermoluminescence dating of coarse-grain quartz from the Malan Loess at Zhaitang section, China: Quaternary Research, v. 28, p. 356–363.

Lu Yanchou, Prescott, J. R., Robertson, G. M., and Hutton, J. T., 1987b, Thermoluminescence dating of the Malan Loess at Zhaitang, China: Geology, v. 15, p. 603–605.

Mann, H. and Fyfe, W. S., 1985, Uranium uptake by algae; Experimental and natural environments: Canadian Journal of Earth Sciences, v. 22, p. 1899–1903.

May, R. J., and Machette, M., 1984, Thermoluminescence dating of soil carbonate: U.S. Geological Survey Open-File Report 84–083, 25 p.

McDougall, D. J., ed., 1968, Thermoluminescence of geological materials: New York, Academic Press, 678 p.

McKeever, S.W.S., 1984, Thermoluminescence in quartz and silica: Radiation Protection Dosimetry, v. 8, p. 81–98.

—— , 1985. Thermoluminescence of solids: London, Cambridge University Press, 370 p.

McKinnon, W. R., and Hurd, C. M., 1983, Tunnelling and the temperature dependence of hydrogen transfer reactions: Journal of Physical Chemistry, v. 87, p. 1283–1285.

Mejdahl, V., 1983, Feldspar inclusion dating of ceramics and burnt stone: PACT Journal, v. 9, p. 351–364.

—— , 1985, Thermoluminescence dating of partially bleached sediments: Nuclear Tracks, v. 10, p. 711–715.

—— , 1986, Thermoluminescence dating of sediments: Radiation Protection

Dosimetry, v. 17, p. 219–227.

Mejdahl, V., and Wintle, A. G., 1984, Thermoluminescence applied to age determination in archaeology and geology, *in* Horowitz, Y. S., ed., Thermoluminescence and thermoluminescent dosimetry: Boca Raton, CRC Press, v. III, p. 133–190.

Mejdahl, V., Kronborg, C., and Strickertsson, K., 1984, TL dating of Scandinavian Quaternary sediments, *in* Proceedings, National Symposium on Thermally Stimulated Luminescence and Related Phenomena: Ahmedabad, India, Physical Research Laboratory, p. R46–R51.

Miller, T. P. and Smith, R. L., 1987, Late Quaternary caldera-forming eruptions in the eastern Aleutian arc, Alaska: Geology, v. 15, p. 434–438.

Mott, N. F., and Gurney, R. W., 1948, Electronic processes in ionic crystals: New York, Dover Publications Inc., 275 p.

Naeser, C. W., Briggs, N. D., Obradovich, J. D., and Izett, G. A., 1981, Geochronology of Quaternary tephra deposits, *in* Self, S., and Sparks, R.S.J., eds., Tephra studies: Boston, D. Reidel, p. 13–47.

Nakajima, T., 1984, The effect of pulverization of irradiated materials on their thermoluminescence emission: Radiation Protection Dosimetry, v. 6, p. 356–358.

Nambi, K.S.V. and Aitken, M. J., 1986, Annual-dose conversion factors for TL and ESR dating: Archaeometry, v. 28, p. 202–205.

Nanson, G. C., and Young, R. W., 1987, Comparison of thermoluminescence and radiocarbon age-determinations from late-Pleistocene alluvial deposits near Sydney, Australia: Quaternary Research, v. 27, p. 263–269.

Northcliffe, L. C., and Schilling, R. F., 1970, Range and stopping-power tables for heavy ions: Nuclear Data Tables, v. A7, p. 233–463.

Norton, L. D., and Bradford, J. M., 1985, Thermoluminescence dating of loess from western Iowa: Soil Science Society of America Journal, v. 49, p. 708–712.

Prescott, J. R., 1983, Thermoluminescence dating of sand dunes at Roonka, South Australia: PACT Journal, v. 9, p. 505–512.

Prescott, J. R., and Stephan, L. G., 1982, The contribution of cosmic radiation to the environmental dose for thermoluminescence dating; Latitude, altitude and depth dependences: PACT Journal, v. 6, p. 17–25.

Proszynska, H., 1983, TL dating of some subaerial sediments from Poland: PACT Journal, v. 9, p. 539–546.

Readhead, M. L., 1984, Thermoluminescence dating of some Australian sedimentary deposits [Ph.D. thesis]: Canberra, Australian National University, 696 p.

Rendell, H. M., 1983, Problems of estimation of water content history of loesses: PACT Journal, v. 9, p. 523–530.

Rendell, H. M., Gamble, I.J.A. and Townsend, P. D., 1983, Thermoluminescence dating of loess from the Potwar Plateau, northern Pakistan: PACT Journal, v. 9, p. 555–562.

Saro, S., and Pikna, M., 1987, Routine alpha-spectrometry of environmental samples: Applied Radiation and Isotopes, v. 38, p. 399–405.

Singhvi, A. K., and Mejdahl, V., 1985, Thermoluminescence dating of sediments: Nuclear Tracks, v. 10, p. 137–161.

Singhvi, A. K. and Wagner, G. A., 1986, Thermoluminescence dating and its applications to young sedimentary deposits, *in* Hurford, A. J., Jäger, E., and Tencate, I.A.M., eds., Dating young sediments: Bangkok, Thailand, UN CCOP Technical Secretariat, p. 159–197.

Singhvi, A. K., Sharma, Y. P., and Agrawal, D. P., 1982, Thermoluminescence dating of sand dunes in Rajasthan, India: Nature, v. 295, p. 313–315.

Singhvi, A. K., Sharma, Y. P., Agrawal, D. P., and Dhir, R. P., 1983, Thermoluminescence dating of dune sands; Some refinements: PACT Journal, v. 9, p. 498–504.

Singhvi, A. K., Deraniyagal, S. U. and Sengupta, D., 1986a, Thermoluminescence dating of Quaternary red-sand beds; A case study of coastal dunes in Sri Lanka: Earth and Planetary Science Letters, v. 80, p. 139–144.

Singhvi, A. K., Sauer, W., and Wagner, G. A., 1986b, Thermoluminescence dating of loess deposits at Plaidter Hummerich and its imlications for the chronology of Neanderthal Man: Naturwissenschaften, v. 73, p. 205–207.

Singhvi, A. K., Bronger, A., Pant, R. K., and Sauer, W., 1987, Thermoluminescence dating and its implications for the chronostratigraphy of loess-paleosol sequences in the Kashmir Valley (India): Isotope Geoscience, v. 6, p. 45–56.

Smith, B. W., 1983, New applications of thermoluminescence dating and comparisons with other methods [Ph.D. thesis]: University of Adelaide, Australia, 226 p.

Smith, B. W., Aitken, M. J., Rhodes, E. J., Robinson, P. D., and Geldard, D. M., 1986, Optical dating; Methodological aspects: Radiation Protection Dosimetry, v. 17, p. 229–233.

Southgate, G. A., 1985, Thermoluminescence dating of beach and dune sands; Potential of single-grain measurements: Nuclear Tracks, v. 10, p. 743–747.

Spanne, P., 1984, TL readout instrumentation, *in* Horowitz, Y., ed., Thermoluminescence and thermoluminescent dosimetry: CRC Press, Inc., v. III, p. 1–48

Strickertsson, K., 1985, The thermoluminescence of potassium feldspars; Glow curve characteristics and initial rise measurements: Nuclear Tracks, v. 10, p. 613–617.

Stuiver, M., Heusser, C. J., and Yang, I. C., 1978, North American glacial history back to 75,000 years B P.: Science, v. 200, p. 16–21.

Sutton, S. R., 1985a, Thermoluminescence measurements on shock-metamorphosed sandstone and dolomite from Meteor Crater, Arizona; 1, Shock dependence of thermoluminescence properties: Journal of Geophysical Research, v. 90, p. 3683–3689.

——, 1985b, Thermoluminescence measurements on shock-metamorphosed sandstone and dolomite from Meteor Crater, Arizona; 2, Thermoluminescence age of Meteor Crater: Journal of Geophysical Research, v. 90, p. 3690–3700.

Takashima, I., 1979, Preliminary study on the determination of alteration age by a thermoluminescence method: Bulletin of the Geological Survey of Japan, v. 30, p. 285–296.

——, 1985, Thermoluminescence dating of volcanic rocks and alteration minerals and their application to geothermal history: Bulletin of the Geological Survey of Japan, v. 36, p. 321–366.

Takashima, I., and Honda, S., 1986, Application of thermoluminescence dating to geothermal history of the Tamagawa–Kowase area, Hachimantai geothermal field, northeast Japan: Journal of the Japanese Association of Mineralogists, Petrologists, and Economic Geologists, v. 81, p. 458–465.

Tanner, A. B., 1964, Radon migration in the ground, *in* Adams, J.A.S., and Lowder, W.M., eds., The natural radiation environment: Chicago, University of Chicago Press, p. 161–190.

Templer, R. H., 1985, The removal of anomalous fading in zircon: Nuclear Tracks, v. 10, p. 531–538.

——, 1986, The localized transition model of anomalous fading: Radiation Protection Dosimetry, v. 17, p. 493–497.

Townsend, T. E., 1987, Discrimination of iron alteration minerals in visible and near-infrared reflectance data: Journal of Geophysical Research, v. 92, p. 1441–1454.

Valladas, H., Geneste, J. M., Joron, J. L., and Chadelle, J. P., 1986, Thermoluminescence dating of Le Moustier (Dordogne, France): Nature, v. 322, p. 452–454.

Visocekas, R., 1979, Miscellaneous aspects of artificial thermoluminescence of calcite; Emission spectra, athermal detrapping, and anomalous fading: PACT Journal, v. 3, p. 258–266.

——, 1985, Tunnelling radiative recombination in labradorite; its association with anomalous fading of thermoluminscence: Nuclear Tracks, v. 10, p. 521–529.

Visocekas, R., Ceva, T., Marti, C., Lefaucheux, F., and Robert, M. C., 1976, Tunnelling processes in afterglow of calcite: Physica Status Solidi (A), v. 35, p. 315–327.

Westgate, J. A., and Briggs, N. D., 1980, Dating methods of Pleistocene deposits and their problems; V, Tephrochronology and fission-track dating: Geoscience Canada, v. 7, p. 3–10.

Westgate, J. A., Easterbrook, D. J., Naeser, N. D., and Carson, R. J., 1987, Lake Tapps tephra; An early Pleistocene stratigraphic marker in the Puget Lowland, Washington: Quaternary Research, v. 28, p. 340–355.

Wintle, A. G., 1973, Anomalous fading of thermoluminescence in mineral samples: Nature, v. 245, p. 143–144.

—— , 1977a, Detailed study of a thermoluminescence mineral exhibiting anomalous fading: Journal of Luminescence, v. 15, p. 385–393.

—— , 1977b, Thermoluminescence dating of minerals; Traps for the unwary: Journal of Electrostatics, v. 3, p. 281–288.

—— , 1978, A thermoluminescence dating study of some Quaternary calcite; Potential and problems: Canadian Journal of Earth Sciences, v. 15, p. 1977–1986.

—— , 1981, Thermoluminescence dating of the late Devensian loesses in southern England: Nature, v. 289, p. 479–480.

—— , 1982, Thermoluminescence properties of fine grain minerals in loess: Soil Science, v. 134, p. 164–170.

—— , 1985a, Stability of the TL signal in fine grains from loess: Nuclear Tracks, v. 10, p. 725–730.

—— , 1985b, Sensitization of TL signals by exposure to light: Ancient TL, v. 3, p. 17–21.

Wintle, A. G., and Brunnacker, K., 1982, Ages of volcanic tuff in Rheinhessen obtained by thermoluminescence dating of loess: Naturwissenschaften, v. 69, p. 181–183.

Wintle, A. G, and Catt, J. A., 1985a, Thermoluminescence dating of Dimlington Stadial deposits in eastern England: Boreas, v. 14, p. 231–234.

—— , 1985b, Thermoluminescence dating of soils developed in late Devensian loess at Pegwell Bay, Kent: Journal of Soil Science, v. 36, p. 293–298.

Wintle, A. G., and Huntley, D. J., 1980, Thermoluminescence dating of ocean sediments: Canadian Journal of Earth Sciences, v. 17, p. 348–360.

—— , 1982, Thermoluminescence dating of sediments: Quaternary Science Reviews, v. 1, p. 31–53.

—— , 1983, Comment on 'ESR dating of planktonic foraminifera': Nature, v. 305, p. 161–162.

Wintle, A. G. and Westgate, J. A., 1986, Thermoluminescence age of Old Crow tephra in Alaska: Geology, v. 14, p. 594–597.

Wintle, A. G., Shackleton, N. J., and Lautridou, J. P., 1984, Thermoluminescence dating of periods of loess deposition and soil formation in Normandy: Nature, v. 310, p. 491–493.

Zeigler, J. F., 1977, Helium; Stopping powers and ranges in all elemental matter; Vol. 4, The stopping and ranges of ions in matter: New York, Pergamon Press.

MANUSCRIPT ACCEPTED BY THE SOCIETY FEBRUARY 16, 1988

Geological Society of America
Special Paper 227
1988

Amino acid racemization kinetics in wood; Applications to geochronology and geothermometry

N. W. Rutter
Department of Geology, University of Alberta, Edmonton, Alberta T6G 2E3, Canada
C. K. Vlahos
Petro Canada Resources, Products Research and Development, 2928 16 Street N.E., Calgary, Alberta T2E 7K7, Canada

ABSTRACT

The geochemistry of amino acids in fossil wood materials appears to be applicable to geological problems such as correlation, relative-age dating, and paleothermometry of sedimentary deposits in the northern Yukon Territory, Canada (Rutter and Crawford, 1984). Activation energies and Arrhenius frequency factors were calculated for the racemization reaction of several bound amino acids (asp, ala, glu, leu). These parameters were obtained by determining elevated temperature rate constants for the bound amino acids isolated from modern and fossil *Picea glauca* (white spruce). The ratios of dextro to levro stereoisomers (D/L ratio) obtained for bound aspartic acid were found to be the most reliable and yielded values of 18.4 ±2.4 Kcal/mol and 27.6 ±3.0 yr^{-1} for activation energy and Arrhenius frequency factor (ρnA), respectively. Slight differences in kinetic parameters were obtained between fossil and modern wood replicates. Aspartic acid also yielded results correlatable to studies performed on *Sequoiadendron giganteum* (Engel and others, 1977). These findings suggested that species specific effects may not be significant for proteinaceous material found within wood matrices.

Extrapolation of a first-order rate constant for bound aspartic acid in the fossil *Picea* sp. yielded a value of $9.75 \times 10^{-7} yr^{-1}$. This constant was derived from the extent of racemization of the dated sample (>53,000 yr B.P.). Rate constants were similarly determined for various fossil localities in the northern Yukon. These rate constants ranged from $9.75 \times 10^{-7} yr^{-1}$ to $3.24 \pm 0.2 \times 10^{-6} yr^{-1}$. As this reaction is temperature dependent, estimations of paleotemperatures that the fossil samples had experienced were calculated. The values obtained (–49°C ± 30°C) were unrealistic since the racemization does not appear to follow simple reversible first-order kinetics. More reasonable results were obtained (–19°C) if the assumption used for calculation was based on the presence of free aspartic acid (complete protein hydrolysis). The apparent rate of racemization of free aspartic acid is characteristically lower than the apparent rates of racemization of protein-bound amino acids.

INTRODUCTION

Amino acid analysis of fossil remains began more than 30 years ago when Abelson (1954) noted amino acid and protein remains in fossil bones and shells. Since then, many workers (Bada and Protsch, 1973; Bada and Schroeder, 1975; Hare, 1969, 1971, 1974; Kvenvolden, 1975; Miller and Hare, 1975; Rutter and others, 1980; Wehmiller and Belknap, 1982) have promoted the use of amino acid racemization as a viable geochronological method. Analyses of the amino acid geochemistry of fossil wood

have been limited to investigations by Zumberge (1979), Engel and others (1977), Lee and others (1976), Rutter and others (1980), and Rutter and Crawford (1984). Their studies suggest that amino acids have strong applicability to correlation studies, geochronology, and geothermometry. However, the details of amino acid geochemical reactions in fossil wood have not been well identified.

This study was designed to investigate amino acid racemization kinetics in the fossil wood of *Picea glauca* (white spruce), to compare the natural diagenesis of amino acids with the thermally induced diagenesis of amino acids in the laboratory using modern *Picea* sp., and to show how those details relate to geochronological and geothermometric applications in the northern Yukon Territory, Canada. Further, this study was also implemented to elucidate species specific effects on the racemization kinetics of amino acids in fossil wood samples.

Each of the 20 common amino acids has its own characteristic reaction rate constant, k, and energy of activation for racemization. The rate of racemization for free amino acids follows reversible pseudo-first-order reaction kinetics (Bada and Schroeder, 1972). This process can be represented as follows:

$$L - amino\ acid \underset{k_D}{\overset{k_L}{\rightleftarrows}} D - amino\ acid$$

where k_L and k_D are the rate constants. An integrated rate expression for the determination of these constants can be derived:

$$\ell n \left(\frac{1 + \frac{D}{L}}{1 - K^1 \frac{D}{L}} \right) - \ell n \left(\frac{1 + \frac{D}{L}}{1 - K^1 \frac{D}{L}t_0} \right) = (1 + K^1)\, k_1\, t \qquad (1)$$

where D/L is the ratio of the enantiomers derived from thermally induced racemization, and t_0 is the value of induced racemization resulting from the analytical procedure. Plots of natural logarithm of the rate constants determined from elevated-temperature kinetic experiments versus the reciprocal of the absolute temperature provide the Arrhenius frequency factor and the activation energy for the racemization of each amino acid:

$$\ell n k_1 = \ell n\ A - E^*/RT \qquad (2)$$

where k_1 is the forward rate constant, A is the Arrhenius frequency factor, E^* is the activation energy, R is the gas constant, and T is temperature in degrees Kelvin (Engel and others, 1977; Masterton and Slowinski, 1973).

Hare and Mitterer (1969) observed that heating of samples simulated diagenetic processes of proteins in nature. The rate of this chemical reaction is more highly dependent upon temperature than time. Other factors of importance in controlling the reaction are: (1) the type of matrix in which the proteins are embedded, (2) whether the amino acids are protein bound or free, (3) the moisture content of the matrix, (4) the acidity of the matrix, (5) oxidation-reduction conditions, (6) trace metal concentration, and (7) the presence of glucose (Engel and others, 1977; Hare and others, 1980).

Under optimum conditions, protein-bound amino acids should be used in dating analysis because they are less susceptible to contamination. However, if only free amino acids are available, they can be employed, if sources of contamination can be eliminated.

The details of amino acid geochemical reactions in fossil wood are not well known. Racemization reactions in wood were first examined by Lee and others (1976). They investigated racemization reactions in wood through thermal diagenetic simulations using proline and hydroxyproline chemistry. Their investigations led to the possibility of using amino acid racemization chemistry for age dating.

Lee and others (1976) used two different techniques; one uncalibrated and one calibrated. The uncalibrated method is more involved, but widely used today for applications of geothermometry and geochronology. This method involves thermal kinetic studies, enantiomer ratios, rate constants and precise temperature histories. The calibrated method, on the other hand, requires only the enantiomer ratios of the amino acids in the fossil material and another fossil found in the same vicinity that has been dated by another method. With this data, a rate constant can be derived that can be applied to date other fossils. This method requires precise knowledge of the past temperature history of the strata in which the remains are found and also implies that fossils in the area have been subjected to the same paleoenvironment.

Other studies involving fossil woods were done by Engel and others (1977), who investigated aspartic acid racemization in *Sequoia giganteum* heartwood (from which they calculated paleotemperatures), and Zumberge (1979), Zumberge and others (1980), and Engel and others (1977), who investigated the potential effects of glucose on the racemization kinetics of aspartic acid, as well as the effects of moisture content, taxonomic and anatomical site dependency, and the presence of amino sugars, terpenes, and melanoidin substances on racemization kinetics in fossil woods.

Studies by Rutter and others (1980) and Rutter and Crawford (1984) indicate that amino acid racemization of aspartic acid can be used for correlation studies of the northern Yukon sedimentary deposits. Aspartic acid has the greatest applicability to such studies in northern latitudes as it undergoes racemization relatively rapidly.

Wood samples

The northern Yukon provides a unique geologic environment for examining amino acid racemization in fossil wood samples. In this region, thick sequences of Quaternary sediments have been exposed by downcutting of the major rivers and tributaries. These sediments contain a plethora of wood samples and thus provide abundant data on amino acid racemization ratios. The ratios of dextro to leuro amino acid stereoisomers (D/L ratios) have been used successfully for correlation of the lithostratigraphic units present in this region (Rutter and Crawford, 1984). The D/L ratios for aspartic acid of wood samples identified as being *Picea* sp. were used for paleothermometric determinations. These samples were collected from the Old Crow, Bluefish, and Bonnet-Plume Basins in the northern Yukon that are presently within the Continuous Permafrost Zone (see Rutter and Crawford, 1984). The permafrost has contributed to the preservation of the wood remnants and provides an opportunity for comparisons of fossil wood to be made on racemization kinetics with modern wood.

The fossil wood sample used in this kinetic study comes from the 25-m-thick Bluefish section, located about 40 km southwest of Old Crow, on the west bank of the Bluefish River in the Bluefish Basin. The lowermost unit, Tertiary sediment with coal, is overlain by alluvial silt and sand, with gravel near the base, containing wood pieces throughout the entire alluvial sequence. Overlying the alluvial silt and sand is an alluvial gravel and sand unit, which underlies a lacustrine silt unit. The fossil wood sample, taken from a spruce log, ^{14}C dated as older than 53,000 yr B.P. (GSC—2373-3), came from the alluvial gravel and sand under the lacustrine silt.

METHODS

Analytical

The basic technique employed in this study involved the thermal treatment of wood samples, extraction of amino acids from those samples, and analysis of ratios of their enantiomers. The results were compared to data available in the literature in order to establish whether or not the laboratory treatment gave an accurate simulation of natural amino acid diagenesis.

The enantiomers of a given amino acid display the same physical properties, and therefore, cannot be isolated by conventional chromatographic techniques. Diastereomers generally display different chemical and physical properties and can be isolated by simple chromatographic methods (Rutter and others, 1979). If the amino acid enantiomers are mixed with enantiomers of a different compound, a mixture of diastereomers can be produced, provided the configuration of all of the chiral carbons is retained. This is the principle used in many laboratory studies to isolate enantiomers and analyze their enantiomeric ratios during kinetic investigations (Rutter and others, 1980; Hare, 1975; Bada and Shroeder, 1972), and the principle upon which this study was based.

In order to determine the rate of amino acid racemization in the wood of *Picea* sp., two wood samples were collected from two separate localities: (1) the fossil wood sample mentioned above and (2) a fresh, living wood sample obtained from Eckville, Alberta, Canada.

Each sample was thoroughly cleaned with double-distilled water and then air-dried in a plastic weighing dish. The sample was then held in a pair of locking pliers and drawn over a rasp blade, which was fitted into a carpenter's plane (both of these instruments had previously been cleaned of all oils). All of the wood shavings were collected in aluminum foil and passed through a clean 20-mesh sieve. This method minimized the possibility of heat-induced racemization.

The sieved wood shavings were washed twice in disposable centrifuge tubes with 2N HCl to remove surface bacterial contamination and free amino acids, and then washed twice with double distilled water. Between washings, each sample was placed in an ultrasonic bath, centrifuged, and decanted. The suspended wood shavings were then vacuum filtered on a Buchner funnel fitted with Whatman glass fibre paper (GF/A—4.25 cm) and washed several times with distilled water. The filtrate was discarded and the washed particles were collected in small plastic vials. These vials were covered with tissue and placed in a vacuum dessicator to be dried.

Each sample consisted of approximately 100 to 200 mg of wood weighed out into a long-necked glass reaction vessel fitted with a 14/20 glass joint. Care was required to keep the vessel neck free of shavings. Three to 4 ml of distilled water were added to the vessel and degassed as follows. The sample was frozen with a dry ice/acetone mixture ($\approx-78°C$), placed under high vacuum, then thawed; the procedure was repeated from 6 to 10 times to ensure that the sample was evacuated so that all dissolved gases were removed. Once the gases were removed, the reaction vessel was sealed. The sealed reaction vessel was next immersed in a temperature-controlled oil bath, which was fitted with a continuous mechanical stirring mechanism to ensure that temperature fluctuations were minimized.

Forty replicate samples were prepared. One replicate was exposed to each of the following temperatures for one of the following time periods: (1) temperatures—110°C, 120°C, 130°C, 140°C and 150°C; (2) times—1 hr, 3 hr, 6 hr, 9 hr, 12 hr, 1 day, 2 days, and 3 days. A duplicate sample was run for each replicate for the time period of 1 day.

At the conclusion of the heating cycle the wood particles from each vessel were filtered with a Buchner funnel and then transferred to an open, screw-top, glass culture tube (13×100 mm). The tube was covered with tissue and placed in a vacuum dessicator for further drying. Once the replicate had dried, 6 to 8 ml of 5.5N HCl (constant boiling) were added to the tube. This mixture was allowed to reflux at 108°C for 24 hr in an aluminum heating block, and then allowed to cool to room temperature. Next, the replicate was centrifuged and the supernatant liquid collected using a Pasteur pipette and transferred to another clean tube.

Each replicate was then dried by evaporation in a Speed Vac Concentrator and the residue resuspended in 1 to 2 ml of double-distilled water. This resulting solution was placed on top of a freshly generated cation exchange column (Dowex AG 50W-X8k 50-100 mesh) in order to desalt the amino acids. After a series of washings with bed volumes of 2N NaOH, followed by 4 bed volumes of double-distilled water, followed by 2 bed volumes of 2N HCl, and finally 4 bed volumes of double-distilled water, the amino acids were eluted from the column using 2 bed volumes of 3N NH₄OH. Only a portion of the final washing elution was retained for study. Approximately 10 ml of the amino acid eluate were collected in a screw-top culture tube after the solvent front was 1.5 to 2.0 cm from the bottom of the column (the solvent front progression is easily monitored as a substantial amount of heat is generated from the reaction of NH₄OH with the resin). The replicate was then evaporated to dryness again using a Speed Vac Concentrator.

Volatile derivatives of the extracted enantiomers were then made in order to allow their identification via gas chromatography. The enantiomers were first esterified by adding 0.1 ml of acidified isopropanol to the dried eluate. This mixture was placed in a sonic bath until a homogeneous suspension was obtained, and then heated to 100°C for 15 min in an oil bath. The mixture was then evaporated to dryness in the Speed Vac Concentrator.

The enantiomers were then acylated by the addition of 0.1 ml pentafluorproprionic anhydride (PFPA) and 0.3 ml distilled methylene chloride (CH_2Cl_2) to the dried residue. Once again each replicate was placed in a sonic bath until the residue dissolved, and then heated at 100°C for 5 min. The excess PFPA was then removed on a rotary evaporator.

Each replicate was now redissolved in 0.1 to 0.5 ml CH_2Cl_2, and sufficient K_2CO_3 was added to remove any excess water as well as neutralize the solution if any acid was present. The solution was then allowed to stand for a minimum of 8 hr, after which it was filtered through a Gelman Alpha-200, 0.20-mm metricel filter. The derivatized amino acids were injected into a Hewlett Packard Model 5840A gas chromatograph equipped with FID detector, an optically active Chirasil-val capillary column (25 m), and controlled by a digital microprocessor terminal (which reports peak areas by integration). Approximately 0.2 to 1.0 μm of replicate was used in the chromatograph.

Figure 1 shows a typical chromatogram of an amino acid standard solution used to determine the retention times of each individual amino acid being studied. A temperature program history was used for the elution of the enantiomers off the chirasil-val column.

The raw D/L ratios are derived from peak area integration of the resultant amino acid peaks of the chromatograms. These data were manipulated according to equation 1 in order to obtain the rate constant of the forward racemization reaction. A necessary correction is the t=0 term, which allows for racemization due to the procedure.

Rate constants for each amino acid are thus generated by plots of $\ln(1+D/L/1-K'D/L)$ versus time (in hours). The slope of the initial portion of the curves generated is considered to be indicative of the racemization rate of the amino acids within the protein. The slope was determined using linear regression analysis.

Plots of the natural logarithm of the rate constant versus the reciprocal of the absolute temperature provide the Arrhenius frequency factor (A) and the energy of activation (E*) for each amino acid (equation 2). The intercept of the resulting straight line is defined as $\ln(A)$, whereas the slope of the line is defined as E*/R, where R is the gas constant. These data were again subjected to linear regression analysis in order to determine estimates of error and correlation coefficients. Measurement of the activation parameters allows an estimate of paleotemperatures to be calculated.

RESULTS

The D/L ratios obtained were derived from peak area integration of the resultant gas chromatograms. Examples of such chromatograms are shown in Figures 2 through 4. In most replicates only five of the seven amino acids being sought were found: alanine, valine, leucine, aspartic acid, and glutamic acid. The

presence of proline and phenylalanine could not be discerned as illustrated in the aforementioned figures.

The D-glutamic acid peak was difficult to isolate for the experiments conducted on fossil wood replicates, as it was always obscured by an unidentified compound. The resultant double peak is illustrated in Figure 3.

A decrease in the total amino acid concentration in each sample was observed as the temperatures to which the replicates were exposed increased. This decomposition (possibly due to hydrolysis) was most noticeable in the fossil wood experiments. The decomposition effect was inferred from need for a larger sample or the need for concentration of the sample so that GC peaks would be observed. The final 3-day run at 150°C for fossil *Picea* sp. could not be analyzed because of this effect. Problems with peak resolution occurred prior to this, as shown in Figure 4.

Table 1 shows the results obtained for the modern *Picea* sp. sample and Table 2 for the fossil *Picea* sp. sample after correction to remove the contribution of racemization induced by the analytical procedure effects. First-order plots are illustrated in Figures 5 through 11.

Reaction time was restricted to less than 72 hr, compared to those used by other investigators, some of whom extended their studies to more than 300 hr (Engel and others, 1977; Zumberge, 1979). The intent here was to study only the first-order kinetics of the racemization reactions occurring early on in the process.

As shown from the curves in Figures 5 through 11, the enantiomer ratios generally increase with time and temperature. The raw enantiomeric ratios obtained never exceeded 0.460, a value that was obtained for aspartic acid using the fossil wood. In general, the data for both fossil and fresh wood show a trend of increasing rates of racemization in the order valine, leucine, glutamic acid, alanine, and aspartic acid. Each amino acid appeared to have its own unique set of curves. All curves show an initial rapid increase and then a change of slope signifying a change in the rate of racemization (due to the liberation of free amino acids). These rate curves generated from these elevated temperature experiments are qualitatively in agreement with the model studies conducted by Smith and de Sol (1980) and Kriausakul and Mitterer (1980).

From examination of the data listed in Tables 1 and 2 and the curves generated in Figures 5 through 11, apparently aspartic acid provided the most reliable and reproducible results in that the curves generated at the various temperatures were similar. Further, the rate constants obtained from the data fell on well-defined straight lines whose correlation coefficient value was .99 for the fresh wood and .94 for the fossil wood samples.

The rate constants obtained from this study are listed in Table 3. These constants were derived from the slope of the initial linear portion of the rate curves generated previously (i.e., Figures 5 to 11). The rate constant obtained for aspartic acid racemization in modern wood at 120°C is $6.53\pm0.1\times10^{-3}$ hr^{-1}, whereas, the rate constant for aspartic acid in the fossil wood sample at 120°C is $1.06\pm0.06\times10^{-2}$ hr^{-1}.

A rate constant of $2.1\pm.3\times10^{-5}$ yr^{-1} has been reported by

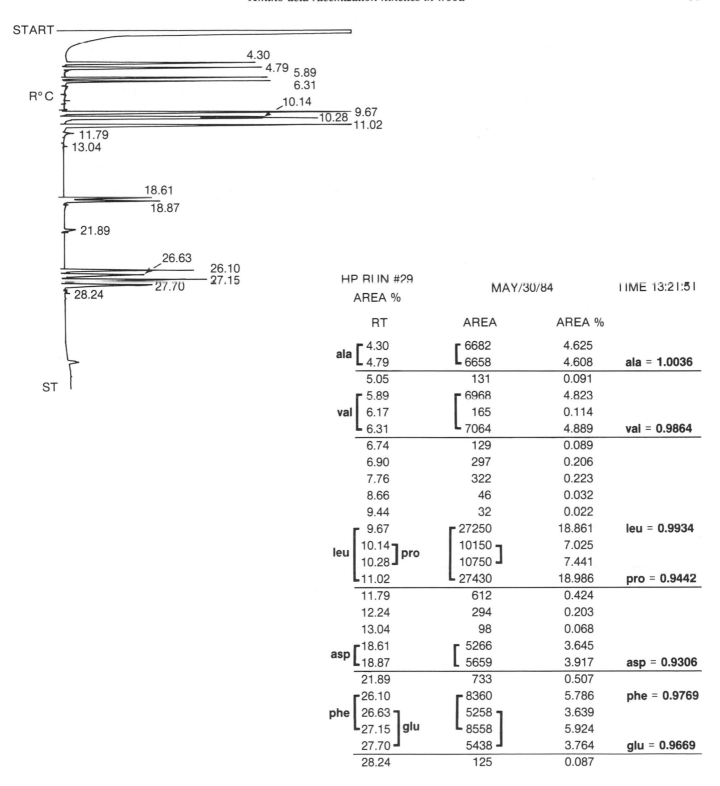

Figure 1. A typical gas chromatogram of the chemically prepared amino acid reference solution, identifying the retention time (RT) of each amino acid sought in sample chromatograms for this study. RT is the time in minutes from the point of sample injection to the time of peak detection and is printed at the top of each peak in the chromatogram. Each amino acid peak in a sample can be located by its retention time.

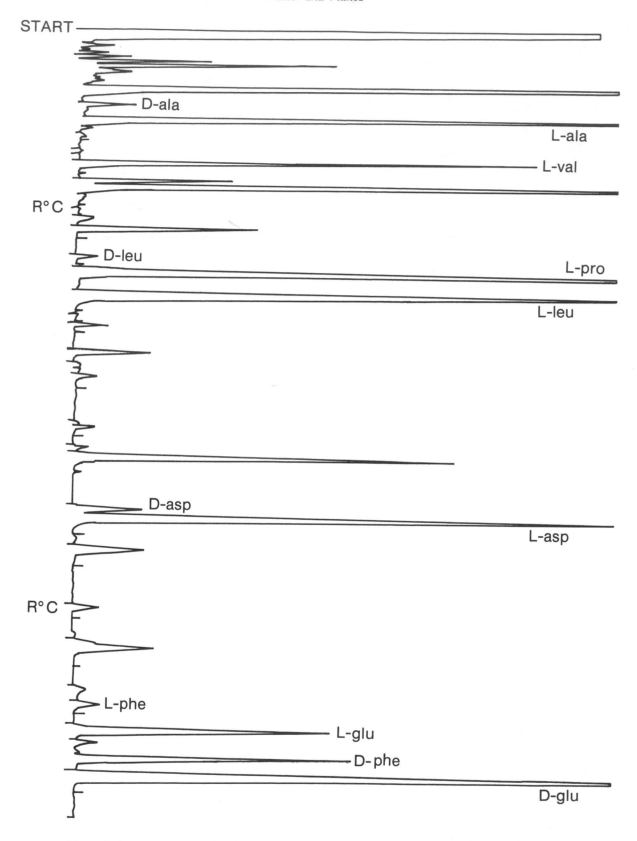

Figure 2. A typical gas chromatogram of the amino acid enantiomers in a word sample. Note the absence of the D-proline peak which should occur between the D-leucine and L-proline peaks. The D-phenylalanine peak is also absent from just before (above) the D-glutamic acid peak.

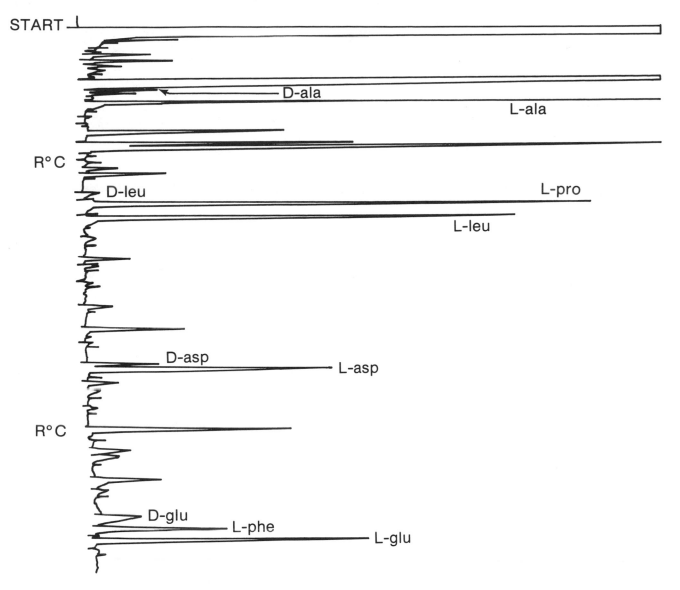

Figure 3. This chromatogram illustrates the double peak observed for D-glutamic acid (labeled as D-glu) in the fossil *Picea* sp. wood sample.

Engel and others (1977) for aspartic acid racemization in *Sequoiadendron* sp. This rate constant was obtained by determining the D/L ratios of various samples of *Sequoiadendron* sp., i.e., 210 B.C., 375 A.D., 1000 A.D., and was not determined by elevated temperature experiments. This value represents an estimation of the rate at ambient temperature conditions (≈5°C). A similar rate constant can be derived for the fossil wood *Picea* sp., as this sample has been radiocarbon dated as greater than 53,000 yr B.P., This rate constant is calculated by rearranging equation 1.

After correcting for time zero and procedure-induced racemization and assuming t of 53,000 yr, k is found to be 8.5×10^{-3} hr^{-1}. The ambient temperature for which this derivation is made is presumably much lower than 5°C as the fossil remnant was found in an area of continuous permafrost. Continuous permafrost is defined as the area north of the −5°C ground-temperature isotherm (Ritchie, 1984).

The rate constant derived by Engel and others (1977) calculated at 110°C (using the Arrhenius equation) is 1.8×10^{-4} hr^{-1}. This rate constant differs by an order of magnitude from the one measured in this work.

A rate constant for free aspartic acid in an aqueous solution has been reported by Zumberge (1979) and is $3.29 \pm .3 \times 10^{-5}$ hr^{-1} at 80°C or 9.95×10^{-3} hr^{-1} at 100°C. Comparison of the data in Table 3 and the published data shows significant differences in rate of racemization of the various amino acids studied between the fresh and fossil wood samples. A faster rate is generally ob-

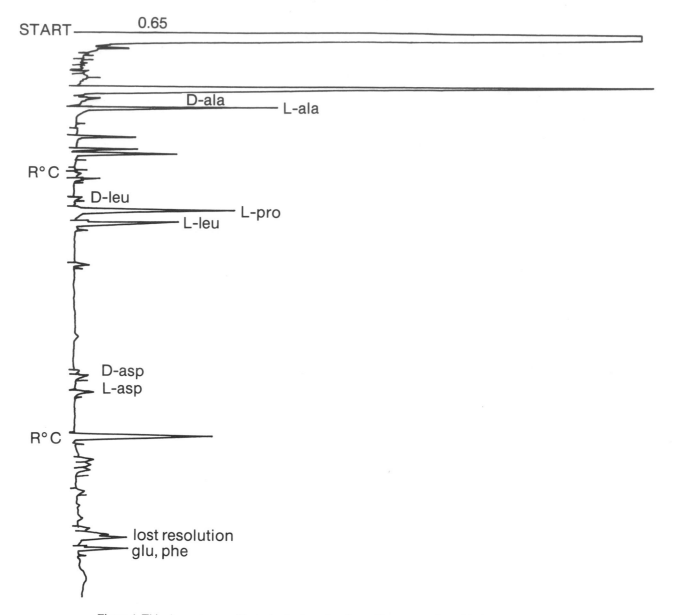

Figure 4. This chromatogram illustrates the loss of amino acid concentration with time and temperature for the fossil *Picea* sp. wood sample. The height and area under each peak show a substantial decrease when compared with the peaks in Figure 3.

served for the fossil wood amino acids, although the rates are of the same order of magnitude. Hence, thermal diagenesis may not truly simulate the natural diagenetic conditions. Because of the paucity of published data concernig the kinetics of racemization in wood samples, proper comparisons cannot be made.

Kinetic studies by Zumberge and others (1980) and Engel and others (1977) show that aspartic acid racemization does not follow reversible first-order kinetics above 150°C for wood samples, while proline does. As proline was not resolved well in this study, no such comparison could be made.

The kinetic constants, such as activation energy (E*) and the

Arrhenius frequency factor (ℓnA), derived from the rate curves are listed in Table 4 and are calculated from the Arrhenius plots illustrated in Figures 12 and 13. These figures show that for those amino acids for which there is sufficient data, the initial slope of the rate curves follows first-order kinetics. However, the kinetics of amino acid racemization in general for fossil and modern wood do not follow simple reversible first-order kinetics.

Activation parameters for the racemization reactions obtained from these experiments are given in Table 4. The energy of activation (E*) for aspartic acid in the fresh wood sample was calculated to be 18.4±2.4 kcal/mol. This value is considerably

TABLE 1. ℓn (1+D/L?1 - D/L) VERSUS TIME FOR THE THERMAL KINETIC EXPERIMENTS CONDUCTED FOR *PICEA* SP. OBTAINED FROM ECKVILLE, ALBERTA

Amino Acid	110°C								120°C							
	1 hr	3 hr	6 hr	9 hr	12 hr	1 day	2 days	3 days	1 hr	3 hr	6 hr	9 hr	12 hr	1 day	2 days	3 days
ala	.0038	.0068	.0250	0	.0162	.0508	.0783	.1090	.0058	.0304	.0302	.0818	—	.0174	.0928	.0454
val	0	0	0	0	.0154	.0156	.0242	.0245	0	0	0	0	0	0	.0122	.0278
leu	0	0	0	0	—	—	—	—	0	.0010	0	0	0	.0046	.0066	.0042
asp	.0173	.0242	.0869	.0302	.0750	.0879	.1463	.2875	.0361	.0649	.0802	.1483	.2842	.2209	.2618	.2890
glu	0	.0190	.0332	.0144	.0200	.0328	.0428	.0651	—	.0062	.0140	.0494	.0356	.0633	.0737	.0974

Amino Acid	130°C								140°C							
	1 hr	3 hr	6 hr	9 hr	12 hr	1 day	2 days	3 days	1 hr	3 hr	6 hr	9 hr	12 hr	1 day	2 days	3 days
ala	.0152	.0282	.0276	.0336	.0338	.0486	.0811	.1091	.0354	—	.0515	.0498	.0414	.0607	.0709	.0797
val	0	0	0	0	0	0	0	0	0	—	.0182	.0200	.0208	.0720	.0744	.0536
leu	0	0	0	.0016	.0048	.0188	.0267	.0267	.0052	—	.0210	.0224	.0148	.0226	.0256	.0266
asp	.0246	.0992	.2030	.2301	.2589	.2731	.3310	.2875	.1150	—	.3156	.3453	.2858	.4661	.3343	.2821
glu	.0008	.0094	.0274	.0386	.0877	.0755	.1648	.1555	.0098	—	.0979	.1275	.1261	.2232	.1842	.2059

Amino Acid	150°C							
	1 hr	3 hr	6 hr	9 hr	12 hr	1 day	2 days	3 days
ala	.0104	.0322	.0374	.0426	.0729	.0825	.0880	.2020
val	0	—	—	.0292	—	.0426	.0522	.0708
leu	0	.0022	.0046	.0178	.0244	.0288	.0370	.1075
asp	.0828	.2682	.2727	.3073	.1940	.0203	.2543	.3346
glu	.0368	.1078	.0915	.0853	.0966	.2465	.2246	.3594

Notes:

1. Data derived from raw D/L ratio values obtained from the gas chromatograms as listed in Appendix III.

2. $\ell n \left(\frac{1+D/L}{1-D/L}\right)_{t=0}$ for the amino acids are as follows:

ala = .0368	asp = .1142	leu = .0428
val = 0	glu = .0326	

lower than the value of 27.4 kcal/mol reported by Zumberge (1979). However, it is in reasonable agreement with the energy of activation determined for *Sequoiadendron* sp. as determined by Engel and others (1977). This value is 23.1 kcal/mol and is within the estimates of error.

A large difference exists between the energy of activation values determined for glutamic acid racemization. The experimentally determined E* for glutamic acid in modern *Picea* sp. is 8.65±0.7, compared to 23.2±0.4 for *Sequoiadendron* sp. This suggests that species and anatomical factors may well influence the rate of racemization for amino acids in wood samples. This has been suggested in the case with molluscan and foraminiferal samples (Rutter and others, 1980; Kvenvolden and others, 1979).

The Arrehenius frequency factor generated from the modern wood experiments for aspartic acid and glutamic acid are $18.5±3.0$ hr^{-1} (or 27.6 yr^{-1}) and $4.37±0.9$ hr^{-1} (or 13.4 yr^{-1}), respectively. The value obtained for aspartic acid is in reasonable agreement with the value of 30.8 yr^{-1} (or 21.7 hr^{-1}) published by Engel and others (1977), whereas the values obtained for glu-

tamic acid are not in agreement. These values also do not compare well with those reported by Zumberge (1979) for free aspartic acid, i.e., 31.39 hr^{-1} (ℓnA).

Other discrepancies also exist in the data in Table 5. The most noticeable is the difference in the activation parameters obtained for alanine and leucine in the fossil and modern wood samples.

DISCUSSION

The rate of racemization for the five amino acids investigated in ascending order are: valine, leucine, alanine, glutamic acid, and aspartic acid. This sequence of rates does not correspond to those observed by Lee and others (1976) for *Pinus aristata* (bristlecone pine), where proline was indicated to have the fastest rate of racemization and aspartic acid a slower rate than glutamic acid. Lee's data is also in disagreement with the data obtained by Engel and others (1977) for *Sequoiadendron* sp., which compares most closely with the data presented in this study.

TABLE 2. ℓn(1+D/L?1 - D/L) VERSUS TIME FOR THE THERMAL KINETIC EXPERIMENTS CONDUCTED FOR *PICEA* SP. FROM THE BLUEFISH BASIN, YUKON (> 53,000 B.P.)

Amino Acid	110°C								120°C							
	1 hr	3 hr	6 hr	9 hr	12 hr	1 day	2 days	3 days	1 hr	3 hr	6 hr	9 hr	12 hr	1 day	2 days	3 days
ala	.0046	—	.0118	—	.0353	.0096	.0636	.0056	.0019	.0066	.0177	.0042	.0046	.0130	.0429	.0196
leu	.0012	—	.0011	—	.1164	.0270	.0172	.0246	.0090	.0314	.0542	.0368	.0170	.0583	.0366	.0462
asp	.0384	—	.0720	—	.1216	.2491	.7631	.3088	.0384	.0717	.1191	.1981	.2696	.2767	.4151	.1402

Amino Acid	130°C								140°C							
	1 hr	3 hr	6 hr	9 hr	12 hr	1 day	2 days	3 days	1 hr	3 hr	6 hr	9 hr	12 hr	1 day	2 days	3 days
ala	0084	0000	0064	0118	0020	0161	—	—	0311	0425	0259	0259	0608	1948	0660	0546
leu	0014	0268	0326	0324	0528	0438	—	—	0448	0534	0350	0787	0713	0949	—	0546
asp	1808	2939	4385	4296	5339	5839	—	—	2478	4182	2079	6271	5818	5834	5722	3656

Amino Acid	150°C							
	1 hr	3 hr	6 hr	9 hr	12 hr	1 day	2 days	3 days
ala	0231	0110	0774	0974	0936	1402	2739	—
leu	0626	0707	0903	0953	0831	1138	1230	—
asp	3554	5001	6107	6515	4400	4741	3487	—

Notes:

1. Values derived from raw D/L ratios obtained from the gas chromatograms listed in Appendix III.

2. Initial $\ln \left(\frac{1+D/L}{1-D/L} \right)_{t=0}$ for the three amino acids are as follows: ala = .0750; leu = .0196; asp = .2176.

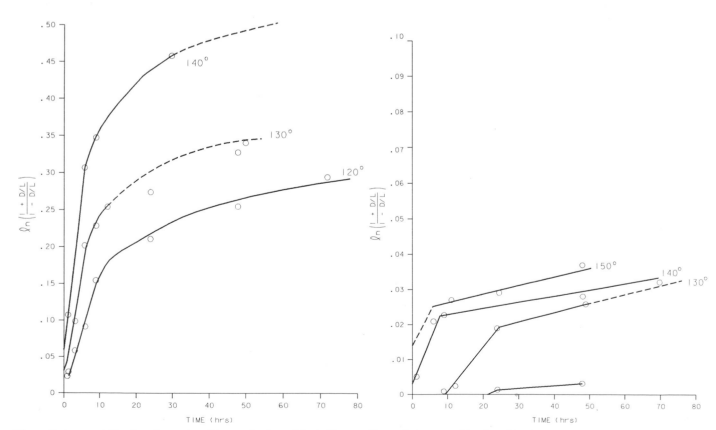

Figure 5. Apparent kinetics of aspartic acid racemization of fresh *Picea* sp. at 120°C, 130°C, 140°C.

Figure 6. Apparent kinetics of leucine racemization in fresh *Picea* sp. at 130°C, 140°C, 150°C.

TABLE 3. EXPERIMENTALLY DETERMINED RATE CONSTANTS FOR ASPARITIC ACID,
GLUTAMIC ACID, LEUCINE, AND ALANINE

Temperature (°C)	Amino Acid	k, [$\ln(\frac{1+D/L}{1-DL})$ versus time]	r = correlation coefficient	Wood Sample
120	aspartic	$6.53 \pm 1.3 \times 10^{-3}$ hr^{-1}	0.9635	fresh
130	aspartic	$1.31 \pm 0.2 \times 10^{-2}$ hr^{-1}	0.9683	fresh
140	aspartic	$2.00 \pm 0.1 \times 10^{-2}$ hr^{-1}	0.9660	fresh
110	aspartic	$3.79 \pm 0.2 \times 10^{-3}$ hr^{-1}	0.9983	fossil
120	aspartic	$1.06 \pm 0.06 \times 10^{-2}$ hr^{-1}	0.9944	fossil
130	aspartic	$1.55 \pm 0.3 \times 10^{-2}$ hr^{-1}	0.9987	fossil
150	aspartic	$2.50 \pm 0.9 \times 10^{-2}$ hr^{-1}	0.9816	fossil
130	leucine	$5.75 \pm 0.1 \times 10^{-4}$ hr^{-1}	0.9999	fresh
140	leucine	$1.13 \pm 0.4 \times 10^{-3}$ hr^{-1}	0.9533	fresh
130	leucine	$1.48 \pm 0.1 \times 10^{-4}$ hr^{-1}	0.9929	fossil
140	leucine	$2.11 \pm 0.4 \times 10^{-3}$ hr^{-1}	0.9999	fossil
150	leucine	$2.81 \pm 0.4 \times 10^{-3}$ hr^{-1}	0.9928	fossil
110	alanine	$1.00 \pm 0.3 \times 10^{-3}$ hr^{-1}	0.9742	fresh
150	alanine	$2.39 \pm 0.5 \times 10^{-3}$ hr^{-1}	0.9441	fresh
110	alanine	$5.94 \pm 3.0 \times 10^{-4}$ hr^{-1}	0.9516	fossil
150	alanine	$4.72 \pm 0.5 \times 10^{-3}$ hr^{-1}	0.9931	fossil
120	glutamic	$1.28 \pm 0.5 \times 10^{-3}$ hr^{-1}	0.9382	fresh
130	glutamic	$1.57 \pm 0.4 \times 10^{-3}$ hr^{-1}	0.9808	fresh
150	glutamic	$2.78 \pm 0.3 \times 10^{-3}$ hr^{-1}	0.9961	fresh

TABLE 4. KINETIC CONSTANTS FOR AMINO ACID RACEMIZATION
IN *PICEA* SP.

Amino Acid	r = correlation coefficient	E* (Kcal/mole)	A*(hr^{-1})
		Fresh	
asparatic	0.9912	18.4 ± 2.4	18.5 ± 3.0
glutamic	0.9967	8.65 ± 0.7	4.37 ± 0.9
alanine	1.000	6.9 ± 1.0	2.19 ± 0.9
leucine	1.000	22.4 ± 3.4	20.5 ± 3.3
		Fossil	
aspartic	0.9435	14.3 ± 3.7	13.4 ± 4.6
alanine	1.000	16.5 ± 2.1	14.2 ± 4.2
leucine	0.9095	9.03 ± 1.3	4.87 ± 0.9

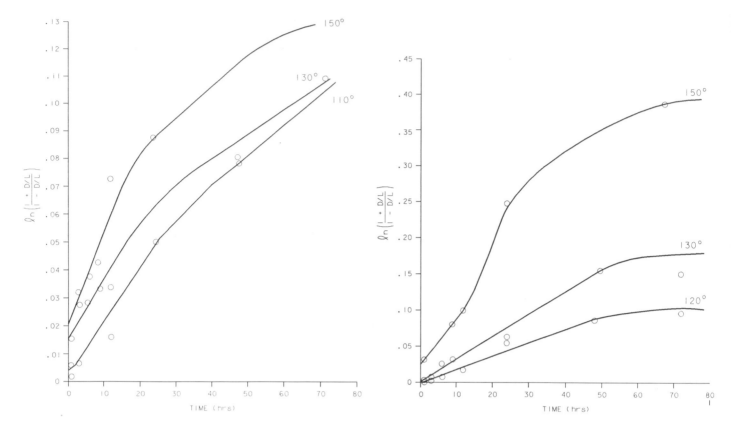

Figure 7. Apparent kinetics of alanine racemization in fresh *Picea* sp. at 110°C, 130°C, 150°C.

Figure 8. Apparent kinetics of glutamic acid racemization in fresh *Picea* sp. at 120°C, 130°C, 150°C.

TABLE 5. D/L RATIOS OBTAINED FROM
THE HUNGRY CREEK AND BLUEFISH SECTIONS

Location	D/L Ratio				Average D/L	k_1 (yr^{-1})
Bonnet Plume Basin	0.15	0.16	0.19	0.18	0.17 ± 0.02	4.06 ± 0.7x10^{-6} yr^{-1}
Hungry Creek	0.16	0.18	0.16	0.12		
	0.20	0.19	0.19			
Bluefish Basin	0.29	0.25	0.19		0.25 ± 0.05	3.24 ± 0.2x10^{-6} yr^{-1}
Bluefish-Section	0.28	0.25				
(approx. 70,000 yr)	0.17	0.31				
Bluefish Basin	0.12				0.12 ± 0.1	9.75 ± 0.4x10^{-7} yr^{-1}
12 Mile Bluff-						
fossil						

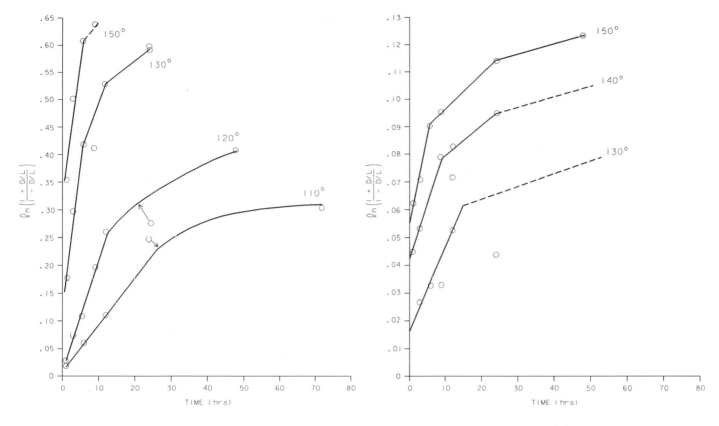

Figure 9. Apparent kinetics of aspartic acid racemization in fossil *Picea* sp. at 110°C, 120°C, 130°C, 150°C.

Figure 10. Apparent kinetics of leucine racemization in fossil *Picea* sp. at 130°C, 140°C, 150°C.

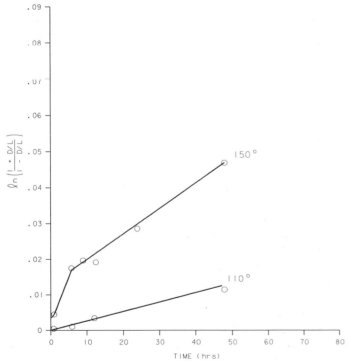

Figure 11. Apparent kinetics of alanine racemization in fossil *Picea* sp. at 110°C, 150°C.

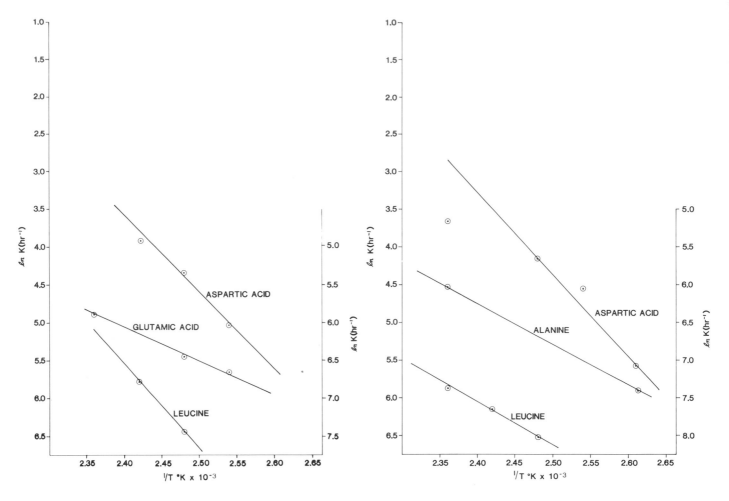

Figure 12. Arrhenius plots of aspartic acid, glutamic acid, and leucine racemization determined from elevated temperature experiments in modern *Picea* sp.

Figure 13. Arrhenius plots of aspartic acid, leucine, and alanine racemization determined from elevated temperature experiments in fossil *Picea* sp.

The rate of racemization for valine was generally very slow and was not a useful investigative tool for the time and temperatures used in this series of experiments. Leucine also has a slow reaction rate, although it proved reasonably useful. Of the amino acids investigated, aspartic acid provided the most reliable and reproducible results for both the fresh and fossil wood samples. Glutamic acid provided reliable information only in the case of the fresh *Picea* sp. sample. Engel and others (1977) suggested that aspartic acid and glutamic acid are less stable than other amino acids due to thermal decomposition. Proline and phenylalanine are more stable and would be more useful for kinetic studies.

The energy of activation found here for aspartic acid in the modern and fossil wood of *Picea* sp. are 18.4±2.4 kcal/mol and 14.3±3.7 kcal/mol, respectively. A slightly lower value for activation energy is obtained from the fossil sample, although the data do agree within experimental error. The slightly lower values

obtained may be attributed to slight differences in natural and thermal diagenesis as protein decomposition in the fossil sample has proceeded to a greater extent. Hence, several factors may have influenced the apparent rate of racemization of aspartic acid in the fossil wood.

The energy of activation and Arrhenius frequency factor for aspartic acid racemization in modern *Sequoiadendron* sp. were determined to be 21.3±0.5 kcal/mol and 30.8±0.3 yr[-1], respectively (Engel and others, 1977). These values do compare well with the value sof 18.4±2.4 kcal/mol and 27.6±3.0 yr[-1] (18.5±3.0 hr[-1]), obtained here for modern *Picea* sp. Again the differences in activation energy suggest that possible species specific and anatomical site effects may be important. This conclusion is less certain when one considers the difficulty in extrapolating rate constants from the rate curves and also the fact that only two wood genera are being compared. In both instances, the trunk of the tree was analyzed, minimizing the effect of anatomical site on the kinetic parameters obtained.

The energy of activation obtained for glutamic acid in the modern *Picea* sp. is 8.65±0.7 kcal/mol. This value does not compare well with the value of 23.2±0.4 kcal/mol for *Sequoiadendron* sp. (Engel and others, 1977). This discrepancy further implies that species specific effects are important in racemization kinetic studies of amino acid residues in wood.

The fresh and fossil wood curves for alanine, leucine, and glutamic acid yield significantly different activation parameters. The different racemization rates may be the result of a number of contributing factors: alanine being created as a reaction product in the decomposition of other amino acids, such as serine (Bada and others, 1978); the introduction of amino acids into the fossil wood matrix by fungi and bacteria (Zumberge and others, 1980); or the nature of the adjacent peptide units and their stereochemical relationships, which may allow the formation of intermediate diketopiperazene structures during chemical reactions that effect the stereochemical L-to-D interconversion as the products may be L- or D-enantiomers (Steinberg and Bada, 1981, 1983).

The apparent disparity in the activation parameters observed for alanine for the fossil and modern wood materials may be related to the diagenetic decomposition of serine and other amino acids. Through the process of dehydration (loss of the OH functional group), serine may produce alanine, as has been observed in the fossil foraminifera (Bada and others, 1978). This dehydration reaction process proceeds more rapidly than the racemization reaction, as do deamination reactions. Hence, the very low activation energy observed for alanine, in modern *Picea* sp. (6.9±1.0 kcal/mol) may well be an indication of such reaction processes. The energy of activation observed for the fossil sample was 16.5±2.1 kcal/mol. This higher value is most probably a truer reflection of the alanine interconversion reaction since most of the simple decomposition reactions (loss of functional groups) should have proceeded to completion over the 53,000-yr interval.

The introduction of secondary amino acids into the wood matrix as contamination is generally identified by the presence of non-protein amino acids, such as β-alanine and λ-amino butyric acid, so that it could have been measured by detailed amino acid analysis. The effect of the secondary amino acid introduction into the matrix may be very pronounced, as unknown mixtures of L- and D-enantiomers may be contaminants as shown in *Pinus aristata* (bristlecone pine) by Zumberge and others (1980).

The stereochemistry of amino acids in proteins is very complex. Proteins are characterized as having primary (linear sequence), secondary (recurring arrangement along one dimension), or tertiary (three-dimensional folded structure), and quaternary (arrangement of two or more chains) structures (Lehninger, 1970). All these structures must influence the rate of hydrolysis, decomposition, and racemization due to neighboring group and steric effects.

Investigations have since revealed an alternate mechanism for the protein diagenesis in fossil materials, the formation of an intermediate diketopiperazene:

The formation of this intermediate is accompanied by extensive racemization of the amino acid residues (Steinberg and Bada, 1983). The formation of these diketopiperazenes during protein decomposition and the rapid rate of racemization that accompanies the formation of these cyclic intermediary structures may help explain the complex amino acid racemization kinetics observed.

Another factor that could explain why discrepancies exist between the fossil wood and fresh wood kinetic parameters is the loss of amino acid concentration through decomposition, most noticeable in the fossil wood of *Picea* sp.

Amino acid decomposition and its effect on the L- to D-racemization of amino acids has been observed by several authors (Hare, 1975; Von Endt, 1982; Bada and others, 1978). Hare (1975) suggested that the effects of hydrolysis and decomposition may be an important factor to consider in racemization studies. The later investigations of Smith and de Sol (1980) suggest that as hydrolysis increases, racemization rate decreases. With continued heating, the composition of the hydrolysate increases in free amino acid concentration. Consequently, the observed rate changes and tends toward the lower value characteristic of free amino acids. A trend toward hydrolysis and amino acid decomposition was generally observed in this work.

Similarly, differences between fossil and fresh wood samples may be related to another reaction, which may have proceeded significantly in the fossil sample. This is the interaction of carbohydrates with amino acids, such as aspartic acid, to produce melanoidins (Engel and others, 1978). This reaction occurs early on in diagenesis and has been demonstrated to increase the apparent rate of racemization of aspartic acid (Engel and others, 1978). The interaction of carbohydrates, glucose for example, with aspartic acid produces an intermediary imine complex (Schiff base). Tautomerization of this intermediate complex often leads to loss of asymmetry. As a result, the reverse hydrolysis reaction can yield racemic-free aspartic acid.

Significant differences have been shown to exist in the rate of racemization of aspartic acid, glutamic acid, leucine, valine, and alanine for fresh and fossil *Picea* sp. samples. The best data for kinetic studies are likely those for fresh wood, so that many of the external factors—such as contamination—are eliminated and only first-order kinetics are examined.

The results of this experiment show that aspartic acid may be used as a geochronological indicator for fossil wood. The presence of an unidentified peak excludes the use of glutamic acid. The slow rate of interconversion of leucine and valine also precludes their use. The possible formation of alanine by the decomposition of serine and other amino acids also precludes the use of alanine. Further, apparently species-specific effects, matrix composition, and anatomical site designations may not be important factors in the use of amino acid dating techniques in fossil wood samples. The differences observed for the racemization rates and rate constants between fresh and fossil woods may be attributed to several factors, such as the introduction of secondary amino acids, amino acid decomposition to produce varied other compounds, and most importantly, the production of various intermediates that enhance the apparent rate of racemization.

APPLICATION: PALEOTHERMOMETRY

One of the objectives of this investigation was to determine the paleotemperature history of various *Picea* sp. samples collected from the three basins in the northern Yukon. The amino acid kinetic data derived for aspartic acid in *Picea* sp. was used for this purpose. This dependency is:

$$\ell n\, k_l = n\ell A - \frac{E^*}{RT}$$

In the case of aspartic acid racemization for *Picea* sp., the Arrhenius equation (equation 2) can abe written:

$$\ell n\, k_l = 27.6 \pm 3.0\ (yr^{-1}) - \frac{18,400\ cal/mol}{1.9872\ T}$$

The values for ρ n A and E^* were derived from the elevated temperature experiments.

The rate equation for the L- to D-interconversion of aspartic acid is based on the assumption that racemization follows linear first-order kinetics. Substituting of the derived Arrhenius equation into the rate equation (equation 1) and solving for $T(°K)$, yields the following temperature prediction equation (Miller and others, 1983):

$$\frac{1}{T} = \frac{1.9872\ cal/mol}{18400\ cal/mol} \cdot \ell n\left(\frac{e^{27.6}\ yr^{-1}}{k_l}\right)$$

This temperature relationship, however, can only be solved if the age of the sediment in which the fossil was found is known (i.e., k_l).

The aspartic acid D/L ratios obtained for this investigation were collected from several fossil localities in the northern Yukon: the Bluefish Basin, the Old Crow Basin, and the Bonnet Plume Basin. This area was selected because many of the wood samples collected were identified as being *Picea* sp. Further, the geologic framework of the area has been identified, and a number of absolute age estimates have been made. For the purpose of illustration, only those values obtained from Hungry Creek and the fossil wood obtained from the Bluefish section are used. The ages of the units for which D/L ratios have been obtained are: Holocene (10,000 yr B.P.), middle to late Wisconsinan (10,000 to 50,000 yr B.P.) and early Wisconsinan/Sangamon (50,000 to 100,000 yr B.P.) (Rutter and Crawford, 1984).

The raw D/L ratios are given in Table 5, as are the associated rate constants. For these units the average temperatures calculated (based on the average D/L ratio obtained and an average age estimate) are:

(a) mid to late Wisconsinan: Bluefish Basin, –49°C±30°C; Bonnet Plume Basin, –40°C±15°C; and –49°C±30°C

(b) early Wisconsinan/Sangamon; Bluefish Basin, –34°C ±13°C

The values obtained through the above calculation are quite unrealistic. Nevertheless, they indicate that the samples have been subjected to extensive periods of permafrost. The continuous permafrost zone is delineated as being the area north of the –5°C ground-temperature isotherm (Ritchie, 1984).

The discrepancy may be related to the fact that the activation parameters were derived for the initial racemization and protein diagenesis occurring early in sample history. As diagenesis and protein decomposition continues, a rate change is observed. This change in rate signifies an increase in the total amount of free amino acid in the hydrolysate. The rate of racemization of the free amino acid is characteristically higher than the rates observed for protein- and peptide-bound amino acids. Aspartic acid was observed to change rate at a ratio of approximately 0.12 to 0.15. The D/L ratios obtained for the fossil samples from the northern Yukon exceeded this limit. Hence, the majority of the aspartic acid present in the wood matrix is non-protein bound. If the calculation is repeated using the kinetic parameters obtained by Zumberge (1979) for an aqueous solution of free aspartic acid (E^* - 27.4 kcal/mol and ℓn A- 31–39 hr^{-1} [40.5 yr^{-1}]), a paleotemperature of –19°C is obtained for the fossil *Picea* sp. sample and –13°C for the Hungry Creek section of the Bonnet Plume Basin. These values are more reasonable.

Clearly, the assumption of simple reversible first-order kinetics is highly suspect. The mechanism of amino acid racemization during protein diagenesis is very complex, as many factors influence the rate of racemization.

ACKNOWLEDGMENTS

The authors acknowledge the technical assistance of Mrs. Z. Kipparisides and Mrs. I. Moffat. Their comments and discussions on experimental technique and problems were most valuable.

REFERENCES CITED

Abelson, P. H., 1954, Organic constituents of fossils: Carnegie Institute of Washington Yearbook, v. 53, p. 97–101.

Bada, J. L., and Protsch, R., 1973, Racemization of aspartic acid and its use in dating fossil bones: National Academy of Science Proceedings, v. 70, no. 5, p. 1331–1334.

Bada, J. L., and Schroeder, R. A., 1972, Racemization of isoleucine in calcareous marine sediments; Kinetics and mechanisms: Earth and Planetary Science Letters, v. 15, p. 1–11.

—— , 1975, Amino acid racemization reactions and their geological implications: Naturwissenschaften, v. 62, p. 71–79.

Bada, J. L., Shou, M. Y., Man, E. H., and Schroeder, R. A., 1978, Decomposition of hydroxy amino acids in foraminiferal tests; Kinetics, mechanisms, and geochronological implications: Earth and Planetary Science Letters, v. 41, p. 67–76.

Engel, M. H., Zumberge, J. E., and Nagy, B., 1977, Kinetics of amino acid racemization in *Sequoiadendron giganteum* heartwood: Annals of Biochemistry, v. 82, p. 415–422.

Engel, M. H., Zumberge, J. E., Nagy, B., and van Devender, T. R., 1978, Variations in aspartic acid racemization in uniformly preserved plants about 11,000 years old: Phytochemistry, v. 17, p. 1559–1562.

Hare, P. E., 1969, Geochemistry of proteins, peptides, and amino acids, *in* Eglington, G., and Murphy, M.T.J., eds., Organic geochemistry; Methods and results: New York, Springer-Verlag, p. 438–463.

—— , 1971, Effect of hydrolysis on the racemization rate of amino acids: Carnegie Institute of Washington Yearbook, v. 70, p. 256–258.

—— , 1974, Amino acid dating of bone; The influence of water: Carnegie Institute of Washington Yearbook, v. 73, p. 576–581.

—— , 1975, Amino acid composition by column chromatography, *in* Needleman, S. B., ed., Molecular biology, biochemistry, and biophysics: Berlin, Springer-Verlag, chapter 7, p. 204–231.

Hare, P. E., and Mitterer, R. M., 1969, Laboratory simulation of amino acid diagenesis in fossils: Carnegie Institute of Washington Yearbook, v. 67, p. 205–208.

Hare, P. E., Hoering, T. C., and King, K., Jr., 1980, Biogeochemistry of amino acids: New York, John Wiley and Sons, 558 p.

Kriausakul, N., and Mitterer, R. M., 1980, Comparison of isoleucine epimerization in a model dipeptide and fossil protein: Geochimica Cosmochimica Acta, v. 44, p. 765.

Kvenvolden, K. A., 1975, Advances in the geochemistry of amino acids: Earth and Planetary Science Annual Review, v. 3, p. 182–212.

Kvenvolden, K. A., Blunt, D. J., McMenamin, M. A., and Straham, S. E., 1979, Geochemistry of amino acids in shells of the clam *Saxodomus* sp., *in* Douglas, A. G., and Maxwell, J. R, eds., Advances in organic geochemistry: Oxford, Pergamon Press, p. 321–332.

Lee, C., Bada, J. L., and Peterson, L., 1976, Amino acids in modern and fossil woods: Nature, v. 259, p. 183–186.

Lehninger, A. L., 1970, Biogeochemistry; The molecular basis of cell structure and function: New York, Worth Publishers, Inc., 1104 p.

Masterton, W. L., and Slowinski, E. J., 1973, Chemical principles: Toronto, W. B. Saunders Company, 709 p.

Miller, G. M., and Hare, P. E., 1975, Use of amino acid reactions in some marine fossils as stratigraphic and geochronologic indicators: Carnegie Institute of Washington Yearbook, v. 74, p. 612–617.

Miller, G. H., Sejrup, H. P., Mangerud, J., and Anderson, B., 1983, Amino acid ratios in Quaternary mollusks and foraminifera from western Norway; Correlation, geochronology and paleotemperature estimates: Boreas, v. 12, p. 107–124.

Ritchie, J. C., 1984, Past and present vegetation of the far northwest of Canada: Toronto, University of Toronto Press, 22 p.

Rutter, N. W., and Crawford, R. J., 1984, Utilizing wood in amino acid dating, *in* Mahaney, W. C., ed., Quaternary dating methods: Amsterdam, Elsevier Publishers, p. 195–209.

Rutter, N. W., Crawford, R. J., and Hamilton, R. D., 1979, Dating methods of Pleistocene deposits and their problems; IV, Amino acid racemization dating: Geoscience Canada, v. 6, no. 3, p. 121–128.

—— , 1980, Correlation and relative age dating of Quaternary strata in the continuous permafrost zone of the northern Yukon with D/L ratios of aspartic acid of wood, freshwater mollusks, and bone, *in* Hare, P. E., Hoering, T. C., and King, K., Jr., eds., Biogeochemistry of amino acids: Toronto, John Wiley and Sons, p. 463–475.

Smith, G. C., and de Sol, B. S., 1980, Racemization of amino acids in dipeptides shows COOH NH₂ for non-sterically hindered residues: Science, v. 207, p. 765.

Steinberg, S., and Bada, J. L., 1981, Diketopiperazene formation during investigation of amino acid racemization in dipeptides: Science, v. 213, p. 544.

—— , 1983, Peptide decomposition in the neutral pH region via the formation of diketopiperazenes: Journal of Organic Chemistry, v. 48, p. 2295.

Von Endt, D. W., 1982, Changes in the amino acids of bone from a well documented and stratified site: Geological Society of America Abstracts with Programs, v. 14, p. 638–639.

Wehmiller, J. F., and Belknap, D. F., 1982, Amino acid age estimates, Quaternary Atlantic Coastal Plain; Comparison with U-series dates, biostratigraphy, and paleomagnetic control: Quaternary Research, v. 18, p. 311–336.

Zumberge, J. E., 1979, Effects of glucose on aspartic acid racemization: Geochimica Cosmochimica Acta, v. 43, p. 1443–1448.

Zumberge, J. E., Engel, M. H., and Nagy, B., 1980, Amino acids in Bristlecone Pine; An evaluation of factors affecting racemization rates and paleothermometry, *in* Hare, P. E., Hoering, T. C., and King, K., Jr., eds., Biogeochemistry of amino acids: New York, John Wiley and Sons, p. 503–525.

MANUSCRIPT ACCEPTED BY THE SOCIETY FEBRUARY 16, 1988

Printed in U.S.A.

Geological Society of America
Special Paper 227
1988

A review of the aminostratigraphy of Quaternary mollusks from United States Atlantic Coastal Plain sites

John F. Wehmiller, Daniel F. Belknap*, Brian S. Boutin, June E. Mirecki, Stephen D. Rahaim, and Linda L. York
Department of Geology, University of Delaware, Newark, Delaware 19716

ABSTRACT

The aminostratigraphic relationships of approximately 150 coastal Quaternary sites from Nova Scotia to Florida and the Bahamas Islands are reviewed. The broad latitudinal range of the sites provides a useful perspective on the relative kinetics of racemization at substantially different temperatures. Local aminostratigraphic sections are presented for five regions in which present mean annual temperatures differ by 3°C or less. Correlation of these individual aminostratigraphies is accomplished by qualitative comparison of results for overlapping sections and by quantitative kinetic modeling. Correlations based on kinetic modeling with local calibration are compared with available U-series data for coastal plain sites. Using basic aminostratigraphic assumptions about the relationship of present and past temperature gradients, the amino acid data from most of the calibration sites follow logical trends. However, significant conflicts between U-series dates and aminostratigraphic age estimates are recognized for sites in South Carolina and for a group of sites in eastern Virginia (central Chesapeake Bay). Reconciliation of the aminostratigraphic data with all of the Atlantic Coastal Plain U-series coral dates is not possible without invoking extreme (and latitudinally variable) thermal effects on the racemization kinetics.

INTRODUCTION

Amino acid racemization (AAR) studies of mollusks from Quaternary marine and nonmarine sites have proven of value in defining relative age relationship and in estimating absolute ages of fossiliferous units (Wehmiller, 1982; G. H. Miller, 1985; Bowen and others, 1985; Miller and Mangerud, 1985; Hearty and others, 1986). The term "aminostratigraphy," first defined by Miller and Hare (1980), involves the direct application of amino acid enantiomeric (D/L) ratios to localities for which identical or similar effective temperatures can be assumed ("effective Quaternary temperatures" or "effective diagenetic temperatures" are the terms most often used to describe these integrated thermal histories). Often such local aminostratigraphies are developed for closely spaced sites with the same current mean annual temperature (CMAT). Ages can be estimated for the different observed groups of D/L values (aminozones) if independent chronologic data are available for calibration or by application of suitable models for the racemization kinetics in the analyzed samples.

In this chapter, we summarize aminostratigraphic studies of mollusks from late Cenozoic marginal marine deposits along the U.S. Atlantic coast. Some of these results have been discussed in previous publications on specific geographic regions. Here we include new data, review analytical methods and interpretive strategies, present several local aminostratigraphic sequences, and provide an overall aminostratigraphic correlation of sites from Nova Scotia (44°N) to Florida (23°N). Our research is part of a large, on-going effort by a number of workers to understand both regional and local stratigraphic sequences in the coastal plain (see Szabo, 1985, and Cronin and others, 1984, for references). This work also has been, and continues to be, an effort to evaluate the reliability of AAR dating methods and to understand the geochemistry, mechanism, and kinetics of diagenetic racemization in mollusks. In a previous paper, Wehmiller and Belknap (1982) noted some serious conflicts between aminostratigraphic age estimates and U-series geochronology. Szabo (1985) has presented some additional isotopic results relevant to these conflicts. These particular issues, which appear to be nearly unique to a few sites on the U.S. Atlantic coast, are discussed in detail in this

---

*Present address: Department of Geological Sciences, University of Maine at Orono, Orono, Maine 04469

paper. Although the conflicts between aminostratigraphy and independent age estimates occur infrequently, they provide useful perspectives on both the assumptions and geochemistry of aminostratigraphy. Unfortunately, a geochemical resolution for these issues is not yet apparent. Therefore, at this time, we choose only to explain the nature of the conflicts and to quantify them using what we consider to be the most appropriate kinetic model for racemization. We also discuss some constraints that can be placed on the thermal histories that might be invoked to reconcile the amino acid and U-series data.

LOCALITIES AND SAMPLES

The localities from which mollusk samples have been obtained are summarized in Table 1 and in Figures 1 through 5. Most of the localities have been sampled at surface outcrops, but some subsurface sites are also included. Samples from higher latitudes (40° to 44°N) have been affected by glaciation: most of those from the southwest coast of Nova Scotia and from Long Island are transported shell fragments contained in glacial drift; samples from Maine are found in latest Pleistocene glaciomarine silt and clay; other unfragmented (and occasionally articulated) samples from Long Island and from eastern Massachusetts appear to have been transported by ice within large blocks of sediment. South of the limit of glaciation, most of the samples have been collected from sediments deposited in lagoonal, estuarine, or back-barrier environments. Articulated and growth position bivalves have been selected for analysis whenever possible, but the majority of samples available at most sites are disarticulated and/or out of life position, so the results presented here are usually limited by the constraint that nontransported samples are rarely available. Fragmented or otherwise visibly altered shells are avoided wherever possible, but sometimes no other sample types are available.

Figure 1 shows the limits of the study area, from Nova Scotia to Florida and the Bahamas Islands. Figures 2 through 5 are larger-scale maps of the "aminostratigraphic regions" that are discussed in more detail in following sections. Current mean annual temperatures are also shown on these maps. The range of temperatures for each region is significant, but small enough so that, to a first approximation, the aminostratigraphy of each region can be considered independently of the effects of major thermal gradients. However, rigorous application of any kinetic model to long-distance correlation and/or age estimation requires the estimation of effective thermal gradients both within and between the regions shown in Figures 2 through 5.

During the course of our work on the Atlantic Coastal Plain, we have analyzed samples collected by many workers. A locality filing system using microcomputer data-base software was used to standardize the nomenclature of these collections. Each locality was given an informal name, and each collection made at that site was given a specific locality number (UDAMS#, for University of Delaware AMinoStratigraphy). Different collections made at what is considered to be the same location received different UDAMS numbers. Appropriate information about latitude, lon-

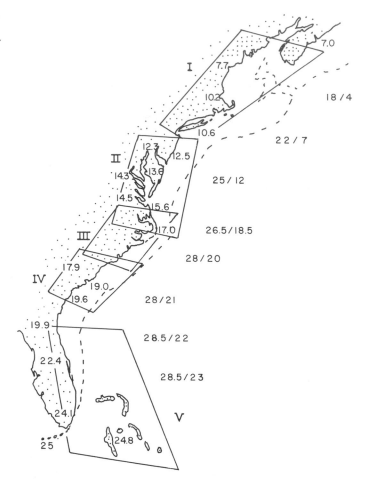

Figure 1. Map of the Atlantic coast of the United States and adjacent regions showing the aminostratigraphic regions discussed in this paper. Regions I through V are shown in Figures 2 through 5. Dashed line is the 200-m isobath. Numerical data shown for coastal sites are mean annual air temperatures (compiled by Belknap, 1979). Numerical data shown for oceanic regions (east of the 200-m isobath) are summer and winter (e.g., 28/20) surface water temperatures (from Imbrie and others, 1983).

gitude, elevation, formation name, collector, date of collection, etc. was also recorded in each file. Because effective or ambient temperature (which is a function of latitude) is a major controlling variable in aminostratigraphic correlation, we find it particularly valuable to keep the developing locality file arranged from north to south by latitude.

Table 1 summarizes the data for the sites discussed here. The abbreviated informal locality names, UDAMS numbers, and locations by latitude/longitude are shown. Figures 2 through 5 show the positions of these collection sites. In the case of subsurface sections with multiple sampling depths (augers, cores, etc.), the site was given one UDAMS number and sample depths were recorded in association with specific data.

Enantiomeric ratios for up to six amino acids in the most

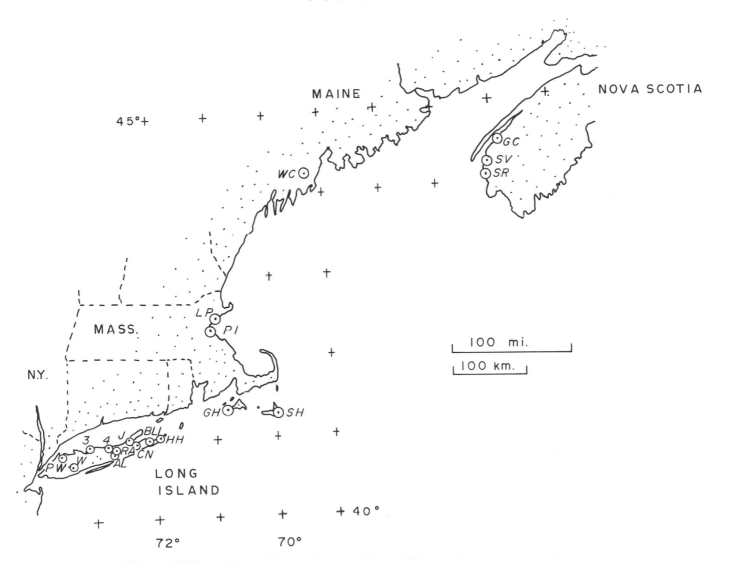

Figure 2. Aminostratigraphic region I, including Nova Scotia, Maine, Massachusetts, and Long Island. Current mean annual temperatures for the region range between 7.5° and 11°C. Locality abbreviations as in Table 1. Holocene localities are Whitney Corner (WC) and Lynn Pit (LP) in Maine and Massachusetts, respectively. Sankaty Head (SH) on Nantucket Island serves as the late Pleistocene calibration locality for this region.

representative genus (genera) for each of the localities considered here are given in Table 1. The number of samples analyzed from each locality is shown in column four of Table 1. Often the number of available samples has been limited. The enantiomeric ratios given in Table 1 are the mean values of multiple sample analyses or, in most of those cases where only one sample was available, the mean of two analyses of that single sample. The chromatographic resolution of alanine, valine, proline, and glutamic acid enantiomers often (roughly 30% of the time) is complicated by interfering peaks, so the mean values for these amino acids shown in Table 1 do not include ratios based on chromatographic peaks with obvious interferences. Data in Table 1 are

grouped according to regional aminozone nomenclature, which is described under the heading "Regional Discussions."

The molluscan genera used for aminostratigraphic studies are the bivalves *Mercenaria, Crassostrea, Anadara, Mulinia, Chione, Rangia, Macoma,* and *Spisula,* and the gastropod *Busycon. Mercenaria* was the genus most often used because it is usually a thick-shelled and mechanically robust sample. In some cases we studied the effects of shell alteration on amino acid D/L values by systematically analyzing different portions of the same shell (with variable degrees of alteration) or by analyzing different shells with variable preservation from the same outcrop. Lower D/L values are often observed in the altered portions of these

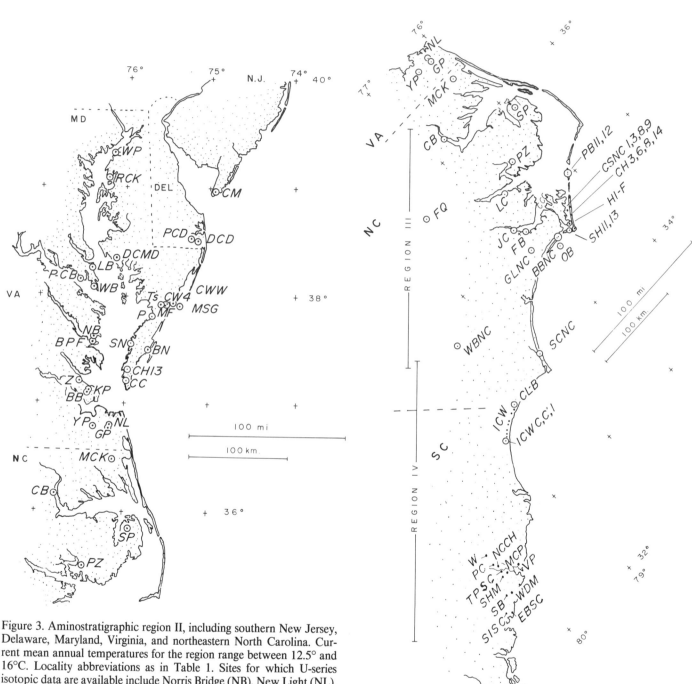

Figure 3. Aminostratigraphic region II, including southern New Jersey, Delaware, Maryland, Virginia, and northeastern North Carolina. Current mean annual temperatures for the region range between 12.5° and 16°C. Locality abbreviations as in Table 1. Sites for which U-series isotopic data are available include Norris Bridge (NB), New Light (NL), Gomez Pit (GP), Moyock (MCK), Stetson Pit (SP), and Ponzer (PZ).

Figure 4. Aminostratigraphic regions III and IV for North Carolina and South Carolina. Current mean annual temperatures for region III range between 16° and 18°C; for region IV, between 18° and 19°C. Locality abbreviations as in Table 1. Region III localities receiving most attention here are Stetson Pit (SP), Ponzer (PZ), and Flanner Beach (FB). Region IV localities of importance include all the Intracoastal Waterway Sites (ICW) near Myrtle Beach, South Carolina, and, in particular, Mark Clark Pit (MCP) and Scanawah Island I (SI), both near Charleston, South Carolina.

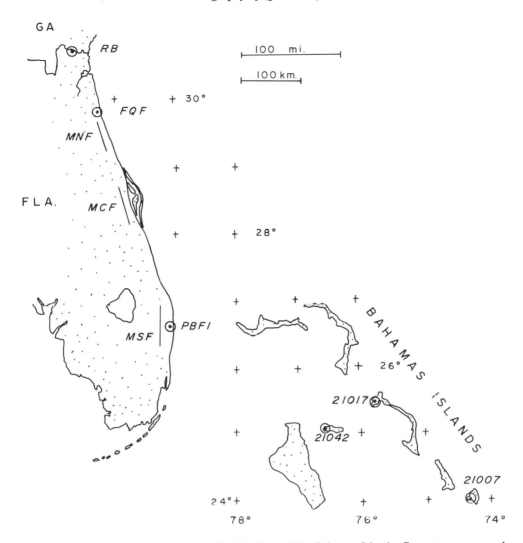

Figure 5. Aminostratigraphic region V for Florida and the Bahamas Islands. Current mean annual temperatures for northern and central Florida range between 20° and 23°C; for southern Florida and the Bahamas, between 23° and 25°C. Locality abbreviations as in Table 1. MNF, MCF, and MSF are abbreviations for "Mitterer North Florida," "Mitterer Central Florida," and "Mitterer Southern Florida," respectively. These regions represent the collections for which amino acid data were published by Mitterer (1975). PBF1 and San Salvador Island (21007) are the two sites in this region at which both mollusks and U-series-dated corals have been collected simultaneously.

shells. In a few cases we analyzed shells that have less than ideal preservation (fragmented, chalky or leached appearance) simply because these were the only samples available. Even in these cases, however, little or no recrystallization (calcite <2 percent) was detected by x-ray diffraction.

Although *Mercenaria* is the preferred sample type, other genera were also studied because: (1) *Mercenaria* is not always present at sites of interest, so the aminostratigraphic value of other genera must be established, and (2) intergeneric relative racemization rates between *Mercenaria* and other common genera must be known so that data for each can be converted into common

values for direct aminostratigraphic comparison. In order to accomplish this task with field samples, several sites (with different ages) should be located, each with a large number of samples of different genera. Analyses of co-existing genera then permit documentation of intergeneric relative racemization rates. Alternatively, these relative rates can be modeled using laboratory pyrolysis experiments. Available data show that the relative rates of racemization in different molluscan genera can be duplicated in this manner (Keenan, 1983). From the results obtained from either laboratory experiments or from field sites, we infer the following kinetic groups for the mollusk genera that we studied:

TABLE 1. ATLANTIC COASTAL PLAIN AMINOSTRATIGRAPHY DATA

UDAMSLOC	Locality	Genus	No. of Samples	Latitude	Longitude	Region	Aminozone	D/L Leu	Allo/Iso	D/L Val	D/L Glu	D/L Ala	D/L Pro
Nova Scotia													
UD00001	GC	Merc	6	44.49	65.91	I	c	0.32			0.25	0.45	0.39
UD00014	SV	Merc	4	44.28	66.14	I	b	0.24			0.20	0.39	0.28
UD01004	WC	Hiat	2	44.16	69.31	I	(a)	0.08		0.05	0.08	0.13	0.12
UD00010	SR	Merc	4	44.03	66.13	I	b	0.24			0.20	0.39	0.28
Massachusetts													
UD03011	LP	Hiat	2	42.46	70.99	I	(a)	0.08		0.06	0.09	0.15	0.12
UD03012	PI	Merc	2	42.30	70.92	I	b	0.21		0.22	0.20	0.42	0.23
UD03033	GH	Merc	1	41.35	70.83	I	d	0.71		0.79	0.64	0.83	0.73
UD03008 - 10	SH (u & l)	Merc	11	41.25	69.96	I	b	0.24		0.17	0.20	0.43	0.34
	SH (u)	Merc	4	41.25	69.96	I	b		0.16				
	SH (l)	Merc	4	41.25	69.96	I	b		0.15				
	SH (u)	Anad	2	41.25	69.96	I	b		0.18				
	SH (l)	Anad	3	41.25	69.96	I	b		0.13				
New York													
UD04000	HH	Unkn	4	41.04	72.98	I	(c)	0.31		0.27	0.24	0.38	0.50
UD04002	BLI	Merc	3	40.99	72.31	I	b	0.23		0.20	0.42	0.46	
UD04003	J	Merc	2	40.95	72.58	I	c	0.43		0.26	0.29	0.65	
UD04001	CN	Merc	2	40.93	72.45	I	b	0.21		0.16	0.19	0.38	0.31
UD04009	3	Merc	1	40.92	73.18	I	b	0.18		0.17	0.18	0.25	
UD04005	RA	Merc	3	40.92	72.96	I	c	0.35		0.26	0.30	0.59	0.34
UD04006	AL	Merc	2	40.85	73.03	I	b	0.19		0.15	0.15	0.29	
UD04004	4	Astar	1	40.84	72.82	I	(b)	0.19		0.14	0.15	0.41	
UD04010 - 13	PW	Crass	7	40.83	73.68	I	(c)	0.28		0.18	0.15	0.66	0.35
UD04015	W	Spis	1	40.58	73.38	I	(b)	0.14		0.10	0.14	0.24	0.30
New Jersey, Maryland, Delaware, Virginia													
UD05009	WP	Rang	3	39.31	76.18	II	(b)	0.35		0.32	0.24	0.60	0.48
UD05008	RCK	Rang	1	39.13	76.17	II	(c)	0.44		0.23	0.42	0.68	
UD05003	CM	Merc	3	38.96	75.25	II	a	0.28	0.21	0.20	0.29	0.45	
UD05012	PCD	Merc	3	38.53	75.25	II	b	0.33	0.28	0.25	0.28	0.47	0.48
UD05007	DCD	Merc	8	38.50	75.12	II	d	0.53	0.44	0.39	0.38	0.72	0.60
UD05007	DCMD	Rang	3	38.48	76.25	II	(d)	0.65		0.44	0.45	0.89	0.90
UD05002	LB	Merc	2	38.24	76.39	II	a	0.28		0.31	0.24	0.56	0.36
UD05001	PCB	Rang	1	38.21	76.60	II	(d)	0.59		0.68	0.44	0.73	0.64
UD05000	WB	Merc	12	38.06	76.36	II	a	0.29	0.25	0.27	0.23	0.45	0.39
UD06009	CW4	Merc	2	37.96	75.49	II	d	0.59		0.55	0.47	0.79	0.70
UD06007	CWW	Merc	2	37.95	75.46	II	e	0.88		0.99	0.90	0.99	0.91
UD06004	MF	Merc	3	37.95	75.50	II	d	0.58		0.47	0.46	0.78	0.72
UD06002	TS	Merc	20	37.95	75.54	II	d	0.54		0.54	0.41	0.70	0.65
UD06006	MSG	Merc	1	37.90	75.36	II	d	0.54		0.39	0.35		0.72
UD06008	P	Merc	2	37.79	75.67	II	d	0.55		0.60	0.44	0.80	0.65
UD06000, 06017	NB	Merc	9	37.63	75.50	II	d	0.55	0.49	0.42	0.39	0.74	0.69
UD06000	NB	Anad	4	37.63	75.50	II	d	0.51		0.39	0.37	0.67	
UD06001	BPF	Merc	1	37.58	75.39	II	e	0.81		0.25	0.23	0.49	0.31
UD06012	SN	Merc	1	37.57	75.90	II	a	0.25		0.25	0.23	0.49	0.31

TABLE 1. (Continued)

New Jersey, Maryland, Delaware, Virginia (continued)

UDAMSLOC	Locality	Genus	No. of Samples	Latitude	Longitude	Region	Aminozone	D/L Leu	Allo/Iso	D/L Val	D/L Glu	D/L Ala	D/L Pro
UD06000	NB	Anad	4	37.63	75.50	II	d	0.51		0.39	0.37	0.67	
UD06001	BPF	Merc	1	37.58	75.39	II	e	0.81		0.90	0.99	0.90	0.89
UD06012	SN	Merc	1	37.57	75.90	II	a	0.25		0.25	0.23	0.49	0.31
UD06013	BN	Mul	2	37.54	75.77	II	(a)	0.24		0.15	0.16	0.42	
UD06011	CH13	Mul	2	37.30	75.98	II	(a)	0.23		0.16	0.21	0.35	0.25
UD06019	Z	Merc	2	37.24	76.53	II	e	0.89		1.00	0.90	0.99	0.89
UD06014	CC	Merc	2	37.21	76.97	II	a	0.25		0.24	0.24	0.45	0.24
UD06021	KP	Merc	5	37.10	76.43	II	c	0.41		0.35	0.31	0.60	0.51
UD06040, 06020	BB	Merc	4	37.08	76.43	II	c	0.44		0.36	0.29	0.60	0.60
UD06023	NL	Merc	7	36.80	76.19	II	a	0.25		0.21	0.21	0.48	0.37
UD06024	NL	Merc	3	36.80	76.19	II	c	0.42		0.37	0.31	0.65	0.53
UD06029	GP	Merc	4	36.79	76.17	II	d	0.52		0.44	0.37	0.71	0.53
UD06030	GP	Merc	8	36.79	76.17	II	a	0.22		0.15	0.15	0.39	
UD06030	GP	Anad	3	36.79	76.17	II	a	0.26		0.16	0.17	0.41	
UD06056	GP	Merc	11	36.79	76.17	II	c		0.33				
UD06045	GP	Merc	15	36.79	76.17	II	a		0.15				
UD06054	GP	Merc	3	36.79	76.17	II	e		1.08				
UD06027	WP (1)	Merc	2	36.79	76.17	II	c	0.36		0.35	0.26	0.58	0.47
UD06027	WP (1)	Merc	1	36.79	76.17	II	a	0.18		0.15	0.19	0.34	0.21
UD06041	YP	Merc	2	36.76	76.38	II	c	0.37		0.34	0.28	0.50	0.46
UD06034 - 06036	YP	Anad	2	36.76	76.38	II	e	0.88		0.98	0.84	0.94	0.88
North Carolina													
UD07000	MCK	Merc	4	36.52	76.17	II, III	b	0.33		0.23	0.23	0.49	0.39
UD07023	CB	Merc	1	36.19	76.75	II, III	e	0.88		0.95	0.99	0.97	0.95
UD07002	SPIT	Merc	6	35.86	75.86	II, III	b		0.23				
UD07002	SPIT	Merc	1	35.86	75.86	II, III	b	0.33		0.20	0.21	0.50	
UD07002	SPIT	Mul	6	35.86	75.86	III	A (2)	0.28		0.24	0.20	0.41	0.37
UD07002	SPIT	Mul	5	35.86	75.86	III	B (2)	0.51		0.66		0.74	0.65
UD07065	FQNC	Noet	1	35.67	77.64	III	(e)	0.89			0.77	0.89	0.91
UD07006	PZ	Merc	2	35.55	76.44	III	c	0.41	0.33	0.27	0.29	0.60	0.56
UD07007	PZ	Mul	3	35.55	76.44	III	A (2)	0.27		0.26		0.37	0.36
UD07007	PZ	Mul	4	35.55	76.44	III	C (2)	0.63		0.66		0.76	0.74
UD07066	LCNC	Merc	1	35.38	76.79	III	e	0.90		0.93	0.93	0.97	0.94
UD07012	JC	Mul	1	35.08	77.02	III	C (2)	0.69		0.78	0.85	0.85	0.85
UD07012	JC	Merc	8	35.08	77.02	III	e	0.88	0.91	0.80	0.75	0.93	0.89
UD07075	PB12	Merc	1	35.05	76.05	III	d	0.78		0.29	0.34	0.95	
UD07074	PB11	Merc	1	35.03	76.08	III	c	0.47		0.16	0.24	0.64	
UD07074	PB11	Mul	1	35.03	76.08	III	A (2)	0.21		0.31	0.30	0.41	0.57
UD07008 - 0711	FB	Merc	6	34.98	76.95	III	c	0.46		0.29	0.31	0.62	
UD07026	NRE2	Merc	1	34.98	76.95	III	c	0.43		0.18	0.23	0.56	
UD07026	NRE2	Mul	2	34.98	76.95	III	A (2)	0.27		0.16	0.17	0.44	
UD07008	FB	Mul	6	34.98	76.95	III	A (2)	0.27				0.40	0.33
UD07008	FB	Mul	30	34.98	76.95	III	A (2)		0.152 ± 0.028				

TABLE 1. (Continued)

UDAMSLOC	Locality	Genus	No. of Samples	Latitude	Longitude	Region	Aminozone	D/L Leu	Allo/Iso	D/L Val	D/L Glu	D/L Ala	D/L Pro
North Carolina (continued)													
UD07042	CH14	Mul	1	34.86	76.31	III	A (2)	0.19		0.12	0.17	0.35	
UD07034	CSNC1	Merc	1	34.75	76.33	III	c	0.42		0.28	0.33	0.63	
UD07033	CSNC3	Merc	1	34.75	76.33	III	d	0.54		0.40	0.37	0.80	
UD07032	CSNC8	Merc	1	34.75	76.33	III		0.08		0.06	0.09	0.15	
UD07031	CSNC9	Merc	3	34.75	76.33	III	b	0.30		0.21	0.30	0.52	
UD07031	CSNC9	Merc	1	34.75	76.33	III	c	0.47		0.42	0.38	0.71	
UD07031	CSNC9	Merc	1	34.75	76.33	III	d	0.61		0.55	0.41	0.82	
UD07044	CH-8	Merc	1	34.73	76.44	III	d	0.62		0.50	0.49	0.90	
UD07036	GLNC	Mul	1	34.73	77.03	III		0.49		0.52	0.50	0.76	
UD07049	HI-F	Merc	1	34.71	76.58	III	b	0.31		0.22	0.24	0.52	
UD07051	BBNC35	Merc	1	34.71	76.78	III	d	0.73		0.54	0.64	0.91	
UD07050	BBNC3	Merc	2	34.70	76.72	III	d	0.59		0.42	0.45	0.82	
UD07045	CH-6	Merc	1	34.68	76.48	III	c	0.38		0.34	0.32	0.61	
UD07057	SH13	Mul	1	34.65	76.56	III	B (2)	0.51		0.48	0.48	0.67	0.62
UD07058	SH11	Merc	2	34.63	76.53	III		0.18		0.17	0.15	0.42	
UD07046	CH-3	Mul	1	34.60	76.53	III	A (2)	0.26		0.19	0.26	0.46	
UD07017	WBNC	Mul	1	34.56	78.48	III	C (2)	0.76		0.73		0.85	0.79
UD07064	OB86	Merc	1	34.50	76.80	III	d	0.66		0.53	0.57	0.85	
UD07064	OB86	Merc	3	34.50	76.80	III	d	0.57		0.49	0.47	0.83	
UD07061	OB84	Merc	3	34.65	76.82	III	d	0.64		0.55	0.54	0.84	
UD07060	OB85	Merc	4	34.65	76.80	III	d	0.58		0.49	0.48	0.83	
UD07059	OB88	Merc	2	34.68	76.80	III	d	0.68		0.60	0.59	0.90	
UD07018	SCNC	Merc	2	34.08	77.92	III	d	0.52		0.35	0.45	0.65	0.67
UD07019	CLB	Merc	4	33.89	78.56	III, IV	e	0.89	0.97	0.90	0.87	0.95	0.87
South Carolina													
UD08011	ICW11	Merc	2	33.86	78.62	IV	a	0.55	0.52	0.41	0.41	0.74	0.64
UD08010	ICW10	Merc	5	33.83	78.69	IV	a	0.57	0.59	0.42	0.38	0.79	0.66
UD08044	MBAP3	Merc	2	33.82	78.75	IV	d	0.88		0.90	0.85	0.96	0.88
UD08009	ICW9	Merc	5	33.82	78.72	IV	b	0.64	0.54	0.47	0.45	0.82	0.73
UD08008	ICW8	Merc	4	33.80	78.76	IV	a	0.54	0.48	0.37	0.40	0.76	0.62
UD08007	ICW7	Merc	3	33.79	78.76	IV	b	0.63	0.52	0.44	0.45	0.80	0.73
UD08024	ICWJ1	Merc	1	33.77	78.81	IV	b	0.66			0.51	0.85	
UD08024	ICWJ2	Merc	2	33.77	78.81	IV	c	0.78		0.70	0.80	0.95	
UD08024	ICWJ4	Merc	1	33.77	78.81	IV	d	0.88		0.84	0.87	0.96	0.87
UD08006	ICW6	Merc	4	33.77	78.81	IV	d	0.93	0.90	0.87	0.85	0.99	0.91
UD08005	ICW5	Merc	3	33.76	78.82	IV	c	0.79	0.86	0.73	0.71	0.96	0.81
UD08019	ICWF'2 - 4	Merc	3	33.76	78.82	IV	c	0.75			0.69	0.95	
UD08019	ICWF'1	Merc	1	33.76	78.82	IV	b	0.66			0.56	0.85	
UD08003	ICW3	Merc	3	33.74	78.87	IV	b	0.66	0.48	0.41	0.46	0.81	0.79
UD08015	ICWC'4	Merc	1	33.71	78.92	IV	a	0.52		0.35	0.37	0.72	
UD08015	ICWC'1, 2	Merc	4	33.71	78.92	IV	b	0.67		0.42	0.44	0.75	0.78
UD08001	ICW1	Merc	1	33.71	78.92	IV	a	0.59	0.54	0.38	0.42	0.78	0.69
UD08014	ICWC	Merc	3	33.71	78.92	IV	a	0.53		0.35	0.35	0.69	
UD08050	W	Merc	1	33.13	80.04	IV	d	0.88		0.92	0.90	0.90	0.90

TABLE 1. (Continued)

UDAMSLOC	Locality	Genus	No. of Samples	Latitude	Longitude	Region	Aminozone	D/L Leu	Allo/Iso	D/L Val	D/L Glu	D/L Ala	D/L Pro
South Carolina (continued)													
UD08001	ICW1	Merc	1	33.71	78.72	IV	a	0.59	0.54	0.68	0.42	0.78	0.69
UD08014	ICWC	Merc	3	33.71	78.92	IV	a	0.53		0.35	0.35	0.69	
UD08050	W	Merc	1	33.13	80.04	IV	d	0.88		0.92	0.90	0.90	0.90
UD08051	PC	Merc	1	32.92	80.08	IV	a	0.57	0.55	0.37	0.46	0.79	0.67
UD08030	NCCH	Merc	3	32.88	80.03	IV	a	0.58	0.77	0.49	0.45	0.81	0.68
UD08032	TPSC	Merc	2	32.87	80.05	IV	c	0.78	1.05	0.65	0.65	0.95	0.85
UD08033	TPSC	Merc	1	32.87	80.05	IV	d	0.96	0.51	0.92	0.91	0.99	
UD08031	SHM	Merc	3	32.85	80.08	IV	b	0.67		0.52	0.52	0.88	0.77
UD08038	MCP	Merc	6	32.83	80.03	IV	a	0.57		0.42	0.42	0.81	0.69
UD08037	VP	Merc	3	32.81	79.84	IV	a	0.55		0.37	0.42	0.78	0.69
UD08052	SB	Merc	1	32.67	80.25	IV	a	0.61		0.60	0.49	0.80	0.71
UD08041	WDM	Merc	1	32.65	80.18	IV	d	0.88			0.88	0.90	0.90
UD08035	SISC	Merc	3	32.56	80.36	IV	a	0.59	0.60	0.45	0.45	0.80	0.70
UD08042	EBSC	Merc	3	32.53	80.27	IV	a	0.51		0.35	0.34	0.72	0.55
Florida													
UD10000	RB	Merc	1	30.72	81.61	V	d	0.82		0.84	0.64	0.84	0.83
UD10076	FQF	Merc	1	29.87	81.25	V	b	0.58		0.36	0.48	0.79	0.69
MNF	MNF(3)	Merc	n/a	29.67	81.25	V	a		0.42				
MNF	MNF(3)	Merc	n/a	29.67	81.25	V	b		0.56				
MNF	MNF(3)	Merc	n/a	29.67	81.25	V	c		0.71				
MNF	MNF(3)	Merc	n/a	29.67	81.25	V	d		0.88				
MCF	MCF(3)	Merc	n/a	28.50	81.00	V	a		0.56				
MCF	MCF(3)	Merc	n/a	28.50	81.00	V	b		0.71				
MCF	MCF(3)	Merc	n/a	28.50	81.00	V	c		0.92				
MSF	MSF(3)	Merc	n/a	26.75	30.25	V	a		0.71				
MSF	MSF(3)	Merc	n/a	26.75	30.25	V	b		0.88				
MSF	MSF(3)	Merc	n/a	26.75	30.25	V	c		0.99				
UD10069	PBF-1	Chione	3	26.71	30.18	V	(a)		0.68				
Bahamas													
UD21017	NELTHRA	Strombus	1	25.47	75.77	V	(a)	0.57		0.52	0.44	0.66	0.56
UD21042	CLFTNPT	Lucina	1	25.00	76.53	V	(a)	0.78			0.73		0.76
UD21007	CKBRNTWN	Chione	3	24.03	74.42	V	(a)	0.62		0.50	0.48	0.89	

Notes:

1. WP samples are poorly documented and ratios are slightly low due to analytical conditions.
2. Capitalized aminozones for Region III Mulinia refer to Figure 12.
3. Localities MNF, MCF, and MSF represent groups of sites from Mitterer (1975); n/a = number of samples not known.
4. A/I data from Regions IV and V from McCartan and others (1982) or Mitterer (1975) unless noted in text.
5. Aminozones are those for Mercenaria; aminozones indicated in parentheses are probable correlative zones derived from other genera.

"fast-racemizing—*Macoma, Busycon;* intermediate—*Mercenaria, Chione;* intermediate to slow—*Rangia, Anadara, Spisula;* slow-racemizing—*Mulinia, Crassostrea.*

Similar categories have been recognized for Pacific coast mollusks (Lajoie and others, 1980; Atwater and others, 1981), and Miller and Mangerud (1985) have recognized different rates of epimerization in samples of European mollusks. A simple chemical model to explain these kinetic groups in terms of their amino acid composition has been proposed (Wehmiller, 1980), and the groups identified above are generally consistent with this model. For any given amino acid, the difference in D/L values within a group is typically less than 10 percent; the difference between groups is usually 15 to 20 percent. The distinction between the groups is usually recognizable in the enantiomeric ratio data for all the amino acids reported in Table 1, although each amino acid is somewhat limited in its resolving power in the high and low ends of the range of enantiomeric ratios. In the case of several genera (particularly *Busycon, Chione, Anadara,* and *Spisula*), we have not had appropriate field samples to confirm that relative intergeneric trends persist over the entire range of observable enantiomeric ratios, so the kinetic groups outlined above require further evaluation.

The kinetic groups above are identified at the generic level. Species-level differences in apparent racemization rates are not well documented. We rarely collected two species of the same genus at the same site, so this question is difficult to answer using analyses of field samples. We often used either *Mercenaria mercenaria* or *Mercenaria campechiensis.* The two can be distinguished easily when living, but identification of fossil samples at the species level is often difficult, particularly for fragmented specimens. Some of the aminostratigraphic issues encountered using *Mercenaria* could be interpreted as being the result of species effects on racemization. However, we satisfied ourselves that no significant species effect exists on racemization kinetics in *Mercenaria* by performing laboratory pyrolysis experiments on both *M. mercenaria* and *M. campechiensis*—the observed kinetics of racemization for different amino acids are identical within experimental error for the two species (Rahaim, 1986).

Several sites have associated biostratigraphic, paleomagnetic, or isotopic data that offer potential calibration for local or regional aminostratigraphies. Szabo (1985) has presented the most recent review of the U-Th data for Atlantic Coastal Plain solitary coral samples, supplementing the discussion of Wehmiller and Belknap (1982) regarding the conflicts between aminostratigraphic age estimates and U-Th dates. Table 2 is a summary of U-series isotopic data (Szabo, 1985) for the sites from which we have specimens.

Two of the U-Th analyses in Table 2 were done at the Lamont-Doherty Geological Observatory on samples obtained by us, but the majority of samples listed in Table 2 were collected by other workers. Consequently, the U-Th-dated corals have only rarely been collected in direct association with the mollusks used for aminostratigraphic studies. Therefore, the control that these dates provide for the aminostratigraphy often depends on in-

ferred correlations between different sites. This issue may be significant for some of the current conflicts between aminostratigraphic and U-Th dates, although other geochemical factors must be considered in the evaluation of both dating methods (Wehmiller and Belknap, 1982).

METHODS AND RESULTS: GENERAL DISCUSSION

The analytical methods that we used have been described in several previous publications (Wehmiller and Belknap, 1978; 1982; Wehmiller and Emerson, 1980; Wehmiller, 1984a; 1984b). These procedures are reviewed here because the data presented were not all obtained using identical methods, and we are currently involved in a program of re-analyzing many samples that were previously analyzed by less sophisticated methods nearly ten years ago. In the course of our work, we usually found little difference between early and later results, with a few notable exceptions. Miller and others (1983) also noted the effects of different analytical procedures on results obtained over long time intervals.

The results produced in the laboratory at the University of Delaware (and also at the University of South Florida and the University of Maine—D. F. Belknap, post–1980) were obtained by one of several gas chromatographic methods. Early phases of our work, from 1974 until 1978, depended almost entirely on the preparation of the NTFA-(+)-2-butyl derivatives of total hydrolyzate amino acid mixtures. Chromatographic separation was achieved using stainless steel capillary columns, coated with either Carbowax 20M or OV225. Beginning in 1978, glass capillary OV225 columns were routinely used along with the stainless steel columns for all (+)-2-butyl derivatives. Although analysis time is a bit longer for the glass capillary systems, the high resolution offered by this method makes it advantageous. Comparison of D/L values obtained by glass capillary and stainless steel columns usually indicated that the two methods agreed to within 1 or 2 percent for those amino acids that are fully resolved by both columns. Also in 1978, we began to use the chromatographic scheme of Frank and others (1977), in which NTFA-isopropyl derivatives separated on 25-m glass capillary Chirasil-val column because this method was somewhat more rapid and employed much less expensive reagents.

Until late in 1980, all enantiomeric ratios derived from these methods were based on peak height ratios measured by ruler. Column overloading, which can happen frequently with glass capillary columns, can cause asymmetric peaks that result in distorted peak height ratios, so multiple chromatograms (between three and five in most cases, employing different sample volumes) were used in order to obtain the most reliable D/L values. In those cases where asymmetric peaks were used in ratio calculations, peak widths at half-heights were used to factor the peak height ratios to the best value. Since late 1980, we have had electronic integration available for the direct calculation of both peak areas and peak heights, from which appropriate ratios have been calculated either manually or by various data reduction

TABLE 2. U-SERIES CORAL DATES FROM ATLANTIC COASTAL PLAIN
AND THEIR POSSIBLE RELATION TO AMINOSTRATIGRAPHY SITES
(from Szabo, 1985)

Aminostratigraphy Site	Nearest U-Th Site*	U-Th Date (ka)
Sankaty Head (SH)	same	133
Norris Bridge (NB)	NB	190
Mathews Field (MF)	MF	190 to >340
New Light (NL)	NL	74
Gomez Pit (GP)	GP	74
Womack Pit (WP)	WP	62
Moyock (MCK)	Moyock spoil pile	74
Moyock: (coral attached to shell; coral U-Th, shell AAR)		108 (LDGO)
Stetson Pit (SP)	Stetson pit	75
Ponzer (PZ)	Ponzer	190
James City (JC)	same	>750
Calabash (CLB)	same	>750
ICW 9	same	>750 [†]
ICW 6	same	>750
ICW 5	same	460
ICW 1	ICW 1	212? [†]
Venning Pit (VP)	VP	94
Mark Clark Pit (MCP)	MC	120
Scanawah Island (SI)	same	89 (LDGO)
Scanawah Island (SI)	same	90
Trailwood Pit (TP)	Ravenel	220
Shadowmoss (SH)	SHMoss	>79
PBF-1	same	159 (McM; see below)

* "Same" indicates that U-series dated sample and aminostratigraphy samples were collected at the same time in direct association with each other. If the mollusks and U-series coral samples were collected at different times, by different workers, at sites that are thought to be equivalent, then the same locality name is used. The quoted U-Th date is the apparent age (majority of dates in Szabo, 1985) derived from $^{230}Th/^{234}U$ isotopic ratios alone. LDGO refers to analyses done at Lamont Doherty Geological Observatory on samples collected by us. McM refers to an analysis done at McMaster University on a sample collected by us. The isotopic data for these samples are given below:

Moyock Coral, LDGO 1618A, Sample wt. 8.46 gm

U (ppm)	3.23 ± 0.08	$^{234}U/^{238}U$	1.093 ± 0.018
Th (ppm)	0.72 ± 0.06	$^{230}Th/^{232}Th$	9.11 ± 0.07
$^{230}Th/^{234}U$	0.6372 ± 0.02	Date = 108,000 ± 6,000	

SISC Coral (same sample as that analyzed by Szabo, 1985) LDGO 1557

U (ppm)	2.89 ± 0.09	$^{234}U/^{238}U$	1.15 ± 0.03
Th (ppm)	0.409 ± 0.04	$^{230}Th/^{232}Th$	13.41 ± 1.39
$^{230}Th/^{234}U$	0.5668 ± 0.023	Date = 89,000 ± 6,000	

PBF-1 Coral: solitary coral collected with mollusks, 10/85
Sample JW85-39B; data from H. Schwarcz, personal communication, 1987

U (ppm)	1.97	$^{234}U/^{238}U$	1.153 ± 0.023
Th (ppm)	0.21	$^{230}Th/^{232}Th$	26.6
$^{230}Th/^{234}U$	0.789 ± 0.021	Date = 159,000 ± 9,000	

[†] Note that Corrado and others (1986) have chosen to emphasize the significance of some stage 5 (ca. 80,000 to 130,000 yr) U-Th dates reported by Szabo (1985) from sections near these sites. Szabo (1985) noted that these particular U-series dates were probably suspect.

computer programs that we employ. Comparison of results derived by electronic integration with those derived by ruler measurements indicates that ruler measurements generally match (within 2 to 3 percent) those obtained electronically. However, we encountered a few results, obtained by electronic integration, that differ significantly from those obtained by simple ruler measurements. In these cases, we consider the integrator results to be the more reliable.

The gas chromatographic methods employed generally yield enantiomeric ratio data for the following amino acids: leucine, glutamic acid, valine, alanine, proline, aspartic acid, and phenylalanine. D-alloisoleucine/L-isoleucine (allo/iso) values have been obtained under conditions of optimum chromatographic resolution (see Wehmiller, 1984a, 1986). Allo/iso values are the most commonly reported enantiomeric ratio (Wehmiller, 1984b) because most laboratories use ion-exchange chromatographic methods, which resolve the D- and L-forms of isoleucine but do not resolve the D/L values of other amino acids. Since 1985 we have performed HPLC ion-exchange analyses on selected samples, making it possible for us to compare directly (using the same shell or sample extract) enantiomeric ratios determined by gas and liquid chromatographic methods. Allo/iso values are included in Table 1 where available.

The amino acids that we have relied on most for aminostratigraphic resolution have been in order of preference: leucine, glutamic acid, valine, proline, and alanine. D/L data for these amino acids are given in Table 1. Aspartic acid and phenylalanine usually can be used for relative age distinctions, but the value of these amino acids is usually diminished because they are either too extensively racemized (in Pleistocene samples) to have much resolving power or their chromatographic resolution is not satisfactory. Alanine, and also proline in those cases where it is measured, can be useful in age resolution of relatively young samples or those from higher latitudes (greater than 40°N). The different chromatographic columns used in the course of our work have somewhat different elution characteristics for the amino acid mixtures being studied, so some enantiomeric ratios appear to be more reliable than others simply because the two peaks (for D- and L-forms) are well resolved and because the same enantiomeric ratio is determined by different chromatographic methods. Leucine is the amino acid for which we have the greatest confidence because D/L leucine values obtained for both (+)-2-butyl and isopropyl derivatives of the same mixtures have been nearly identical. For most other amino acids, differences of up to 10 percent in D/L values determined by the different analytical methods are common. Because of these analytical artifacts, we rely on leucine for establishing a relative aminostratigraphy, and then use the other amino acid enantiomeric ratios as tests of internal consistency. Because the methods employed in our lab have changed over the past ten years, systematic shifts in results for some amino acids may occur, but these shifts do not preclude the use of multiple enantiomeric ratios for tests of consistency. In some cases, faster racemizing amino acids (proline, alanine) imply age differences that are not apparent in the data for slower racemizing amino acids (valine, glutamic acid, leucine). In a few cases, major inconsistencies among the D/L values of a single sample indicate the sample is unreliable. Examples of the relationships (intrageneric relative rates) between the different amino acid enantiomeric ratios are found in Belknap (1979) and Lajoie and others (1980). Blank areas in Table 1 indicate that the D/L value for a particular amino acid was not determined. The sample analyses that do not include any D/L value for proline were obtained using the isopropyl alcohol derivative method, whereas all other analyses were obtained using the butanol derivative (OV225 or Carbowax columns). Direct comparisons of D/L values should be made only for those samples analyzed by the same method.

An effort to compare enantiomeric ratio results obtained in different laboratories (Wehmiller, 1984b) has recently been made. This effort employed three homogeneously powdered Pleistocene mollusk samples, each with different degrees of racemization. Two of the sites (MCSC and CLB) involved are also discussed here, as they are from the southeastern U.S. coastal plain. These interlaboratory samples have been repeatedly analyzed in our laboratory as a check on long-term trends in analytical results. A summary of these results is given in Appendix IV. The results obtained in the initial phase of the interlaboratory comparison (Wehmiller, 1984b) indicate that some very large discrepancies can exist between laboratories or between analytical methods, but that most laboratories can reproduce their own measurements with a precision of 10 percent or better, depending on particular amino acids. Comparisons of results obtained at Delaware (J. F. Wehmiller and colleagues) with those obtained by D. F. Belknap at either the University of South Florida or the University of Maine have always shown good agreement (ca. 3%).

We consider here the D/L values for the total amino acid population, those amino acids released by acid hydrolysis of the dissolved sample. Free amino acids (those determined after sample dissolution without acid hydrolysis) have not been measured routinely in this laboratory, primarily because they are so extensively racemized in most Pleistocene Atlantic coast samples that they have little age resolution capability. However, quantitative (HPLC) free amino acid data are useful for assessing the extent of diagenetic hydrolysis of our samples. Aminostratigraphic studies of high-latitude regions have often employed free amino acid determinations (see G. H. Miller, 1985, for example).

In the presentation of data in the following sections, enantiomeric ratios are presented in individual graphs for each aminostratigraphic region. Emphasis is placed on leucine data that are suitably screened for internal consistency with other amino acids; results for other amino acids are presented only for those cases in which additional relative chronologic information is available. In all cases, weight is given to those sites or groups of sites for which multiple analyses are available.

The precision of multiple analyses is probably affected by several factors, many of which are difficult to evaluate objectively. The precision of multiple chromatograms of a single sample extract is usually 1 to 3 percent; the precision of multiple analyses (separate fragments) of a single shell can be as good as 1 to 3 percent or as poor as 20 percent if the shell is not homogenously preserved. *Mercenaria* generally yields D/L values 15 to 20 percent lower in the growth edge region than in the hinge or inner layers portions of the shell (Mirecki, 1985). We have also found that small fragments (less than 20 mg) cut from large *Mercenaria* valves (usually weighing between 30 and 60 g) can also yield highly variable D/L results, probably because of small, local inhomogeneities (of diagenetic or biochemical origin) in the shell structure. For shells of any size, samples taken for analysis must be large enough to be representative of the shell's chemistry, even though instrumental sensitivity permits much smaller sample sizes. Our preferred sample size is about 0.5 to 1.0 g for a typical *Mercenaria* sample.

The question of how many samples are required to establish the resolution of different aminozones in a local area is of constant concern. Often the small number of available samples precludes rigorous distinction of aminozones (see Miller and Hare, 1980, for discussion of one statistical approach to aminozone resolution). We usually sought at least three shells of any genus at a particular site, but for several sites we have analyzed as many as 12 of a particular genus. At T's Corner, Virginia (T'S, 06002 in Table 1), 20 *Mercenaria* samples were analyzed over the course of several years. All but three yielded D/L leucine values between 0.49 and 0.55, but, coincidently, out of the first four samples analyzed from the site, two yielded D/L leucine values of 0.44 and 0.42, and two yielded D/L leucine values of 0.52 and 0.53. At the time of these analyses, we felt that this range of results could represent samples of significantly different ages (all the samples had been excavated by a dragline operation and their stratigraphic relationships were not known), simply because all prior information indicated that the precision for multiple analyses at a single site should be much better than observed with these four samples. Subsequent study indicated that the best D/L leucine value for this site is undoubtedly in the order of 0.53; furthermore, the two samples from this site with the lowest D/L values all appeared more altered (chalky rather than lustrous on interior surfaces) than the samples with higher ratios, even though none of these samples appeared recrystallized when studied by x-ray diffraction. In contrast to this example, results from other sites on the Atlantic Coastal Plain, and from sites in Europe, the Arctic, and the Pacific coast of the U.S. often indicate that "high-resolution" aminostratigraphy can be accomplished with as few as two well-preserved samples per site, although we acknowledge that more samples per site would be desirable. For purposes of discussion in the following sections, and based on the results obtained from the T's Corner site and neighboring correlative localities (Mixon, 1985), we assume that a standard deviation of ±10 percent of the mean D/L leucine value for a particular aminozone is a conservative estimate of the overall uncertainty of that aminozone. Miller and Mangerud (1985) commented that the commonly used criteria for data evaluation are largely empirical because the precision of a particular data set is often used as

the primary test of the reliability of the results. While good precision is necessary for confidence in the results, precision alone is not proof of the accuracy of the data.

REGIONAL DISCUSSIONS

Aminostratigraphy, based on leucine or isoleucine in combination with other amino acids, of specific geographic regions as shown in Figures 2 through 5, is discussed in this section. Qualitative identification of clusters (aminozones) of amino acid enantiomeric ratios in each geographic region is based on the assumption that effective temperature differences within each region have been small. Each region is assigned a roman numeral (I through V), and each apparent aminozone in the region is given a letter. Qualitative correlations of local aminozones are presented, assuming that enantiomeric ratios for coeval units would be similar, or slightly higher, for sites at lower latitudes (higher temperatures). Following these regional presentations, correlations between aminozones from each region are discussed by considering latitudinal gradients of temperature and kinetic models of racemization. The association of each aminozone with formations or outcrops of particular significance is discussed where appropriate. In some sections we refer to our "preferred" kinetic-model age estimate (as presented in Wehmiller and Belknap, 1982), although we consider optional age estimates where appropriate.

Region I: Nova Scotia to Long Island (Fig. 2)

Shells analyzed from this region include samples from shelly drift along the southwest coast of Nova Scotia, from numerous drift and marine clay subsurface and outcrop samples on Long Island, and glaciomarine samples from the latest Pleistocene Presumpscot Formation of Maine. D/L leucine values for samples from Region I are shown in Figure 6. All samples are *Mercenaria* unless otherwise indicated; in the case of *Crassostrea* samples analyzed from several sites on Long Island, the D/L leucine data have been converted into equivalent *Mercenaria* ratios in order to simplify the presentation of data. The method of conversion is discussed in Belknap (1979) and Atwater and others (1981). Although the intergeneric conversion factor for *Mercenaria-Crassostrea* is ambiguous at some localities, we conclude from a majority of our results that *Crassostrea* has D/L values about 15 to 20 percent lower than coeval *Mercenaria*. Figure 7 shows D/L alanine and valine ratios plotted against D/L leucine, in order to delineate various regional aminozones using combinations of enantiomeric ratios.

Four aminozones occur in Region I: Ia, representing the *Hiatella* samples from the Presumpscot Formation (WC), which has ^{14}C dates of about 11,000 to 13,000 yr B.P. (Stuiver and Borns, 1975) and from Lynn Pit (LP), Massachusetts; Ib, the majority of sites from Long Island, Boston Harbor (Peddocks Island), Nova Scotia, and, most importantly, the early stage 5 calibration site at Sankaty Head (SH), Nantucket Island, Massachusetts (data from Oldale and others, 1982, and additional un-

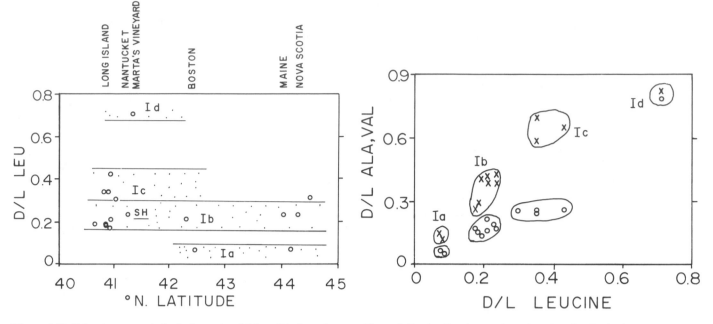

Figure 6. D/L leucine versus latitude for region I, Nova Scotia to Long Island (southeastern New York). Each open circle represents the mean value for multiple samples (or multiple analyses of the same shell) from a single site. Aminozones Ia, Ib, Ic, and Id are shown. Ia is represented by latest Pleistocene (ca. 12,000 yr B.P.) glaciomarine samples; Ib is the best calibrated of the Pleistocene zones because it includes the Sankaty Head site (SH) from which a U-Th coral date of about 130,000 yr has been obtained (Oldale and others, 1982).

Figure 7. Region I aminostratigraphy: plot of D/L alanine (X) and D/L valine (O) versus D/L leucine (mean values for each site plotted), showing how these amino acids conform to the leucine stratigraphic relationships.

published analyses); Ic, one site in Nova Scotia and as many as four subsurface sites and one outcrop from Long Island; and Id, an outcrop of Pliocene or early Pleistocene, glacially deformed, marine sediment at Gay Head, Martha's Vineyard, Massachusetts. The range of D/L values in aminozone Ib potentially represents early to late oxygen isotope stage 5 (130,000 to about 80,000 yr B.P.), but relating the lithostratigraphy of these Ib sites precisely enough to be certain of these apparent age differences is not possible.

Aminozones Ib and Ic include both surface outcrop and subsurface samples from Long Island associated with the Gardiners Clay, implying two different ages for this unit. Significantly, aminozone Ib probably includes a sample (*Spisula*, D/L leucine = 0.14) from the Wantagh Formation of Rampino and Sanders (1981), a unit that is interpreted to overlie the Gardiners Clay in southwestern Long Island. Because aminozone Ib also includes the Sankaty Head calibrated data, the Wantagh Formation appears to correlate with oxygen isotope stage 5. A stage 7 age for at least a portion of the Gardiners Clay would then be consistent with its stratigraphic position in southwest Long Island. A stage 7 age for all the samples of aminozone Ic is consistent with available kinetic models, using the Sankaty Head samples as calibration. *Anadara, Mercenaria,* and *Crassostrea* samples from

outcrops of the Gardiners Clay near the eastern end of Long Island (see Gustavson, 1976) have yielded ratios assigned to aminozone Ib (stage 5). Shell fragments from shelly drift sites along the north shore of Long Island also have ratios in this Ib range. Some of these samples have ratios low enough that they might represent late stage 5 (ca. 80,000 yr), but the resolution of early and late stage 5 requires many more samples than are currently available.

Aminozone Ic is controversial. The majority of samples in this aminozone come from an outcrop of glacially transported marine sediment (described as the Gardiners Clay by at least some workers) at Port Washington, New York (Sirkin and Stuckenrath, 1980), or from two subsurface samples taken from water-wells approximately 10 km west of Port Washington (Ricketts, 1986). All of the samples from northwestern Long Island are *Crassostrea,* which have yielded D/L leucine values between 0.25 and 0.31. Based on probable intergeneric relative racemization kinetics, we interpret the Port Washington *Crassostrea* data to represent *Mercenaria* aminozone Ic. This interpretation is suported by D/L leucine values between 0.17 and 0.21 for *Crassostrea* from other Long Island sites and from Sankaty Head. Based on finite ^{14}C dates for wood and shell samples from the Port Washington beds, Sirkin and Stuckenrath (1980) interpreted the pollen record in these beds to indicate interglacial conditions, and they defined this section as the "Port Washingtonian Interstadial."

Two *Mercenaria* samples from aminozone Ic have come from separate water-well samples near the west end of the North Fork of Long Island. Duplicate analyses of each sample confirm the precision of the enantiomeric ratios, but because the samples are single shell fragments, whether they represent an older, in situ, marine unit or reworked (glacially?) source beds equivalent to the older Gardiners Clay cannot be determined.

The Sankaty Head site consists of two beds (Upper and Lower Sankaty Sand) with distinct microfaunal characteristics (Oldale and others, 1982) suggestive of a transition from full interglacial to early glacial conditions. The Lower Sankaty Sand contains many in-place, articulated, and very well preserved *Mercenaria mercenaria* samples that retain their original luster and coloration. Mollusks from the Upper Sankaty Sand are primarily reworked (disarticulated, not in life position) and are often fragmented or somewhat bleached and chalky in appearance, although many large single valves of *M. campechiensis* are present. The U-Th dated solitary coral that serves to calibrate the Sankaty site was collected from the Upper Sankaty Sand. Although the faunal data, combined with models of interglacial-glacial transitions, imply some age difference between the Lower and Upper Sankaty Sand units, the currently available aminostratigraphic data do not resolve (within about 10,000 to 20,000 yr) any such age difference. Enantiomeric ratios obtained on *Mercenaria* samples from the Sankaty Sand actually revealed a slight (ca. 10 percent) inversion of D/L values in relation to the stratigraphy (Oldale and others, 1982). The cause of this inversion was not evident in the original data set, partly because the samples analyzed were not collected with possible thermal or diagenetic effects in mind. More recently, a detailed study of more than 50 shells from this outcrop (Rahaim, 1986) has shown that this inversion of aminostratigraphic data can probably be explained by two factors: (1) the thermal effects of exposure during reworking of the Upper Sankaty Sand samples, and (2) diagenetic alteration of the shells in the Upper Sankaty Sand due to ground-water percolation to the contact between the porous Upper Sankaty Sand and the indurated Lower Sankaty Sand. Allo/iso data (Rahaim, 1986) for hinge portions of the best-preserved *Mercenaria* samples from the Sankaty Sands (four shells from each unit) are as follows: 0.151 ± 0.012 (Lower Sankaty Sand) and 0.160 ± 0.015 (Upper Sankaty Sand). Allo/iso data for hinge samples from 12 Upper Sankaty Sand and 13 Lower Sankaty Sand less well preserved *Mercenaria* are more indicative of the same aminostratigraphic inversion observed in the leucine data: 0.1715 ± 0.034 (Upper Sankaty Sand) and 0.138 ± 0.029 (Lower Sankaty Sand). We conclude from these data that the age difference between the Upper and Lower Sankaty Sands is not resolvable based on enantiomeric ratios. Alternatively, reworking of the Upper Sankaty Sand coral and *Mercenaria* samples from a unit equivalent to the Lower Sankaty Sand could explain the apparent discrepancy between aminostrat-

igraphic and paleoclimatic interpretations of the Upper and Lower Sankaty Sand.

Nova Scotia. Extrapolation of aminozones Ib and Ic from the Long Island–Sankaty Head region (CMAT about 10°C) to Nova Scotia (CMAT about 7.5°C) is tentative because of the substantial temperature difference between the regions. Nevertheless, the similarity of results from the two temperature regions presents a strong case for the correlations shown. The only option other than the one shown in Figure 6 would be to correlate the lower D/L values in Nova Scotia (leucine 0.23) with the higher ratios (leucine about 0.34) on Long Island, but this option would not be consistent with the kinetic model of leucine racemization (see Fig. 19). The Nova Scotia samples represent seven sites (the three sites for which multiple analyses are available are shown here) between Yarmouth and Digby. Samples were taken from two units, the Red Shelly Drift (with a lithology indicating a source in New Brunswick, across the Bay of Fundy) and the Salmon River Beds (moderately deformed interglacial marine and fluvial clay and sand) (nomenclature after Grant, 1980). Aminozone Ib includes both shell fragments from the Red Shelly Drift and whole shells from the Salmon River Beds; aminozone Ic includes whole shells from one drift site (locality GC) near the north end of St. Mary's Bay. Marine beds of two ages (stages 5 and 7) apparently acted as sources for the shells that were transported (and usually fragmented) various distances by ice advances from the northwest. The Salmon River Beds represent the least transported of these marine units. In both the Nova Scotia and Long Island sequences, at least one phase of ice transport (post-stage 5) is implied by the amino acid age clusters, with marine deposits of two or more ages being incorporated into the resultant drift. This interpretation appears consistent with the correlations summarized by Stone and Borns (1986).

All of the Nova Scotia samples discussed here were analyzed without the aid of electronic integration of chromatographic peaks. Therefore the direct comparison of these results with those from Sankaty Head and Long Island may be biased slightly by these differences in analytical methodology.

Region II: Southern New Jersey, Delaware, Maryland, Virginia, and northeastern North Carolina (Fig. 3).

This region contains over 40 sites that have been studied intensely by several workers (Belknap, 1979; Belknap and Wehmiller, 1980; Mirecki, 1985; York, 1984). The region includes the Norfolk, Virginia, area where on-going borrow-pit excavation has permitted detailed lithostratigraphic studies (Peebles, 1984; Mixon and others, 1982; Oaks and others, 1974). Repeated collection at many of these sites over the past ten years has permitted us to develop an aminostratigraphy for this region based on a large number of well-preserved samples and to address a number of issues related to the resolving power and temperature assumptions of aminostratigraphy. U-Th coral dates (Szabo, 1985; Cronin and others, 1981) are available for several of the sites in this region.

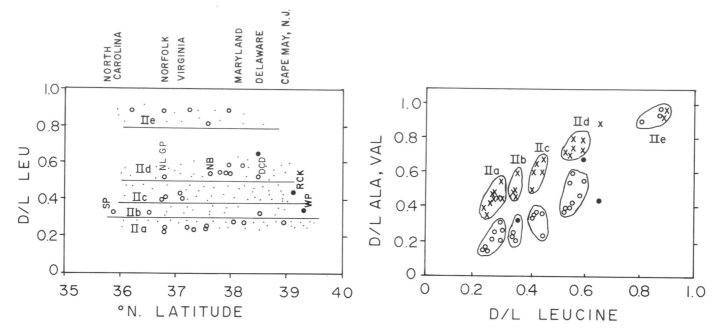

Figure 8. D/L leucine versus latitude for region II, including southern New Jersey, Delaware, Maryland, Virginia (DelMarVa), and northeastern North Carolina. Each open or closed circle represents the mean value for multiple samples from a single site. Aminozones IIa, IIb, IIc, IId, and IIe are shown. *Rangia* data, which are not used in the delineation of aminozones, are shown by the solid circles. IIa includes two sites (NL and GP) with 75,000-yr U-Th dates, and IId includes on site (NB) with a 190,000-yr U-Th date (Szabo, 1985) and the Dirickson Creek (DCD) site with a biostratigraphic age estimate of greater than 300,000 yr (Cronin and others, 1984).

Figure 9. Region II aminostratigraphy: plot of D/L alanine (x) and D/L valine (o) versus D/L leucine (mean values for each site plotted), showing how these other two amino acids conform to the leucine stratigraphic relationships. Notation as in Figures 8 and 6. *Rangia* data not included in aminozone IId.

Figures 8 and 9 present aminostratigraphic data for sites in region II. Five aminozones (IIa, IIb, IIc, IId, and IIe), all based on a minimum of ten *Mercenaria* samples, are recognized. Aminozones IIa through IId include at least two sites from which three or more *Mercenaria* samples were analyzed, with good precision. Other sites from which only a limited number of samples were collected exist, but these sites can be assigned to a particular aminozone with confidence because these aminozones are now well defined by multiple samples. Data for *Rangia* from a few sites are presented in Figure 8. Because of different salinity tolerances, *Rangia* and *Mercenaria* do not coexist; the limited data from *Rangia* and *Mercenaria* samples thought to be the same age suggest that these two genera are members of the same kinetic group. Data for *Crassostrea* at many of the sites shown in Figure 8 were also obtained, but because of the abundance of more reliable *Mercenaria* data from all of these sites, *Crassostrea* data are not discussed here. Multiple *Anadara* samples from Moyock, Gomez Pit, and Norris Bridge were analyzed in order to supplement the *Mercenaria* data from these three critical sites with results for a second genus.

Aminozone IIa includes shell beds at Gomez and New Light

pits (Norfolk) that have multiple U-Th coral dates clustering around 75,000 yr (the Mears and New Light samples of Szabo, 1985). These two sites collectively represent the best U-Th calibration of coastal plain aminostratigraphic data because of the abundant isotopic data that appear to represent closed-system samples (Szabo, 1985; see Wehmiller and Belknap, 1982, for discussion).

Coral samples from Moyock and Stetson Pit have similar, late-stage 5, U-series dates, but the relationship of the coral dates to the molluscan aminostratigraphy is less clear. The Moyock, North Carolina, site is a spoil pile from which coral and shell samples were collected at different times. No stratigraphic section is known for this site (T. M. Cronin, personal communication, 1984), although a lithologic description was provided with the original publication of the U-Th date (Cronin and others, 1981). *Mercenaria* analyzed from the Moyock spoil pile reproducibly fall into aminozone IIb, while *Anadara* from the Moyock spoil pile reproducibly fall into aminozone IIa, implying either that samples of two ages may have been mixed in the excavation process at Moyock, or that the intergeneric relationship between *Anadara* and *Mercenaria* inferred from other sites (see Appendix

V) is not maintained at Moyock (the Moyock *Anadara* are rather chalky and poorly preserved). One of the Moyock *Mercenaria* analyzed (D/L leucine = 0.33, aminozone IIb) was encrusted by a coral that yielded a U-Th date of about 108,000 yr (Table 2), but the Th isotope data imply open-system conditions. D/L data suggest that the mollusks from nearby Womack Pit fit into aminozones IIa and IIc. A U-series coral date from this site is 62,000 yr (Table 2). Although two apparent ages for the Womack Pit section might be consistent with the local stratigraphy (Belknap, 1979), little emphasis is placed on these results, because we did not collect the samples, and the relationship between the coral and mollusk samples is not clear. At Stetson Pit, a U-series-dated coral was obtained from an excavated outcrop at an elevation of ca. −3 m (T. M. Cronin, personal communication, 1982); Stetson Pit *Mercenaria* samples that fall into aminozone IIb have been collected from the same outcrop (T. M. Cronin, personal communication, 1986) and from subsurface sections at depths to 11 m below sea level (Miller, 1982; York and Wehmiller, 1987).

Aminozone IId represents at least five subsurface and one surface outcrop sites in central Delmarva and two surface outcrop sites on the west shore of Chesapeake Bay. At Gomez Pit (Fig. 10), aminozone IId shells are found stratigraphically below aminozone IIa shells (Mirecki, 1985). Aminozone IId includes the Norris Bridge, Virginia, site 9 (the Shirley Formation of Peebles and others, 1984) for which U-series data suggest a stage 7 (ca. 200,000 yr) age (Szabo, 1985) and also the Dirickson Creek, Delaware, site (Omar Formation) for which biostratigraphic data (T. M. Cronin, personal communication, 1980; Cronin and others, 1984) suggest an age of 300,000 yr or more. Prior aminostratigraphic interpretation (Wehmiller and Belknap, 1982) of the Norris Bridge and Delmarva sites implied that units of two ages might be present, but additional analyses have led us to the conclusion that one aminozone, with a mean D/L leucine value of 0.55 (±0.06), represents Norris Bridge, Mathews Field, and Dirickson Creek (all discussed in Wehmiller and Belknap, 1982), as well as shells from the Gomez Pit section and three subsurface sites within 15 km of Mathews Field (Belknap, 1979). Multiple *Rangia* samples from the subsurface (ca. −10 m) of Dorchester Island, Maryland (Jacobs, 1980), and from Poplar Creek Bluff, Maryland also appear to belong to this aminozone.

Aminozone IIc represents four sites (three with multiple shells) that appear older than aminozone IIb and younger than aminozone IId. At NL and WP (Fig. 10), aminozone IIc shells are found stratigraphically below (ca. 2 m) aminozone IIa shells. At Gomez Pit, aminozone IIc is found stratigraphically beneath aminozone IIa along an exposure of ca. 100 m. Two sites (KP, BB) within 2 km of of each other on the York–James Peninsula also fall into aminozone IIc.

Literature on the stratigraphic nomenclature and depositional history of the Norfolk–York–James Peninsula region is extensive (see review in Peebles and others, 1984). Figure 10, redrawn from Peebles and others (1984), shows the complexity of the relationships between aminostratigraphy and lithostratigraphy in the region. At least three aminozones, IIa, IIc, and IId,

have been found superposed in separate outcrops (IIa over IIc, IIa over IId) of the Sedgefield member of the Tabb Formation. Reworking of some of the older (IIc or IId) shells is likely, based on field evidence (Mirecki, in preparation). Only very subtle lithologic breaks are detected between some of these aminozones at GP and NL (Fig. 10). On the basis of aminostratigraphic data, the Sedgefield member appears to record multiple transgressive events that are resolvable by enantiomeric ratio data but which may not be recognizable in the field.

Aminozone IIe includes nearly racemic samples from the Yorktown Formation or its equivalents at outcrop and subsurface sites in the region. These samples provide early Pleistocene/Pliocene calibration but no attempt is made to resolve age differences in such extensively racemized samples.

The age assignments that we propose for aminozones IIa, IIb, IIc, and IId, using selected 75,000-yr U-series dates as calibration and the non-linear leucine racemization kinetic model (Wehmiller and Belknap, 1982), are as follows:

IIa, ca. 80,000 yr B.P. (late stage 5)
IIb, ca. 120,000 to 130,000 yr B.P. (early stage 5)
IIc, ca. 250,000 yr B.P. (stage 7)
IId, ca. 500,000 yr B.P. (stage 13 or perhaps as young as stage 11)

The method of kinetic model age assignment has been discussed elsewhere (Wehmiller and Belknap, 1982) and is reviewed under the heading "Aminostratigraphy, Kinetic Model Dates, and Philosophy of Geochronology" below. Optional age assignments for these four aminozones are also considered therein. The age estimates for region II are based on the standard assumptions of aminostratigraphy, specifically that the differences in effective temperatures for nearby sites was similar to the present temperature differences between the sites.

A major conflict exists between these age estimates and independent chronologic data for the Norris Bridge site, which is estimated to be about 200,000 yr old by U-Th data (Szabo, 1985). Numerous analogies have been presented to justify the aminostratigraphic argument that the Norris Bridge site must be substantially older (more than one glacial cycle) than the New Light–Moyock group (Wehmiller and Belknap, 1982). In addition, the aminostratigraphic data for the Norris Bridge, Mathews Field, and Dirickson Creek sites are consistent with their accepted correlation (Mixon and others, 1982; Mixon, 1985), so the Norris Bridge U-Th date must not only be compared with the aminostratigraphic kinetic model age estimate of ca. 500,000 yr, but also with the biostratigraphic evidence (Cronin and others, 1984) that supports a mid-Pleistocene age for Dirickson Creek. Additional aminostratigraphic evidence for the antiquity of the Norris Bridge site can be derived from recent data for *Anadara* (Mirecki, 1985): this genus shows the same aminostratigraphic relationship between Gomez Pit (mean D/L leucine = 0.21) and Norris Bridge (mean D/L leucine = 0.50) as does *Mercenaria,* so the conflict that centers on the Norris Bridge site does not appear to be resolvable by strange results for a single genus. The Norris Bridge coral was not collected at the same time as any of the

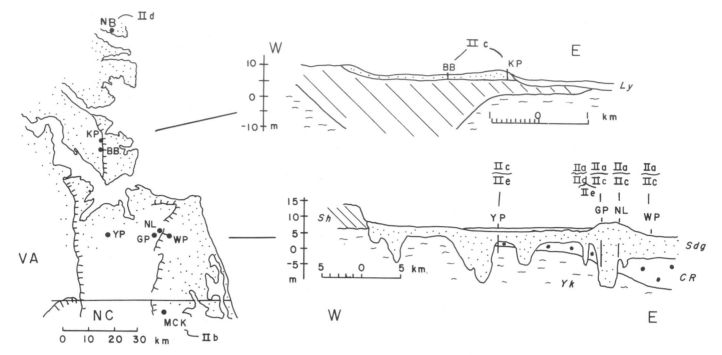

Figure 10. Map and cross sections of the eastern York–James Peninsula and Norfolk, Virginia, areas (redrawn from Peebles and others, 1984). The two major scarps in this region are the Suffolk and Hickory scarps, west and east of locality YP, respectively. Samples were collected from sites labeled KP, BB, YP, GP, NL, WP, NB, and MCK. The formational terminology of Peebles and others (1984) is: Lynnhaven member of the Tabb Formation (Ly); Sedgefield member of the Tabb Formation (Sdg); Chowan River Formation (CR); Shirley Formation (Sh); Yorktown Formation (Yk). Mollusks from the Sedgefield were collected at KP, BB, YP, GP, NL, and WP. Mollusks from the Yorktown Formation were collected at YP and BB. Mollusks from the Chowan River Formation were collected at GP and at CB (Table 1). Mollusks from GP, NL, and WP yielded several apparent aminostratigraphic clusters that imply distinctly separate depositional episodes within the Sedgefield member. Vertical lines beneath each locality symbol indicate the approximate position of sampled sections, from which the indicated aminozone stratigraphies (IIa/IIc, IIa/IId, or IIc/IIe) have been recognized. Not shown is the additional observation that nearly racemic samples were obtained from the base of the section at Gomez Pit (GP), so four aminozones (IIa, IIc, IId, and IIe) were recognized in this one borrow pit, but not all in any one specific outcrop section. The exact positions of the WP samples are not known to us so the relationship of our WP data to those for NL and GP is uncertain.

shells analyzed from this site, yet the apparent simplicity of the stratigraphic section at NB has led all workers to assume that the coral and mollusk samples are all the same age. In light of the subtle unconformities identified by aminostratigraphy in the Gomez Pit (GP) section, the Norris Bridge site requires further study.

The kinetic model age estimate presented here for aminozone IId is slightly lower than that presented previously (Wehmiller and Belknap, 1982). This revised age estimate is not due to any revision in kinetic models or temperature assumptions (each of which could be modified to change an age estimate, if the modifications involved reasonable assumptions). Instead, many additional analyses of *Mercenaria* from the Norris Bridge site have been performed, showing that the earlier Norris Bridge D/L ratios (which had been obtained without the use of electronic

integration) were systematically higher (by 5 to 10%) than all subsequent analyses of either the original shells or more recently collected samples.

The age assignment for the Norris Bridge site and its correlatives remains controversial. However, the aminostratigraphic data can be used in a qualititative manner to describe the development of the southern portion of the Delmarva Peninsula. All of the old sites (aminozone IId) from the peninsula are located north of latitude 37.75°; five sites on the peninsula south of this latitude have yielded shell fragments that fall into either aminozone IIa or IIb. Consequently, the aminostratigraphic data imply that the southern Delmarva Peninsula evolved southward during the late Pleistocene and that the mouth of Chesapeake Bay was much less restricted during the middle Pleistocene (aminozone IId time) (Belknap, 1979, Fig. 45; Belknap and Wehmiller, 1980). This model has been proposed by others, based primarily on litho-stratigraphic and geomorphic evidence (Peebles and others, 1984; Peebles, 1984; Mixon, 1985).

In previous presentations of data from the Delmarva-Chesapeake region (Belknap and Wehmiller, 1980), we proposed (based on kinetic models) that some of the analyzed localities were of early stage 5 and others were of late stage 5 age. As additional data have been obtained, we remain confident that these two ages can be resolved. Unfortunately, however, samples of the two aminozones (IIb and IIa) have not yet been collected in a stratigraphically superposed relationship, and IIb samples are relatively infrequent, so unequivocal field evidence is not available to confirm this implied age difference. This issue has been a classic case of lumping versus splitting for our work in Atlantic Coastal Plain aminostratigraphy, and the most conservative approach would be to lump all these stage 5 aminozones until their age differences are confirmed independently. Similar relative ranges of enantiomeric ratio data have been observed for early- to late-stage 5 stratigraphic sequences elsewhere (California: Lajoie and others, 1979; Kennedy and others, 1982; Spitszbergen: Boulton and others, 1982; Norway: Miller and others, 1983; Mediterranean: Hearty and others, 1986).

Region III: North Carolina

Region III encompasses the state of North Carolina (Fig. 4) and overlaps local aminostratigraphic sections in the Norfolk, Virginia, and Myrtle Beach, South Carolina, areas (Fig. 11). Region III includes two important subsurface sections at Stetson Pit and Ponzer, North Carolina, and an outcrop at Flanner Beach, North Carolina, from which abundant samples were available. Because we obtained a substantial amount of data for the genus *Mulinia*, in addition to many results for *Mercenaria*, separate aminostratigraphic summaries of the data from each of these genera are presented and compared. *Mercenaria* generally has significantly greater D/L values than coexisting *Mulinia* (see Appendix V), but the conversion of results between these two genera is not always internally consistent. In general, the precision of multiple analyses of *Mulinia* is poorer than for *Mercenaria*.

Figure 11 presents the *Mercenaria* data for Region III. The latitude range for this figure permits comparison of region III aminozones with those of regions II and IV. Five aminozones are identified for region III, though aminozone IIIa is really a southward extension of results from region II and could overlap with data from aminozone IIIb at 34.8°N.

The data from Stetson Pit, Ponzer, and Flanner Beach are most useful in establishing a relative aminostratigraphy for this region. Aminozone IIIc includes the Flanner Beach outcrop samples and samples from 3 to 4 m below sea level at Ponzer. The Ponzer site is used as calibration for aminozone IIIc because of U-Th and U-trend dating, which collectively indicate that this site is probably stage 7 in age (ca. 200,000 to 220,000 yr B.P., Szabo, 1985; Cronin, Rosholt, Szabo, personal communication, 1985). *Mercenaria* from Stetson Pit (outcrop and from depths of ca. 7 to

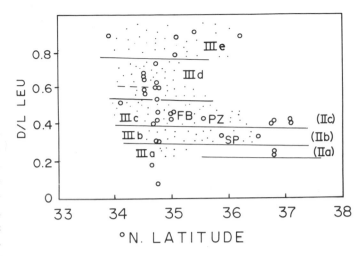

Figure 11. D/L leucine values versus latitude for *Mercenaria* samples from region III, North Carolina. Aminozone IIIb includes the Stetson Pit (SP) stage 5 calibration site. Aminozone IIIc includes the stage 7 Ponzer calibration site (PZ) and Flanner Beach (FB). SP, PZ, and FB are the three most heavily studied sites in this region, hence the two aminozones representing these sites are the best established.

11 m below sea level) fall into aminozone IIIb (York and Wehmiller, 1987; York and others, 1988; Miller, 1982). A coral from the Stetson Pit outcrop has yielded a 75,000-yr date (Szabo, 1985). We interpret the IIIb and IIIc zones to represent stages 5 and 7, respectively.

One vibracore (Snyder and others, 1982) in Core Sound, North Carolina (CSNC-9, UDAMS 07031), appears to have penetrated three aminozones in proper stratigraphic order between 5 and −9 m; IIId, IIIc, and an overlying younger aminozone designated either IIIb or IIIa, with an implied correlation to either the IIb or IIa aminozones discussed for region II (Fig. 8). Preliminary *Mercenaria* data from outcrops of a unit known as the Core Creek Sand along the Core Creek Canal (Snyder and others, 1982), between the Neuse River Estuary and Cape Lookout, also fall into aminozone IIIa or IIIb. The Core Creek Sand (or equivalents) has usually been considered to be younger than the Flanner Beach Formation (Johnson and Peebles, 1986), and the aminostratigraphic data appear entirely consistent with this interpretation. Aminozone IIId is represented by *Mercenaria* samples only for a dense cluster of data points at latitude 34.6°N. These samples are all from subsurface vibracores, primarily from Onslow Bay and Core Sound (Snyder and others, 1982; Belknap, 1984). Aminozone IIId apparently has a range of *Mercenaria* leucine D/L values between about 0.54 and 0.70. Because many of the shell samples obtained from these cores were fragments, this wide range may be caused at least partially by diagenetic effects related to fragmentation. Further detailed study is necessary to refine the aminostratigraphy of this zone. Based on other

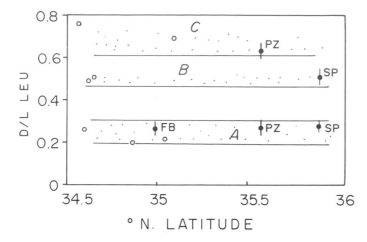

Figure 12. Region III *Mulinia* aminostratigraphy. The solid circles represent mean values of multiple shell samples (at least five) while the open symbols represent mean values for fewer samples analyzed per site (one or two). Standard deviations shown for those sites at which five or more *Mulinia* valves were analyzed. Aminozones A, B, and C are shown for comparison with the *Mercenaria* data from this same region (Fig. 11). These aminozone designations are different from those for *Mercenaria* (IIIa, IIIb, IIIc, etc.) because they are based on a different genus.

observations, we expect that this zone can be resolved into two separate zones (as suggested by the dashed line in Fig. 11F), each stratigraphically consistent. Aminozone IIIe is represented by *Mercenaria* samples from units referred to as the Yorktown, the Chowan River, and Waccamaw Formations, but no attempt is made here to resolve these IIIe units based on the amino acid data.

An un-numbered aminozone, apparently younger than aminozone IIIa, is represented by two *Mercenaria* samples (mean D/L leucine = 0.18) from about 15 m below sea level at the east end of Shackleford Banks, North Carolina. A 29,280 +2000/−2680 ^{14}C date (Susman and Heron, 1979) was obtained from this same core depth (Belknap, 1984) in Susman and Heron's core SH-11. Radiocarbon dates in this same range were obtained from other nearby cores, but preliminary amino acid data on shells from these other cores seem to indicate ages that ought to be well beyond the ^{14}C range (Belknap, 1984). Some of these age differences could be due to sample contamination by young carbon, and some could be due to reworking of older shells into younger deposits. Many of the offshore vibracores examined by Belknap (1984) penetrated the Diamond City Formation, usually interpreted as mid-Wisconsin because of many finite radiocarbon dates. The data of Belknap (1984) and others indicate that the Diamond City Formation contains shells of several ages from latest Pleistocene to early Pleistocene, most likely because of channel cutting and filling and sample reworking during the deposition of this unit.

Figure 12 shows the D/L leucine data for *Mulinia*, plotted versus latitude, between 36.0° and 34.5°N. Three aminozones, designated A, B, and C, are recognized in the *Mulinia* data presented in Figure 12. These aminozone designations are specific for *Mulinia*, and no correlation with aminozones IIIa, IIIb, or IIIc is implied by these letter designations. Aminozones B and C, although they have some overlap, appear to be distinguishable on the basis of several amino acid D/L values (York, 1984), biostratigraphic and U-trend data (T. Cronin, J. Rosholt, B. Szabo, personal communication, 1985). Aminozone B is found at a depth of −17 to −33 m below sea level in the Stetson Pit section and aminozone C is found at a depth of −10 to −15 m below sea level in the Ponzer section (York, 1984; York and Wehmiller, 1987). Aminozone A is problematical because all the *Mulinia* data from the Flanner Beach outcrop and shallow portions of the Ponzer and Stetson Pit sections fall into this zone, contradicting the aminostratigraphic interpretation of the Stetson Pit section based on the *Mercenaria* data. The Stetson Pit *Mulinia* data in aminozone A conflict with all other observations of relative D/L values in *Mercenaria-Mulinia* pairs (Appendix V). Aminozone A represents sedimentary units that, on the basis of *Mercenaria* data, are thought to represent stages 5 and 7, so region III *Mulinia* data cannot be used to resolve age differences of less than about 100,000 yr.

The other *Mulinia* results from region III provide less aminostratigraphic control than those for the Stetson Pit–Ponzer–Flanner Beach sections because fewer samples per site have been analyzed. Nevertheless, aminozones B and C appear to be represented in the data for sites in central and southern North Carolina, including deposits referred to as the James City and Waccamaw Formations (which are included in aminozone C, although they may represent an older aminozone). Additional analyses are required to determine if aminozone C can be split into separate aminozones.

The age relationships for the Stetson Pit, Ponzer, and Flanner Beach sections, based on the combined amino acid and U-series data, are shown in Figure 13. The *Mercenaria* and *Mulinia* data for these sections do not resolve the stratigraphy in the same manner. Although the results for each genus are consistent with the local stratigraphic sequences at Ponzer (PZ) and at Stetson Pit (SP), the results for the two genera imply different correlations when the shallow portions of the PZ and SP sections are compared. Only the *Mercenaria* data at PZ and SP seem consistent with the isotopic data at these sites. Both *Mercenaria* and *Mulinia* imply correlation of Flanner Beach (FB) and the shallow portion of the Ponzer (PZ) section.

Understanding the aminostratigraphy of the region represented in Figures 11 through 13 is critical to many of the issues raised by Wehmiller and Belknap (1982) regarding long-distance latitudinal correlations along the Atlantic Coastal Plain. Corrado and others (1986) and Hearty and Hollin (1986) proposed that the Flanner Beach site should be late stage 5 in age (approximately 75,000 yr) based on new analyses and/or reinterpretations of other amino acid data. Such a young age assignment for

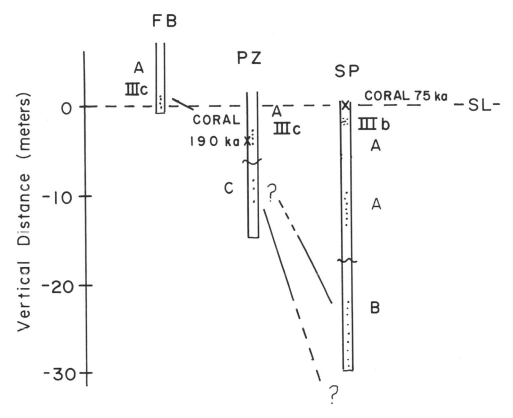

Figure 13. Schematic chronostratigraphic section of northeastern North Carolina from Flanner Beach (FB) northeast to Ponzer (PZ) and Stetson Pit (SP), redrawn from York (1984). Depths below sea level of U-series-dated samples and sampled aminozones indicated by (x) or by (:), respectively. Aminozones with roman numeral III are based on *Mercenaria*; those labeled A, B, or C are based on *Mulinia*. Aminozone IIIb is calibrated by the 75,000-yr U-Th date at Stetson Pit and aminozone IIIc is calibrated by the 190,000-yr U-Th date at Ponzer.

the Flanner Beach Formation has not been proposed in any previous stratigraphic study of this unit (W. Miller, 1985; Johnson and Peebles, 1986). If this age assignment is correct, then the aminozone slopes shown in Figure 11 would have to be steepened significantly (rising from D/L of about 0.25 at 36.8°N to about 0.50 at 34°N). Such steepening of aminozones might appear to resolve some aminostratigraphic conflicts (Corrado and others, 1986), but would not be consistent with the Ponzer U-series isotopic data or the presence of aminozone IIIb samples at the same latitude as Flanner Beach. The slopes of any aminozones are directly proportional to latitudinal gradients of effective temperature. Reconciliation of the present latitudinal temperature gradient in this region with the gradient of effective temperature that would be required by a late-stage 5 age assignment for Flanner Beach is extremely difficult. Corrado and others (1986), in their attempt to separate the Ponzer U-Th date from the aminostratigraphic data for the combined Flanner Beach–Ponzer sections, have suggested that deposits of two ages exist at these sites, one dated by amino acids and one dated by U-Th. We have also considered this possibility (York, 1984) and admit that the apparent inversion of *Mulinia–Mercenaria* (see Appendix V) aminostratigraphic data for the Stetson–Flanner Beach sites forces the consideration of numerous stratigraphic options. However, we prefer to rely on previous, single-age, lithostratigraphic interpretations of the Flanner Beach–Ponzer section (e.g., W. Miller, 1985; Johnson and Peebles, 1986) rather than proposing a multiplicity of units based on amino acid data alone. Assuming that FB is the youngest unit in North Carolina, as Corrado and others (1986) have done, is not necessary because younger units (based on aminostratigraphy) have been found in the region, both offshore and onshore.

Region IV: South Carolina (Fig. 4)

This region includes two local areas from which numerous samples have been analyzed: Charleston, South Carolina, and the Intracoastal Waterway near Myrtle Beach, South Carolina. McCartan and others (1982) and Wehmiller and Belknap (1982) discussed the relation of *Mercenaria* aminostratigraphic data to lithologic, biostratigraphic, and U-series data for sites in this region. In addition, Corrado and Hare (1981) and Corrado and others (1986) have summarized data for *Mulinia* samples in the Charleston–Myrtle Beach region. The Myrtle Beach region has been the focus of several biostratigraphic and lithostratigraphic studies (Oaks and DuBar, 1974; McCartan and others, 1982). Szabo (1985) summarized the available U-series isotopic data bearing on the geochronology of units in the Myrtle Beach and Charleston areas. The formational nomenclature of the Myrtle Beach region is misleading in its apparent simplicity, and different workers have used the same formation name for different units or groups of units. The names associated with the youngest (Pleistocene) marine units in the Myrtle Beach area are Waccamaw, Canepatch, and Socastee Formations (oldest to youngest). McCartan and others (1984) proposed that correlatives of some or all of these units, plus the younger Wando formation, are found in the Charleston area. Some major and fundamental conflicts between the aminostratigraphic and lithostratigraphic correlations exist between units in the Charleston and Myrtle Beach areas (McCartan and others, 1982; Wehmiller and Belknap, 1982). Corrado and others (1986) and Hearty and Hollin (1986) have presented some new amino acid data or reinterpreted the data of Wehmiller and Belknap (1982) in an attempt to reconcile these conflicts with recently reported U-series data (Szabo, 1985) for this region.

Region IV also includes two of the sites from which multiple shells were collected for interlaboratory comparison of amino acid data (Wehmiller, 1984b; ILC-B, Mark Clark Pit, Charleston, South Carolina, and ILC-C, Calabash Pit, North Carolina, along the Intracoastal Waterway just north of the South Carolina–North Carolina border). Consequently, aminostratigraphic data for these two sites have been obtained by at least ten laboratories.

Figure 14 presents the D/L leucine data (*Mercenaria*) for the latitude range 32.5°N to 34.5°N, covering the area from central South Carolina to southeastern North Carolina. The present mean annual temperature range for this region is about 1°C. Four aminozones are recognized for region IV: IVa, IVb, IVc, and IVd. In the simplest interpretation of the region IV data, each of the four aminozones is recognized in the Charleston and Myrtle Beach areas, and the correlation isochrons between the two areas are nearly horizontal, as would be expected given the slight temperature difference between the two regions. Corrado and Hare (1981) also documented a four-fold *Mulinia* aminostratigraphic sequence in each area, but most of the sites they studied were not the same as those discussed here. However, the horizontal correlation trends shown in Figure 14 are not consistent with the stratigraphic model of McCartan and others (1982), who argued that broader ranges of aminostratigraphic data represent each depositional interval (mappable unit) and that correlation isochrons between the two areas should be much steeper. In their model, aminozones IVa and IVb are not considered as separate units, and the IVa–IVb unit in Myrtle Beach is correlated with IVc in the Charleston area. This proposed correlation is shown as a diagonal area (enclosed in dashed lines) of Figure 14. Corrado and others (1986) and Hearty and Hollin (1986) have proposed

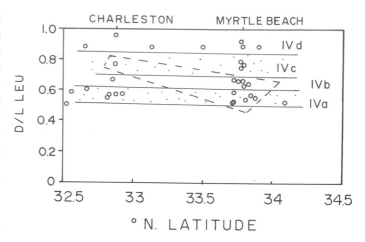

Figure 14. D/L leucine values (mean values for multiple analyses from each site) for *Mercenaria* samples from region IV, South Carolina. Mean annual temperatures for Charleston and Myrtle Beach are about 19° and 18°C, respectively. Four aminozones are shown: IVa, IVb, IVc, and IVd. The simple, near-horizontal, correlation bands shown in this figure are based on the similarity of apparent aminozones from the two areas and their similar temperatures. The area enclosed in dashed lines shows the correlation preferred by McCartan and others (1982).

aminostratigraphic correlations between Charleston and Myrtle Beach that would have slopes very similar to those that we propose here.

Studies by McCartan and others (1982), Wehmiller and Belknap (1982), Szabo (1985), and Corrado and others (1986) raised a number of issues about aminostratigraphic resolving power, choice of U-series calibration data, and expected latitudinal trends of isochrons. These issues are best addressed by careful examination of the leucine data from the Intracoastal Waterway, where abundant samples with stratigraphic control are available.

Figure 15 shows the mean D/L leucine data obtained for multiple *Mercenaria* samples at each of the Intracoastal Waterway sites. Some, but not all, collections were made at the same time that corals or sediment samples were collected by others for U-series, paleomagnetic, or biostratigraphic analysis. Because landmarks are rare and locality description in the waterway is difficult, the relationship of our collection sites to those of other workers is not always clear. Sites identified in the literature as "ICW-(number)" were all collected in March of 1979 in a major joint collection effort by many of the stratigraphers and geochemists involved in this study area (McCartan and others, 1982). Other sites, with designations such as MB- or ICW-(letter), were collected by the Delaware group alone. Based on comparisons of results from our multiple collections in the Myrtle Beach area, the 1979 group collection effort, and site descriptions in the literature (particularly in Szabo, 1985), Figure 15 shows what we believe to be the best relation of the Myrtle Beach area aminostratigraphic data to local formational nomenclature and isotopic and paleomagnetic geochronology.

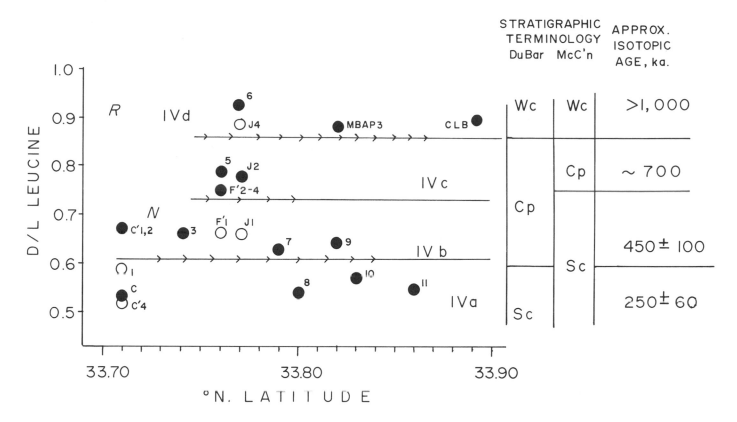

Figure 15. Detailed *Mercenaria* leucine aminostratigraphy for the Myrtle Beach area of region IV. Solid circles represent mean values (c.v. ±6 percent) for multiple shells from each specific outcrop position. Open circles represent data for single shells. Aminozone designation as in Figure 14. Locality designation for each data point is given in Table 1. The formational designations used by DuBar and others (1980) and McCartan and others (1982) for each of our aminozones are shown on the right side of the figure, as are the approximate isotopic ages preferred by Szabo, (1985). Reversed magnetic polarity is associated with aminozone IVd; normal polarity with IVb (McCartan and others, 1982). Formation terms: Sc = Socastee; Cp = Canepatch; Wc = Waccamaw. The Route 501 Bridge section is represented by the points labeled ICW C'-1, 2 (probably equivalent to the "beach level" of Szabo, 1985) and ICW 1, ICW C, and ICW C'-4 (probably equivalent to the "mid-section" of Szabo, 1985).

Figure 15 depicts aminozones IVa, IVb, IVc, and IVd. One of the major and continuing questions relates to the resolving power of these data. McCartan and others (1982) group all of our IVa and IVb data (Fig. 15) into the Socastee Formation, thought to be approximately 200,000 to 250,000 yr old (stage 7) based on U-Th coral dates and U-trend dating (Szabo, 1985). At the Route 501 Bridge section (Szabo, 1985; our sites ICW C and ICW C'), aminozone IVa stratigraphically overlies aminozone IVb. This is the section that has proven controversial in the past. As Szabo (1985) pointed out, McCartan and others (1982) interpret the entire section as the Socastee Formation, whereas DuBar and others (1980) interpret the upper portion of the section (our ICW C) as the Socastee Formation and the lower portion (our ICW C') as the Canepatch Formation. Our data, as well as the U-trend data reported by Szabo (1985), support the two-fold nomenclature of DuBar and others (1980). The D/L leucine values for these two groups lead to kinetic model age estimates of about

240,000 ±50,000 and 350,000 ±75,000, respectively Wehmiller and Belknap, 1982). Although these age ranges overlap slightly, they are consistent with the ages inferred from the U-trend data (Szabo, 1985). Szabo's U-Th coral dates for this section range from >70,000 to 500,000 yr (Szabo, 1985, Table 4). Although any of several pairs of dates selected from these Th/U results could be interpreted to be consistent with the local two-fold stratigraphy at the Route 501 Bridge section, Szabo (1985) appears to prefer those isotopic data that indicate ages of about 220,000 to 250,000 yr and about 400,000 yr for the upper and lower portions of this section. Corrado and others (1986) have chosen instead to emphasize the significance of the apparent stage 5 isotopic dates at this section, thereby using these younger dates as calibration for their aminostratigraphic correlations in the South Carolina region.

The section at the Route 501 Bridge is the only place where aminozones IVa and IVb are in stratigraphic superposition. Elsewhere along the waterway, these two aminozones are exposed separately (five other sites for IVb and three other for IVa). At one site, about 1.5 km southwest of the Route 501 Bridge (not plotted in Fig. 15), a bimodal distribution of results falls into each of the two aminozones. We propose that this bimodal population at one site probably represents reworking of older shells into the younger deposit, because of the abundance of data that implies that IVa and IVb are two separate aminozones. The Route 501 Bridge section clearly plays a pivotal role in understanding the aminostratigraphy and lithostratigraphy of the Myrtle Beach area.

Figure 15 also depicts another aspect of the stratigraphic issues of the area. Two aminozones (IVb and IVc) represent different sites mapped by DuBar and others as the Canepatch Formation. Aminozone IVc most closely represents the type locality of this formation, from which Szabo and others collected corals that have U-series age estimates of 440,000 yr or more (see summary in Wehmiller and Belknap, 1982). Szabo (1985) also reports that U-trend dating of sediments at the type locality of the Canepatch Formation yields age estimates in the range of 600,000 to 700,000 yr. Our kinetic model age estimate for aminozone IVc is approximately 550,000 yr (Wehmiller and Belknap, 1982), which does not disagree with either the U-series or U-trend data or with the available paleomagnetic data for the Canepatch. The aminostratigraphic data lead us to conclude that the deposit mapped as the Canepatch Formation (by DuBar and others) formed during at least two separate transgressive cycles, separated by approximately 200,000 yr. The maximum estimate of the age differences between these aminozones suggests that these two cycles could correspond to stages 9 and 15 of the marine isotope record, but the possibility that they correspond to stages 9 and 11 instead cannot be ruled out. Szabo (1985, Fig. 6) showed the Canepatch (type section) as representing either stage 11 or stage 13, while he showed the ICW-1 site (equivalent to our ICW C' and DuBar's "Canepatch at 501 Bridge") as being equivalent to stage 9. This dual age for the Canepatch, and the age estimates themselves, are internally consistent for the two approaches to dating of the Canepatch sites. At two sites in the waterway (ICW-F' and ICW-J), shells of aminozones IVb and IVc occur in superposed and lithologically distinct units, and at ICW-J, aminozone IVd underlies IVc, so these results are not an artifact of sampling or analysis.

Figure 16, showing D/L data for other amino acids in the Intracoastal Waterway samples, supports the resolution of the four aminozones recognized in the leucine data.

Aminozone IVd corresponds to the unit mapped as the Waccamaw Formation. This aminozone is stratigraphically below either aminozone IVc or IVb at two sites along the waterway. The kinetic model age estimate for aminozone IVd is >1 m.y., which is consistent with biostratigraphic, paleomagnetic, and isotopic information for the Waccamaw Formation (Wehmiller and Belknap, 1982).

For the Charleston area, we rely on three primary references (McCartan and others, 1982, 1984; Szabo, 1985), because our samples from this area have come from some of the outcrops

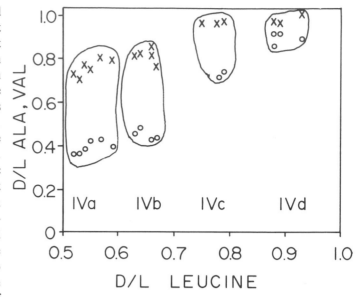

Figure 16. Myrtle Beach area data for alanine (x) and valine (o) (mean values, multiple shells each site) compared with leucine data. See Figure 15 for discussion. In spite of slight overlaps, alanine and valine resolve aminozones IVa and IVb, as does leucine.

studied by these workers. The four aminozones recognized in the Charleston area (Fig. 14) appear, on qualitative grounds, to be numerically equivalent (and geochronologically equivalent) to the four recognized in the Myrtle Beach area. As noted, however, McCartan and others (1982) used U-series isotopic data for sites in the Charleston area to argue that our aminozone IVa (the Wando Formation of McCartan and others, 1982) in Charleston is substantially younger than our aminozone IVa in Myrtle Beach (by approximately 100,000 to 150,000 yr, or one interglacial cycle). Wehmiller and Belknap (1982) acknowledged this conflict but pointed out that the coral samples from the Wando sites all had $^{230}Th/^{232}Th$ values that implied contamination or open-system behavior, thereby suggesting that the apparent precision of multiple coral U-series dates from the Charleston area might be misleading. U-series-dated corals, and mollusks for amino acid analysis, were collected simultaneously at several of the Charleston sites, so these particular conflicts cannot be explained as artifacts of sampling.

The sloping isochron band in Figure 14 represents the interpretation preferred by McCartan and others (1982), showing their Wando Formation to be younger than the Socastee Formation in Myrtle Beach. Our aminostratigraphic correlation implies that the Wando Formation and the Socastee Formation are temporally equivalent. In addition to several corals' dates of 200,000 yr and more, Szabo (1985) obtained three, stage 5, U-series, isotopic dates for corals from "probable Socastee" sites in Myrtle Beach, but rejected these three dates as being too young; we agree with his interpretation and point out that the isotopic data (espe-

TABLE 3. COMPARISON OF LITHOSTRATIGRAPHIC AND GEOMORPHIC NOMENCLATURE OF
THE CHARLESTON AREA QUATERNARY MARINE UNITS WITH RADIOMETRIC DATA AND
APPARENT LEUCINE *MERCENARIA* AMINOSTRATIGRAPHY

McCartan Unit	Colquhuon Terrace	Formation	Average Elevation (m)	Ward: 0 - 18 Stage	Isotope Age (Ma)	Aminozone (this work)
Q1	Recent	no name	0	1		
Q1	Silver Bluff	no name	2.5	5e	0.007	Possibly younger than IVa
Q2	Princess Anne	Upper Wando	5	7	0.075 - 0.10	IVa
Q3	Pamlico	Lower Wando	8	9	0.12 - 0.13	IVb?
Q3	Talbot (lower)	Ten Mile Hill Beds	11	11	0.20 - 0.24	IVc
Q4	Talbot (upper)	Ladson	14	13	0.450	
Q5	Penholoway		23	15?	0.73 - 0.97	IVd

Notes:

1. Stage designations from Ward's (1985) geomorphic aproach.
2. "Terrace" designations from Colquhuon (1974) as summarized in McCartan and others (1984).
3. Unit designations from McCartan and others (1984); ages are averages of dates reported by Szabo (1985), with stratigraphic associations as presented by either McCartan and others (1984) or Corrado and others (1986).
4. Formation nomenclature and elevations as summarized by Corrado and others (1986).
5. Aminozones as developed here from *Mercenaria* leucine data using aminostratigraphic and kinetic model approaches to age estimation. Other options, such as those of McCartan and others (1982) and Corrado and others (1986), are discussed in the text.

cially $^{230}Th/^{232}Th$ values) for the rejected samples are similar to the Wando Formation coral results that were not rejected in earlier studies (McCartan and others, 1982). The ^{232}Th abundances in most of the younger South Carolina coral samples are high enough to require that these results be interpreted with caution (Wehmiller and Belknap, 1982). The $^{234}U/^{238}U$ values, which Szabo has employed for age estimation in some cases, are low enough in both the Wando and Socastee corals to suggest ages of greater than 200,000 yr for these samples.

The stratigraphy in the Charleston region consists of a complex sequence of marginal marine units preserved as terraces underlain by a variety of barrier, shelf, and estuarine deposits. McCarton and others (1984) summarized how their mapped units in the Charleston area correspond to previous nomenclature for terraces and associated terrace deposits in this area. Corrado and others (1986) proposed a slight revision of this terminology. The relation of our aminozones for the Charleston area to the stratigraphic interpretations of these workers is summarized in Table 3. We also show the elevations of the relevant units. The ages assigned to the units by McCartan and others (1984) are based on the Th/U results of Szabo (1985). Ward (1985), using the isotopic record of climatic change as a model for correlation

of coastal sea level records, proposed a chronology for the Charleston area (in particular, the Wando Formation) that is essentially the same as we proposed on the basis of aminostratigraphy and kinetic modeling. Ward's model correlations are also shown in Table 3. Our aminostratigraphic correlations and Ward's chronology imply that the Wando Formation is correlative with isotope stage 7 (approximately 200,000 yr B.P.), in conflict with the stage 5 (mostly late stage 5) U-Th dates for the Wando. In addition, the one Charleston area site from which we obtained aminozone IVc data is not directly associated with either of the coral collection sites used to designate unit Q3 as stage 7 (ca. 200,000 to 250,000 yr in age), although it is interpreted by McCartan and others (1984) to be correlative to both of these dated sites.

The relation of aminozone IVb, which consists of one poorly exposed site from which three shells with only modest preservation characteristics were obtained, to the mapped units of McCartan and others (1984) is uncertain and requires further investigation.

If the U-series coral dates for the Charleston area are incorrect, and if the Wando Formation belongs to stage 7, then it appears that isotope stage 5 is very poorly represented in region

IV. Only one site, Eddingsville Beach, South Carolina, south of Charleston at 32.5°N, might be representative of a stage younger than aminozone IVa; this site is geomorphically younger (seaward) in relation to the Wando sites and is mapped as the Silver Bluff terrace by prior workers (Colquhuon, 1974). The one sample (plotted in Fig. 14) that we analyzed from Eddingsville Beach was not collected by us, but a more recent collection at this locality has produced leucine D/L values of between 0.39 and 0.45 for *Mercenaria* and *Anadara*, consistent with a stage 5 age. If correct, these results would further support our stage 7 age interpretation for the Wando sites.

Corrado and others (1986) and Hearty and Hollin (1986) have offered aminostratigraphic interpretations for this region that are similar to ours in that they do not require steeply sloping correlation lines between Charleston and Myrtle Beach. However, their choice of certain U-series dates from the Charleston and Myrtle Beach regions for calibration points results in some major conflicts with the absolute ages that we propose for sites in both region IV and region III (particularly Flanner Beach). In particular, Corrado and others (1986) proposed a high-resolution *Mulina* aminostratigraphy (recording substages 5e, 5c, and 5a) in the Charleston area and have applied these three substage correlations in a reinterpretation of our results. This reinterpretation requires a much different model for the latitudinal gradient of effective temperature along the Atlantic Coastal Plain.

Corrado and others (1986) performed a large number of alloisoleucine/isoleucine measurements on samples of *Mulinia* from outcrop and subsurface samples. The majority of samples were obtained by augering, a method that Corrado and others acknowledge as probably causing mixing of samples of different ages. Consequently, Corrado and others chose to interpret their data in terms of statistical groups thought to represent specific times of deposition. Their groups (modal peaks) of allo/iso values, and the inferred times of deposition (derived from the U-Th data of Szabo, 1985) are as follows:

    0.35—late stage 5
    0.44—middle stage 5
    0.50—early stage 5
    0.65 and 0.72—late and early stage 7, respectively
    0.92, 0.99, and 1.12—all greater than 700,000 yr

Because *Mulinia* is such a thin-shelled mollusk, and because of our own experience with *Mulinia* in region III (e.g., coefficients of variation of about 17 percent and the inability of *Mulinia* to resolve stage 5 and stage 7), we doubt the ability of this genus to provide the high-resolution stage 5 aminostratigraphy proposed by Corrado and others. Careful analysis of their data, using only results for samples from outcrops, reveals a somewhat different interpretation. Data from three outcrop collections, sites L2, L5, and L22 of Corrado and others (1986), demonstrate the complexity of this issue. Site L5, which we also studied (Mark Clark Pit), falls into our aminozone IVa. Corrado and others have divided the L5 site into an upper and lower unit. The *Mulinia* data for these sites are listed below:

Site	Allo/iso		No. of samples	Elevation in meters
	mean	st. dev.		
L2	0.3809	0.058	11	+1.2
L5a	0.451	0.057	10	+2.4
	0.425	0.019	8	
L5b	0.469	0.032	11	+1.5
L22	0.548	0.034	5	+1.5

Note: The results for L5a are shown for all ten samples and also for the eight that are interpreted (based on the allo/iso data) not to have been reworked.

Corrado and others argue that the L5a–L5b stratigraphy represents an early- to late-stage 5 chronology because a coral that yielded a 120,000-yr U-Th date was collected from "somewhere in the section, most likely in the lower [L5b] unit" and because they reject two allo/iso values of 0.54 and 0.57 from the upper unit (L5a) as being reworked. Rejection of these two allo/iso values results in a lower mean value and smaller standard deviation for L5a, as shown above. The two rejected samples do not appear to be reworked, based on their total amino acid content (data of Corrado and others, 1986, Table 2). Without rejection of these two samples, the problem of two units at L5 is unresolvable. Corrado and others (1986) also forced this high-resolution apparent stratigraphy at L5 into the interpretation of the *Mercenaria* data that we obtained (Wehmiller and Belknap, 1982) at this site. Corrado and others (1986, p. 34) suggested that three of our samples came from the upper bed and three from the lower bed and proposed two very tight clusters of D/L leucine values (0.547 ±0.005 and 0.593 ±0.013) for the upper and lower units. However, one articulated shell sample (JW80-163), from which we analyzed fragments from both valves, yielded the entire range of D/L values (0.55 and 0.61) that Corrado and others interpreted to represent the early- to late-stage 5 sequence at this exposure. Although we acknowledge that in our data (Wehmiller and Belknap, 1982; Table 2), articulated shell samples were not clearly identified, one shell could not represent such a long time span. In addition, we are certain that the *Mercenaria* we collected at Mark Clark Pit all came from the same stratigraphic unit and that only one unit was exposed at our collection site. We conclude that the two-fold aminostratigraphic interpretation of Corrado and others for their site L5 is based on some extremely precarious assumptions about our data and their own.

The allo/iso data from site L2 show a broad range of ratios, which can be interpreted as bimodal. Corrado and others interpret this distribution to represent mixing of two populations (allo/iso values of 0.33 ±0.02 and 0.43 ±0.04). However, we conclude that these results represent the typical distribution of allo/iso values for *Mulinia* at a single outcrop; some of the lower allo/iso values might be rejected as indicative of diagenetic contamination, but the amino acid abundances alone (Corrado and others, 1986) do not serve as unique indicators of either age differences or contamination.

The allo/iso data from site L22 are closely related to a site that was dated as 202,000 ±18,000 yr by Th/U on a coral (Corrado and others, 1986). Therefore, the aminostratigraphy of site L22 should be most rigorously tied to a stage 7 calibration site, yet the mean allo/iso value for L22 of 0.548 is substantially below the modal peak of 0.66 (range 0.60 to 0.68) that Corrado and others proposed as representative of late stage 7 (Corrado and others, 1986, Fig. 3). Only one other locality (L7 of Corrado and others, which yielded five allo/iso values between 0.65 and 0.76) was related to a stage 7 calibration date, but L7 was an auger hole 0.9 km from another auger hole in which the U-Th-dated coral was obtained (depth information not provided in Corrado and others). In light of the complexities inferred from aminostratigraphic data in this region and elsewhere, we suggest avoiding the use of coral dates as calibrations unless all relevant samples were collected from the exact locality and stratigraphic position at the same time.

Three outcrops, each designated as a type locality (using the nomenclature of DuBar, as shown in Fig. 15) for each of the formations mapped in Myrtle Beach were also studied by Carey ([Corrado] 1981). *Mulinia* allo/iso data (Carey [Corrado], 1981) for these three sites are listed below:

Type Socastee	Type Canepatch	Type Waccamaw
0.332 ± 0.033 (6)	0.33 ± 0.036 (3)	
	0.425 ± 0.035 (2)	
0.505 ± 0.035 (2)		
	0.673 ± 0.02 (3)	
		1.006 ± 0.04 (4)

The number in parentheses is the number of shells analyzed. In each case, the allo/iso values are in proper stratigraphic order. In many cases, more than one *Mulinia* aminozone is found in the Myrtle Beach outcrops, which typically range in elevation from sea level to between +3 and +6 m above sea level. The data are arranged here to show the apparent *Mulinia* aminostratigraphy in the Myrtle Beach area—these relationships are similar to those shown in Figure 15.

As Corrado and Hare (1981) noted, numerically equivalent *Mulinia* aminozones are observed in both the Myrtle Beach and Charleston areas, and Corrado and others (1986) indicated that they favor horizontal aminostratigraphic isochrons (i.e., no major effective temperature gradient) between these two areas. From the Myrtle Beach *Mulinia* data given above, we conclude that the Canepatch Formation in Myrtle Beach does have an aminostratigraphic equivalent in the Charleston area (the group of allo/iso values between 0.66 and 0.72, indirectly related to the Ten Mile Hill Beds), and that deposits mapped as Canepatch in Myrtle Beach probably could yield *Mulinia* allo/iso values of ca. 0.45 or 0.67, suggesting (as do the *Mercenaria* data) that the Canepatch represents two depositional events. Nevertheless, Corrado and others (1986, p. 40) argue that the Canepatch Formation in Myrtle Beach has no correlative marine beds in the Charleston area, instead suggesting that the fluvial/deltaic Ladson Formation (pre–Ten Mile Hill in age) is the Charleston-area equivalent of the Canepatch.

Because both *Mulinia* and *Mercenaria* aminostratigraphic correlations imply that the Canepatch equivalent marine sediments exist in Charleston, and because of the assignment of stage 7 ages to modal allo/iso values much higher than those observed at site L22, some serious ambiguities exist in the stratigraphy and absolute age assignments proposed by Corrado and others (1986) for South Carolina. We argue that at least two, and perhaps three, post-Waccamaw *Mulinia* aminozones are identified in both the Charleston and Myrtle Beach areas and that the youngest of these aminozones (lowest D/L values) implies age equivalence of deposits labeled Wando in Charleston and Socastee in Myrtle Beach. Age equivalence for these deposits is not indicated by the majority of the U-series data at these sites. Consequently, as with the *Mercenaria* data, reconciliation of the South Carolina *Mulinia* aminostratigraphy with *all* the U-Th obtained from the Charleston and Myrtle Beach areas is virtually impossible. The suggested age equivalence (stage 5, Corrado and others, 1986; Hearty and Hollin, 1986) for units in Charleston and Flanner Beach with *Mulinia* allo/iso values of ca. 0.43 and 0.16, respectively, is difficult to accept because Flanner Beach is only 2.5°C cooler today than Charleston. A more detailed discussion of these issues is presented elsewhere (Wehmiller and others, 1987).

Region V: Florida and Bahamas Islands (Fig. 5)

Region V (Fig. 5) includes Pleistocene deposits along the east coast of the Florida peninsula and three sites in the Bahamas Islands. D/L leucine data are available for the Bahamas sites and one Florida site; *Mercenaria* isoleucine (D-alloisoleucine/L-isoleucine) data are from Mitterer (1974, 1975). Most of the Florida localities are not well documented in the literature, but Mitterer (personal communication, 1983) has provided descriptions of the sites from which *Mercenaria* were collected and subsequently analyzed for alloisoleucine/isoleucine values. All of Mitterer's sites are apparently above sea-level, an important consideration in any comparison of the Florida chronology to late Pleistocene sea-level records. *Chione* isoleucine data are available from one site in south-central Florida at which both coral and mollusk samples were collected. The alloisoleucine/isoleucine and D/L leucine values for region V are plotted versus latitude in Figure 17.

Four apparent aminozones (Va, Vb, Vc, and Vd, youngest to oldest) are found in northern Florida. In central and southern Florida, three aminozones are apparent in the isoleucine data. Within the experimental uncertainty of conversions between different genera and between leucine and isoleucine data (see Appendices I and III for discussion of leucine-isoleucine comparisons), the three Bahamas sites all appear to be equivalent to Florida aminozone Va. Mitterer (1975) proposed that aminozone Va was equivalent to oxygen isotope stage 5e (or perhaps 5c), based on the probable correlation of samples of this aminozone with U-series (coral) dated sites in central and southern Florida having ages between 110,000 and 130,000 yr. This conclusion is supported by the data for the Bahamas sites, which all have been

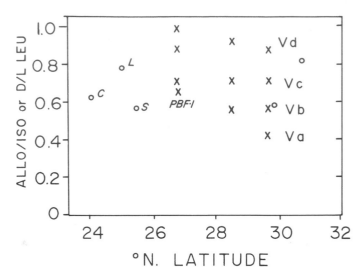

Figure 17. Leucine and isoleucine aminostratigraphy for region V. Mean annual temperatures for the northern portion of this area are about 20°C; they are about 24.5° for the southern portion. The isoleucine data points (x) represent the groups of ratios reported by Mitterer (1975) for *Mercenaria* from northern, central, and southern Florida, respectively, and one *Chione* site (PBF-1) in southern Florida. The leucine data (o) are for *Mercenaria* from two northern Florida sites and *Strombus* (S), *Lucina* (L), and *Chione* (C) from Bahamas sites.

associated with U-series coral dates between 100,000 and 140,000 yr (Neumann and Moore, 1975; Carew and others, 1984; J. S. Carew, personal communication, 1986). The *Chione* isoleucine data (PBF-1) support these comparisons between different genera and different amino acids. A solitary coral sample (weighing about 60 g) from PBF-1 yielded a U-Th date of about 160,000 yr (Table 2). This date is difficult to interpret because it does not correspond to either isotope stage 5 or stage 7. The coral was not attached to any mollusk sample used for amino acid analysis and was the only one found at the collection site (in a 2 × 50 m outcrop), so it could have been reworked from an older deposit. Although the coral sample showed signs of recrystallization in some areas, its large size permitted subsampling the best-preserved portions for U-series analysis.

Leucine data from a single sample from the Anastasia Formation near St. Augustine indicate a late Pleistocene age (aminozone Va or Vb) for this deposit, consistent with the isoleucine data and the radiometric data summarized by Mitterer (1975) for this area. The Reids Bluff site (leucine data) probably fits in aminozone Vd.

We sampled only two of the Florida sites represented in Figure 17. Preliminary data for other sites in Florida are consistent with the four-fold aminostratigraphy depicted in Figure 17. Stratigraphic implications of these data are not discussed here, other than to note the probable correlation of aminozone Va with

isotope stage 5. This age control for aminozone Va is important in understanding the overall latitudinal framework of amino acid data.

AMINOSTRATIGRAPHY, KINETIC MODEL DATES, AND PHILOSOPHY OF GEOCHRONOLOGY

Ward (1985) discussed a variety of approaches used in the application of dating methods to Quaternary stratigraphic problems. By using the marine isotopic record to predict positions of shorelines in selected coastal records, Ward (1985) avoided direct reliance on U-series coral dates, which, he points out, are commonly accepted without question, even when some dates are suspect or when "the geology is fitted to the dates" (Ward, 1985, p. 1156). The range of attitudes to dating methods includes those in which the "dating determines the geology" and those in which the dates only confirm a time-stratigraphic sequence that is well established by field relationships (see Ward, 1985, p. 1158).

Applications of AAR have often represented this same spectrum of approaches. One approach, which is probably most attractive theoretically, is to use amino acid enantiomeric ratios and kinetic models to assign dates directly to individual analyzed samples. This approach has often been used in the application of AAR to bone samples (Bada, 1985) and, in principle, is similar to most isotopic dating methods. The second approach, which becomes more viable in a region with large sampling density, employs enantiomeric ratio data as a simple correlation or stratigraphic tool. Such data can also be interpreted in terms of ages (using kinetic models and/or independent calibration). The basic premise of this approach is that, given suitable samples, enantiometric ratios become valid time-stratigraphic tools with a status at least equal to that of classical lithostratigraphic, biostratigraphic, and geomorphic methods. In this approach, amino acid data form a component of the time-stratigraphic information and their use in determining depositional sequences does not represent a case in which the "dates determine the geology."

The critical ingredient for successful application of aminostratigraphic methods to a region like the Atlantic Coastal Plain is an adequate density of sampling sites over the entire latitude-temperature range of interest. Because many of the current issues in Atlantic Coastal Plain aminostratigraphy involve assumptions in correlations over distances of 3 or 4 degrees of latitude (Wehmiller and Belknap, 1982), having closely spaced (ca. 100 km or less), age-equivalent localities over the entire latitudinal range would be ideal. The availability of such data would permit local aminozones to be assembled into continuous isochrons that represent the latitudinal gradient of effective temperatures. The density of potential collection sites will probably never approach this ideal situation, so use of a combination of independently calibrated data and kinetic modeling is necessary to correlate aminostratigraphic data between regions having a fairly high density of collection sites. The range of Atlantic coast collection-site densities for which aminostratigraphic data are available is from approximately 30 sites/degree of latitude (portions of South

Carolina) to as low as 2 sites/degree of latitude. Given the fragmentary nature of the preserved record of the late Cenozoic marginal marine units of the coastal plain (Cronin and others, 1984), most approaches to correlation are hampered by regions of low sample density.

Two principle approaches to estimation of ages from the Atlantic Coastal Plain aminostratigraphic data are possible: (1) a kinetic model for leucine racemization presented elsewhere (Wehmiller and Belknap, 1982), and (2) a more empirical approach (modified from Corrado and others, 1986) that is dependent on the choice of U-Th calibration sites.

The kinetic model approach relies on the following assumptions:

(a) The present latitudinal temperature gradient is an adequate model for the shape of past temperature gradients. Steeperthan-present late Pleistocene temperature gradients would have been likely, but inversions of thermal gradients would not be considered reasonable. Wehmiller and Belknap (1982) summarized the information on which these late Pleistocene thermal gradients were modeled. Kennedy and others (1982), Miller and Mangerud (1985), and Hearty and others (1986) relied on similar assumptions.

(b) The non-linear kinetic model for racemization (Wehmiller and Belknap, 1982; Wehmiller, 1984a; Miller and Mangerud, 1985; Hearty and others, 1986) is appropriate for the interpretation of these results.

(c) The combination of temperature and kinetic models will predict smooth trends of isochronous D/L values versus latitude, with no inversions or major divergences of these isochrons.

(d) Local aminozones have enough variation in them so that they should be grouped (i.e., put into mappable units) for kinetic model age estimation, rather than estimating ages for individual outcrops. This variation is probably due to a combination of local thermal and diagenetic effects, as well as slight (5–10%) age differences for different deposits within a particular region. These observed local variations are assumed to be due to natural causes and are not an artifact of poor analytical resolving power.

Calibration of model racemization kinetics

Aminostratigraphic data from marine terrace calibration sites on the Pacific coast of North America provide the best available model for the construction of isochrons over a broad latitude-temperature region (Kennedy and others, 1982). Figure 18 (from Wehmiller, 1982, 1984a) presents D/L leucine values in *Protothaca* from four sites, all 120,000 to 130,000 yr B.P., from 36°N (CMAT about 13.3°C) to about 23°N (CMAT about 21.5°C).

Figure 18 shows how aminostratigraphic data from independently dated samples can be used to estimate the integrated thermal history (effective Quaternary temperatures) for these late Pleistocene calibration localities. Model isochrons (Fig. 18) predicted for 120,000-yr samples exposed to effective temperatures equal to, or slightly less than (0° and 5°C) present temperatures at

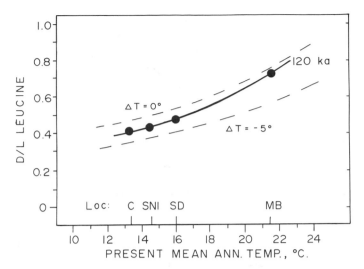

Figure 18. Modified from Wehmiller (1982, 1984a). D/L leucine data in *Protothaca* from Pacific coast calibration sites (all early stage 5), compared with kinetic model isochrons (dashed lines) for 120,000 yr at two possible effective temperatures: identical with present-day ($\Delta T = 0$), and 5°C cooler than present-day ($\Delta T = -5$). Solid line connects calibrated data and shows that the actual effective temperature for these four sites has been about 3° cooler than present for the more northerly sites and about 1° cooler than present for the most southern site. Site Abbreviations: C = Cayucos, California; SNI = San Nicolas Island, California; SD = San Diego, California (Nestor terrace); MB = Magdalena Bay, Baja California, Sur, Mexico.

the sites. These isochrons are specific for leucine in *Protothaca* and are derived from a kinetic model developed by Wehmiller and Belknap (1978, 1982; see Wehmiller, 1982, for further discussion). The calibration isochron, fixed to the calibration samples, represents an effective temperature approximately 2° to 4°C below modern temperatures. Because these effective temperatures represent the integrated thermal history of samples, they do not uniquely define any one temperature during that history. Nevertheless, when compared with modern temperatures for a site and with the oxygen isotope model for the timing of late Pleistocene temperature fluctuations, reasonable estimates of full-glacial maximum-temperature reductions can be derived from these effective temperature values (Wehmiller, 1982). By this approach, the estimated full-glacial (stage 2, 18,000 yr B.P.) temperature reduction for coastal California is between 3° and 6°C, with greater reductions at higher latitudes. Full-glacial temperatures near southern Baja California appear to have been only 2° to 3°C cooler than present. A similar range of sea-surface Pacific fullglacial temperature reductions has been inferred from micropaleontologic data (Imbrie and others, 1983). The marine-dominated nature of coastal climate from 48° to 23°N is reflected in the similar slopes of latitudinal temperature gradients for modern air- and sea-surface-, full-glacial sea-surface-, and effective temperatures inferred from the AAR data at dated sites. The smooth latitudinal temperature gradient and the apparent lack of extensive temperature variation during the late Pleistocene of the Pacific coast simplifies the development of kinetic models of molluscan amino acid racemization, which form the basis for current aminostratigraphic correlations on the Atlantic Coastal Plain.

In previous studies (Wehmiller and Belknap, 1982), evidence derived from the Pacific coast kinetics was used, and information on the relative racemization kinetics of *Protothaca* and *Mercenaria* was also used to argue that the general trend of isochrons versus latitude and/or effective temperature should be similar for both Atlantic and Pacific coast data. However, the required extrapolations or interpolations of either calibrated or theoretical kinetic models have usually involved graphical approaches that are difficult to duplicate by other workers. We present here a more quantitative approach based on equations that translate the Pacific coast model to the Atlantic coast. The fundamental conclusions derived from this approach are the same as earlier ones (Wehmiller and Belknap, 1982), but they now have a more quantitative basis. Appendix III presents the relationship between leucine and isoleucine curves for *Mercenaria* and *Protothaca.*

The model presented here, a quantitative, non-linear, kinetic model, focuses on leucine racemization in the molluscan genera *Protothaca, Mercenaria,* and *Chione.* Evidence from both field and laboratory studies indicates that these genera have similar relative racemization rates (Keenan, 1983; Lajoie and others, 1980). This model differs from that given in Wehmiller and Belknap (1982, Fig. 6 and 8) only in that it describes the kinetic pathway in terms of a continuously decreasing rate according to the equation:

$$y = a + b[\ln t], \text{ where } y = (X_E - X)/X_E \qquad (1)$$

where, X is $D/(D+L)$ at any time t, and X_E is the equilibrium $D/(D+L)$ value (0.50 for leucine).

Appendix II contains further discussion of equation 1 and a set of curves derived from it. The general form of the quantitative model equation indicates that the initial few thousand years of diagenetic history cannot be modeled reliably because equation 1 is asymptotic to the y-axis. Miller and Hare (1980) showed a rapid decrease in the total amino acid content of molluscan fossils inthe early stages of diagenesis, after which the amino acid concentrations decrease only slowly with time. The quantitative model equations appear inadequate for modeling the earliest phases of diagenetic racemization, but they are adequate for modeling racemization in the total amino acid mixture (even though the relative mix of different molecular weight components might be changing) as long as the amount of material in this mixture is relatively constant.

A separate component of any racemization model is the modeled temperature history of any sample or region, and the resulting EQT (effective Quaternary temperature) assigned to a

particular site or region. However, the temperatures themselves are independent of the kinetic model because they can be inferred from independent paleoclimatic data. The Pacific coast results have been particularly useful because both paleoclimatic and kinetic model estimates of EQT agree. The approach outlined here represents a further refinement of that kinetic model.

In order to establish specific kinetic equations for various values of EQT, determination of an equation is necessary for a single EQT (latitude) value at which two calibration sites of different age are available. The pair of terraces on Pt. Loma, near San Diego, California, provide excellent control for this calculation. *Protothaca* samples from the lower (Bird Rock) and upper (Nestor) terraces have mean D/L leucine values of 0.39 and 0.47, respectively (Wehmiller and others, 1977; Kennedy and others, 1982). U-series data for solitary coral samples from the Nestor terrace indicate that it is early-stage 5 age, and the combined geomorphic, isotopic, faunal, and aminostratigraphic evidence indicate that the Bird Rock terrace is late stage 5 in age (Lajoie and others, 1979; Kennedy and others, 1982; Kennedy and Muhs, personal communication, 1987). For the purpose of developing a calibrated kinetic equation, ages of 120,000 and 80,000 yr are assigned to the Nestor and Bird Rock terraces, respectively.

For reasons summarized elsewhere (Wehmiller, 1982, Appendix I), terraces of early- and late-stage 5 age cannot be interpreted in terms of the same EQT, because a greater proportion of the younger terrace's history has been affected by cooler temperatures. If the Bird Rock terrace had the same EQT (13.5°C) as the Nestor terrace, the *Protothaca* D/L leucine values would be approximately 0.415. Substituting the appropriate D/L leucine values and terrace ages into equation 1, one can solve for the constants a and b, as follows:

Nestor terrace—
  $t = 0.12 \times 10^6 \quad \ln t = -2.12$
  $D/L = 0.472 \quad y = 0.3587$

Bird Rock terrace—
  $t = 0.08 \times 10^6 \quad \ln t = -2.526$
  $D/L = 0.415 \quad y = 0.4134$
  t in million yr. units
  $[(a-0.3587)/(2.12 \times b)] = [(a-0.4134)/(2.526 \times b)]$
  $a = 0.0732 \quad b = -0.135$

Using the Magdalena terrace (Baja California Sur) as a higher-EQT, 120,000-yr, calibration site for the equations presented above, assuming an EQT of about 21.0°C for this site (Wehmiller and Emerson, 1980), the constants of equation 1 can be related to the temperature dependence of racemization as shown in Table 4.

The b constant does not vary with temperature (although it would vary with different kinetic groups or taxa), but the a constant varies in a manner indicating a temperature dependence of about 16.5 percent per degree C, quite similar to various estimates of the activation energy of racemization (Bada and

TABLE 4. VALUES OF THE CONSTANTS a AND b OF THE GENERAL KINETIC MODEL EQUATION $y = a + b(\ln t)$ FOR VARIOUS VALUES OF THE EFFECTIVE QUATERNARY TEMPERATURE (EQT)

EQT (°C)	a	b
8.0	0.2066	-0.1349
9.0	0.1821	-0.1352
11.0	0.1336	-0.1348
12.0	0.1090	-0.1352
13.0	0.0851	-0.1347
14.0	0.0614	-0.1352
15.0	0.0378	-0.1346
18.0	-0.0362	-0.1352
21.0	-0.1081	-0.1347
		mean = -0.1350

Schroeder, 1972; Mitterer, 1975). The values for the constants a and b shown above can then be used to generate leucine racemization curves for different temperatures for any genus (i.e., *Mercenaria, Chione,* and probably others) that has kinetics similar to those of *Protothaca*. These curves, depicted in Figure 19, are quite similar to Figures 6 and 8 of Wehmiller and Belknap (1982), except in the region above D/L values of about 0.85. The age estimates derived from these curves are here referred to as "isochron model ages."

Also shown in Figure 19 are mean D/L leucine data points for each of the regional aminozones presented in previous discussions. The choice of EQT values at which to plot each leucine value is critical to the age estimate; the values used in Figure 19 are chosen to be consistent with present-day temperature differences between the sites, with independent paleoclimatic information (Wehmiller and Belknap, 1982), and with stage 5 coral dates for sites in Massachusetts, Virginia, and Florida (SH, NL & GP, and SF, respectively). The slopes required for each regional aminozone (especially the least racemized) to be consistent with the model isochrons increase for the lower latitude regions. Figure 19 can be used to quantify the conflicts between isochron model ages and the U-series data and correlation models presented elsewhere (McCartan and others, 1982; Szabo, 1985; Corrado and others, 1986). Figure 8 of Wehmiller and Belknap (1982) presented age interpretations in the same manner, but did not group data into each regional aminozone for comparison with model isochrons. The approach used here tends to smooth any data variability due to local temperature differences and to consider only the larger temperature differences between regions. Table 5 presents a summary of the isochron model ages and correlations for each regional aminozone.

The aminozones for the Florida sites, which are based mostly on Mitterer's (1975) isoleucine data, have a large uncertainty because of the reported range of isoleucine results and the uncertainty involved in conversion from isoleucine data to leucine data (see Appendix I). Because the model isochrons at D/L values

TABLE 5. SUMMARY OF APPARENT REGIONAL AMINOZONES AND THEIR PROBABLE CORRELATIONS (BASED ON THE KINEMATIC MODEL CORRELATIONS SHOWN IN FIG. 19) WITH THE MARINE ISOTOPIC RECORD

Isotopic Stage	South V	IV	III	II	North I
1	H		H		a
3					
5 (late)	a		a?	a	b
5 (early)	a		b	b	b
7	b	a	c	c	c
9	b/c?	b			
11	c/d?	b	d?	d?	
13	d?		d	d	
15		c	d	d	
17		c			
19		c			
ca. 1 Ma					
	d	e	e	d	

Note: Correlations with the isotopic record are based on probable age ranges for each aminozone, as derived from leucine kinetic model shown in Figure 19. Many aminozones, especially the older ones, could possibly represent more than one isotopic stage. Horizontal correlation lines place limits on the probable age range for particular aminozones.

greater than about 0.50 become rather steep, and because of the large temperature range within region V, it is necessary to plot the data for each region V aminozone as separate points (blocks) rather than in one large block representing the entire latitudinal range for the region.

Selected aminozones in Figure 19 are also identified with solid triangles. The lines connecting these triangles represent the chronostratigraphic correlations proposed by McCartan and co-workers. Wherever one of the lines connecting triangles intersects one of the isochrons, a conflict arises between the two approaches to correlation.

The group represented by the inverted triangle (▼) contains D/L data from the Wando Formation (aminozone IVa near Charleston, South Carolina), and data from aminozones Va, IVb, IIIb, IIa and IIb, and Ib. The data points plotted in Figure 19 for

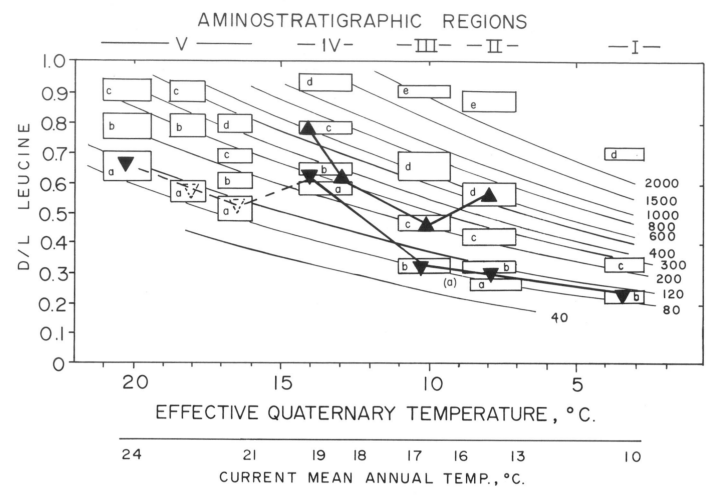

Figure 19. Kinetic model isochrons for the Atlantic Coastal Plain, plotted versus Effective Quaternary Temperature fixed for calibration sites in Massachusetts (SH, aminozone Ib), Virginia (GP and NL, aminozone IIa), and Florida (aminozone Va, derived from data of Mitterer, 1975). Isochrons drawn from the kinetic model derived from equation 1 in text. Ages represented by each isochron, in 1,000-yr units, are shown at the right end of each isochron. Regional aminozones (depicted by roman numerals across the top of the figure and letters within each data cluster) are superimposed on the isochrons for purposes of age estimation. The inverted and upright triangle symbols depict the portions of our aminozones that are thought to be correlative with either stage 5 or stage 7, respectively (McCartan and others, 1982). Wherever correlation lines based on these symbols diverge from, or intersect with, kinetic model isochrons a conflict between the two approaches to relative age estimation is apparent.

region V are "equivalent leucine" D/L values, converted from the isoleucine data shown in Figure 17 where needed. McCartan and co-workers would assign stage 5 ages to these points, based on U-series coral dates from several sites in the Wando Formation (late stage 5); the Norfolk, Virginia, area (late stage 5); and at Sankaty Head, Nantucket Island, Massachusetts (early stage 5). The correlation to aminozone Va is based on the work of Mitterer (1975), who has indicated that the samples from aminozone Va were collected from sites correlative with U-series (coral) dated sites along the Florida coast.

The group represented by a triangle (△) contains data from aminozones IVa and IVb (Myrtle Beach, South Carolina), aminozone IVc (Charleston, South Carolina), and aminozones IIIb and IId. McCartan and co-workers generally consider the localities in this group to represent deposition during isotope stage 7 (ca. 200,000 to 240,000 yr B.P.), based on the U-series coral dates at Norris Bridge and Mathews Field (aminozone IId sites) and at two sites near Charleston, which they correlate to our site TPSC (aminozone IVc).

The proper trends of increasing D/L values with increasing sample age (at a given latitude) are generally observed in Figure 19. In addition, the isochron model ages (given at the right end of each isochron) agree reasonably well with the U-series coral dates from many of the localities. However, significant conflicts between isochron model ages and U-series coral dates are seen in Figure 19 for the following localities:

(1) Wando Formation, Charleston, South Carolina, aminozone IVa: 80,000 to 100,000 yr by U-series but about 200,000 to 250,000 yr by the isochron model. The Wando Formation appears age-equivalent to Socastee Formation near Myrtle Beach by amino acids, yet the Wando Formation is thought to be younger than Socastee by McCartan and others (1982). This conflict is related to the differences between lithostratigraphic and aminostratigraphic resolution in the Myrtle Beach and Charleston areas. Only by interpreting these IVa results in terms of significantly warmer effective temperatures (about 20°C instead of about 14°C) can these results be forced into agreement with the U-Th dates.

(2) Norris Bridge, Mathews Field, and Dirickson Creek Ditch, all represented by aminozone IId. The aminostratigraphic correlation of these sites is consistent with litho-stratigraphic models for the region, but the amino acid age estimate for zone IId is substantially greater than the dates of 190,000 to 210,000 yr quoted by Szabo (1985) for Norris Bridge. The amino acid age estimate for these sites is in much better agreement with the biostratigraphic age estimate of greater than 300,000 yr, proposed by Cronin and others (1984) for the Dirickson Creek Ditch site. Only by interpreting the IId data in terms of an effective temperature of about 14°C (very nearly the present temperature at the sites) can the amino acid data be reconciled with the U-Th date at Norris Bridge.

The issue of the age of aminozone IId is demonstrated in a different manner in Figure 19. The current model of coastal plain stratigraphic correlation (of McCartan and coworkers) maintains that the localities represented by aminozones IIIc and IId are all correlative. The aminostratigraphic data agree with each of these local correlations within specific regions, but the age equivalence of IIIc and IId would require that there be an inversion of the latitudinal trends of D/L values, contradicting one of the fundamental principles of aminostratigraphy. Inversions of this type have not been observed in studies along other north–south-trending coastlines that span a comparable latitude–temperature range (Wehmiller and Emerson, 1980; Kennedy and others, 1982). If such inversions are realities, then their existence challenges some of the fundamental principles of aminostratigraphy, and criteria for their recognition must be developed.

A similar apparent inversion of latitudinal D/L trends is seen when the Wando Formation data are compared with results from lower latitudes (region V). Although some uncertainty exists regarding the conversion of isoleucine data (Fig. 17) into the leucine ratios shown in Figure 19, the Wando Formation samples still appear too racemized to be age equivalent with the localities in aminozone Va. Figure 20 presents isoleucine data for *Mercenaria* and emphasizes the same point. When the trend lines of alloisoleucine/isoleucine (allo/iso) results from Florida are projected northward into the mid-Atlantic region, the lowest allo/iso

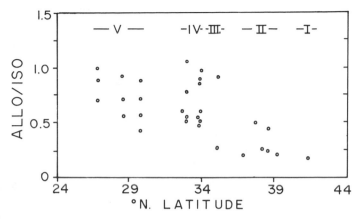

Figure 20. Isoleucine data for *Mercenaria* samples from Massachusetts to Florida, shown for comparison with Figure 19. Data for region V from Mitterer (1975). Data for regions IV and III from McCartan and others (1982). Most of the data for regions II and I have been produced at the University of Delaware. Although the relative paucity of data precludes the kinds of aminozone recognition that has been presented above, any southward projection of the region I and II results (A/I = 0.18 to 0.28 for stage 5 calibration sites) to the Florida region clearly would not include the Wando Formation samples in region IV (A/I = 0.5 to 0.55) that are considered to be correlative with stage 5.

values in South Carolina appear too high to be age equivalent to the youngest (stage 5) allo/iso values in Florida.

Localities in Figure 19 for which the isochron model ages and for which the U-series coral dates are internally consistent are as follows: Sankaty Head, Nantucket Island, Massachusetts; Gomez and New Light Pits, Virginia; Stetson Pit, North Carolina; Ponzer, North Carolina; Moyock, North Carolina; Intracoastal Waterway, Myrtle Beach, South Carolina (most of the coral dates); Miami, Florida (isoleucine data from Mitterer, 1975; consistent with leucine data from the Bahama Islands). At all of these sites, at least three (and in some cases 15 or more) shells were analyzed in order to confirm the observed D/L values for these sites.

An additional issue regarding the relation of latitudinal temperature gradients to aminozone correlation is identified by the Myrtle Beach to Charleston trends. McCartan and others (1982) and Szabo (1985F) indicated that localities in Myrtle Beach (our aminozones IVa and IVb, with a "pooled" mean D/L leucine value of 0.60) should be correlative with a locality in Charleston with a D/L leucine value of 0.78. As the present temperature difference between these localities is about 1°C, such a large difference in D/L values for isochronous samples cannot be explained unless significantly greater (7° to 10°C) temperature differences prevailed over this short latitudinal range during the late Pleistocene. The simplest aminostratigraphic interpretation, based on the assumption that the present temperature differences between these two areas have been maintained throughout the late Pleistocene (Wehmiller and Belknap, 1982), would imply that the Wando and Socastee localities are correlative, but

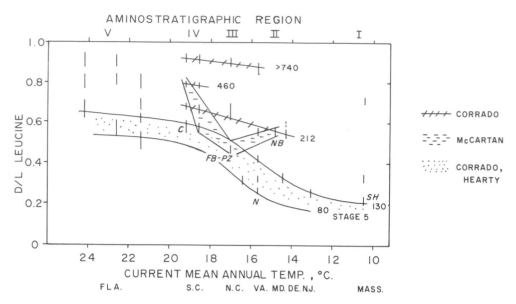

Figure 21. Plot of D/L leucine data from regional aminozones of Figure 19 versus current mean annual temperature. See Figure 19 for comparison. Regional designations and locality identifications given as in Figure 19. Correlations shown are those proposed either by Corrado and others (1986), Hearty and Hollin (1986), or McCartan and others (1982). These correlations are based on the assumption that the stage 5 U-series coral dates obtained at Wando Formation sites (Charleston) and from two outcrops in Myrtle Beach are correct. Stage 5 envelope and specific dates (212 ka, 460 ka, and >740 ka) are from Corrado and others (1986, Fig. 10). Localities: C = Charleston (Wando Formation); FB-PZ = Flanner Beach–Ponzer; N = Norfolk; NB = Norris Bridge; SH = Sankaty Head.

McCartan and co-workers feel that this correlation is inconsistent with the lithostratigraphy. Wehmiller (1982, Fig. 10) discussed the details of the conflicting interpretations, which are also depicted in Figure 6 of McCartan and others (1982). Corrado and others (1986) and Hearty and Hollin (1986) also offered aminostratigraphic interpretations of this region that rely on the assumption that the present temperature gradient between Charleston and Myrtle Beach has changed little during the history of the samples.

In light of the above conflicts between the isochron model ages and U-series data or regional models of coastal plain depositional history, some examples of optional interpretations of the amino acid data need to be considered. One of these interpretations involves the use of a kinetic model that differs significantly from that discussed above. Previously (Wehmiller and Belknap, 1982) we compared an "extended linear" racemization model (in which leucine racemization proceeds by apparent first-order reversible kinetics to a D/L value of about 0.55) to the available U-series data. Several versions of this extended linear model have been proposed (Mitterer, 1975; Kvenvolden and others, 1979). However, the extended linear model is not consistent with known temperature dependencies of the racemization reaction and D/L data for both late Pleistocene and Holocene calibration samples (Wehmiller and Belknap, 1978; Wehmiller, 1981, 1984a), or with deep-sea-sediment data where the racemization reaction has been more fully documented (King, 1980). Hearty and others

(1986) and Miller and Mangerud (1985) have demonstrated that a more non-linear model, like that used here (i.e., linear up to D/L values on the order of 0.35) is appropriate for the interpretation of Mediterranean and northwest European sites.

A second alternative approach to the interpretation of Atlantic Coastal Plain results was presented by Corrado and others (1986). This is an empirical approach that has as its basic assumption that all the U-series coral dates from the Charleston area localities are correct. Figure 21 summarizes this method of data interpretation. This approach, like the model presented above, must also exclude certain U-series dates. This approach does not rely on any assumptions about latitudinal temperature gradients or any particular kinetic model for racemization, and all data are plotted versus current mean annual temperatures. Because Corrado and others incorrectly interpreted our Wando Formation leucine data, we have revised their approach (their Fig. 10) to account for this error. Also included in Figure 21 is the correlation band proposed by McCartan and others (1982) for stage 7 sites in region IV (TP), region III (FB & PZ), and region II (NB). Hearty and Hollin (1986) proposed a correlation of apparent stage 5 data in regions IV and III that is quite similar to that suggested by Corrado and others (1986).

The following conclusions or issues can be derived from Figure 21, assuming that this is the correct approach to aminostratigraphic correlation of coastal plain sites:

(1) An extreme steepening of isochrons is required for the

region between Norfolk and southeastern North Carolina, roughly represented by our region III. Although the possibility of such steepening cannot be ruled out in the absence of detailed paleotemperature data, such variations in latitudinal temperature gradients are fundamental violations of the basic assumptions of aminostratigraphy and have not been observed in other regional studies (Kennedy and others, 1982; Miller and Mangerud, 1985; Hearty and others, 1986).

(2) The implied resolving power (0.25 to 0.42 in D/L values) shown in Figure 21 for stage 5 samples is without precedent and leads to some remarkable predictions about the preservation of detailed sections, particularly in the Norfolk, Virginia, area, but also in the Charleston, South Carolina, region. In the Norfolk area, aminozones IIa, IIb, and IIc would be predicted by the Figure 21 model to be late, middle, and early stage 5 in age, respectively, and aminozone IId would be stage 7. Coral U-Th dates equivalent to all but early stage 5 have been observed in this area (Table 2). Aminozones IIa, IIc, and IId are all preserved in superposition at one or more sites southeast of Norfolk (Fig. 10), all within 5 m of present sea level. Aminozone IIb is preserved at isolated sites in this same region. These deposits, if all stage 5 to stage 7 in age, clearly cannot be interpreted in terms of the same sea-level curves (i.e., Dodge and others, 1983; Chappell and Shackleton, 1986) that have been observed elsewhere without there having been some variable uplift and/or local isostatic effects (Cronin and others, 1981). Figure 21 requires the conclusion that a low-uplift coastline (<0.02 m/1,000 yr) has preserved a late Pleistocene sea level record that is rarely recognized in such detail except in regions with substantially greater (10× or more) uplift rates. The absence of early-stage 5 coral dates in this region, in contrast to the frequent observation of coral dates corresponding to this time elsewhere in the world, is also remarkable. Mixon and others (1982) suggested that the absence of early-stage 5 coral dates on the coastal plain was due to diagenetic factors.

(3) What temperature histories would need to be invoked in order to reconcile the isochrons in Figure 21 with the present temperature differences between region II and region IV? The effective temperatures for a particular aminozone can be derived from the leucine kinetic model (Appendix II or Fig. 19) using assigned ages, derived from Figure 21, as follows:

Aminozone	Assigned Age	EQT (°C)
IIa	75,000	6.5
IIb	105,000	8.5
IIc	125,000	11.0
IId	220,000	13.0
IVa	90,000 or	18.5
	125,000*	17.0
IVb	220,000	17.0

*125,000-yr coral dates not associated with our aminozone IVa, but this age assignment seems reasonable (see Fig. 21).

Because effective temperatures represent the integrated effect of all temperature fluctuations that have affected a sample, no unique temperature history is specified when an EQT value is determined. Two possible temperature histories for the Charleston and Norfolk areas are shown in Figure 22. These temperature histories are constructed to be consistent with the present temperatures at the sites (19° and 16°C, respectively), with the most probable timing of climatic change (based on the 0–18 record of ice volume), and with the EQT values derived as outlined above. The procedure by which a temperature history such as those shown in Figure 22 is calculated is discussed in Wehmiller (1982, Appendix).

The two constraints on the curves in Figure 22 are the present temperatures at the sites and the effective temperature (EQT) derived from the kinetic model for D/L data fixed to specific ages. For the Norfolk area, late-, middle-, and early-stage 5 ages were assigned to aminozones IIa, IIb, and IIc, respectively, and a stage 7 age to aminozone IId. Effective temperatures were then derived from the kinetic model and are plotted as the solid circles. The solid line (labeled N) is one possible temperature history for the Norfolk region that would be consistent with these EQT values.

For the Charleston-area samples, aminozone IVa was assigned a late-stage 5 age, and aminozone IVb was assigned a stage 7 age; EQT values were derived from the kinetic model. The solid squares indicate these EQT values and the line labeled C is one possible temperature history for the Charleston area that would be consistent with these EQT values. If aminozone IVa were assigned an early-stage 5 age, a longer period of temperatures between 15° and 16°C (i.e., from about 100,000 to 20,00 yr B.P.) would be needed to explain the apparently lower EQT value for the aminozone IVa samples.

Figure 22 shows that the Norfolk region must have experienced a much larger and much longer glacial-stage (ca. last 100,000 yr) temperature reduction than the Charleston region in order to explain the effective temperature differences between these regions. If the 125,000-yr age is assigned to the aminozone IVa samples, then a longer period of cooling (between 100,000 and 50,000 yr B.P.) from 19°C to about 16°C is consistent with the apparent EQT inferred from the D/L data. The model temperature histories presented in Figure 22 suggest that the temperature difference between Norfolk and Charleston has been between two and five times greater than at present during the last 100,00 yr, but that temperature differences during isotope stage 6 were similar to those of today. Because of the similar aminostratigraphies for the Charleston and Myrtle Beach areas, these conclusions are actually more specifically relevant to the thermal gradients between Norfolk and Myrtle Beach.

The model temperature histories presented in Figure 22 cannot be rejected or confirmed with paleoclimatic data, although a steepened latitudinal thermal gradient in the mid-Atlantic region is inferred from full-glacial palynological data (Delcourt and Delcourt, 1984) and from marine paleoclimatic data (Imbrie and others, 1983). In all previous attempts to model

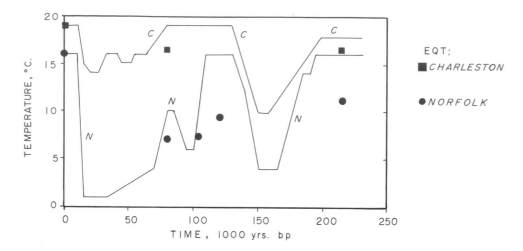

Figure 22. Possible temperature histories for the Charleston and Norfolk regions that could be invoked to reconcile observed U-series dates and D/L leucine values with the leucine kinetic model (Appendix II). See text for discussion.

latitudinal gradients of EQT, it has been necessary to assume that temperatures at specific sites changed roughly synchronously but changes at higher latitudes were greater than at lower latitudes (Wehmiller and Belknap, 1982). Figure 22 shows that invoking non-synchronous temperature changes and ca. 4× steeper temperature gradients is required to reconcile the D/L and U-Th data is required. While these temperature histories are not impossible, the necessity to invoke them represents a contradiction of some of the fundamental temperature assumptions of aminostratigraphy.

(4) The lack of parallelism between the stage 5 and stage 7 isochrons in Figure 21 also contradicts these same temperature assumptions. The near equivalence of EQT values for samples of different ages but similar current mean annual temperatures has long been a component of all aminostratigraphic models.

(5) A major conflict still exists in the interpretation of the aminostratigraphic data for the Ponzer and Flanner Beach sites (PZ and FB) in region III. The smooth stage 5 isochrons (Fig. 21) connecting the Charleston and Norfolk sites overlap the FB-PZ data points, thereby implying that these sites are also correlative with stage 5. This age estimate conflicts with the stage 7 U-Th coral date at Ponzer (Szabo, 1985) and with biostratigraphic age estimates (McCartan and others, 1982; Cronin and others, 1984; W. Miller, 1985), and lithostratigraphic models for the region (summarized in Johnson and Peebles, 1986). Consequently, a large body of stratigraphic data and at least one U-Th date have to be rejected if the FB-PZ sites are assigned a stage 5 age. In addition, this age assignment would then force our aminozone IIIa to be post-stage 5 in age.

The data for the FB-PZ sites also present a problem for the correlation model of McCartan and others (1982), which is also depicted in Figure 21. This correlation model is based on the apparent age equivalence for TP (aminozone IVc), FB-PZ

(aminozone IIIb), and NB (aminozone IId). The inversion of latitude trends of D/L data for "isochronous" samples contradicts all of the basic temperature assumptions of aminostratigraphy. We argue that the Norris Bridge stage 7 U-Th date is incorrect based on the evidence of open-system isotope ratios (Wehmiller and Belknap, 1982), and because of the biostratigraphic data (Cronin and others, 1984) for other samples from the same aminozone as NB. The conclusion that some U-Th dates, either from NB, PZ, or the Wando Formation sites, must be rejected seems inescapable. The isotopic evidence (Szabo, 1985) for open versus closed-system chemistry favors the PZ date as the only correct one.

A comparison of Figures 19 and 21 permits the following conclusions:

(1) If the Wando, PZ, and NB U-series dates for the coastal plain are correct, the latitude gradient of EQT is much steeper than the gradient of modern temperatures, and inversions of EQT gradients apparently have occurred. Furthermore, this particular issue is not model dependent. Locally calibrated aminostratigraphies can be forced to be consistent with local sequences but cannot be extrapolated to other regions.

(2) The kinetics required to produce the isochrons depicted in Figure 21, when interpreted in terms of local aminostratigraphic sequences only, appear to support the extended linear model discussed by Wehmiller and Belknap (1982). An extended linear kinetic model results in a short high-resolution chronology for most coastal plain deposits, but does not identify from the D/L data any deposits with apparent ages between ca. 200,000 yr and ca. 1.0 m.y. The extended linear model cannot be reconciled with calibrated D/L data for both Holocene and late Pleistocene samples, with kinetics observed in other calcareous samples, or with known racemization activation energies.

(3) The non-linear kinetic model (Fig. 19), which is con-

sistent with racemization activation energies and most observations of actual racemization pathways at ambient temperatures, has been combined with the assumption that smooth latitudinal gradients of effective temperature are appropriate for the region. This approach, which is more consistent with the basic assumptions of aminostratigraphy, yields many age estimates that are consistent with U-Th data and other independent age control, but it yields age estimates for the Wando sites (aminozone IVa) and Norris Bridge (aminozone IId) that are two and three times greater, respectively, than the associated U-Th coral dates.

(4) The non-linear model leads to longer chronology for the coastal plain deposits, with several deposits having apparent ages in the range of 200,000 yr to ca. 1.0 m.y. The resolving power for this model is substantially less than that predicted by the model in Figure 21, especially for samples representing isotope stage 5.

SUMMARY AND CONCLUSIONS

Aminostratigraphic data for marine mollusks from nearly 150 Atlantic coastal sites between Nova Scotia and the Bahamas Islands are represented by what we interpret as the most reliable data points, based on analytical criteria, sample documentation, or field relationships. The data presented here can be used to derive stratigraphic and chronologic conclusions and also for an understanding of geochemical issues (analytical precision, sample preservation, etc.).

Several fundamental issues have arisen as a result of the comparisons of aminostratigraphic, isotopic, lithostratigraphic, and biostratigraphic data for Quaternary deposits of the Atlantic Coastal Plain, and a combination of conflicts involving models for regional lithostratigraphy, racemization kinetics, late Pleistocene temperature histories, U-series dating, and local sea-level histories may be observed. The broad regional framework of aminostratigraphic data permits a much clearer delineation of these conflicts than simple, localized, aminostratigraphic studies.

One of the most important questions is the resolving power of the method within any particular temperature region. Conflicting views of this issue have been encountered: McCartan and others (1982) proposed lithostratigraphic depositional models for portions of the coastal plain that require rather large ranges of D/L values to be lumped into the time frame of a single interglacial high sea stand. Corrado and others (1986) argued the opposite case, that aminostratigraphy has the capability of resolving all three substages of stage 5.

Issues of resolving power are linked partly to the choice of kinetic models for racemization. The non-linear model (Fig. 19) that we believe is most consistent with observed kinetic parameters of racemization results in several conflicts with available U-series dates, but requires relatively simple assumptions regarding thermal histories for the coastal plain. The extended linear model that has been proposed by others is applicable to the results presented here, but results in thermal and kinetic conflicts as significant as those conflicts between the non-linear mode and U-series dates. The empirical approach (Fig. 21) also requires

that some U-series dates be rejected. Furthermore, if all the U-series coral dates from the coastal plain are used as rigid calibrations for the amino acid data, one of the fundamental assumptions of aminostratigraphy (that present temperature differences are representative of differences in effective temperatures) is apparently not valid for the Atlantic Coastal Plain. Inverted or steepened latitudinal temperature gradients (as required by the Norris Bridge and Wando Formation U-Th dates) cannot be ruled out solely on the basis of aminostratigraphic assumptions, but unusual temperature trends such as these have not been required in prior aminostratigraphic studies of latitudinally spaced coastal samples.

The density of sampling along the coastal plain appears to be great enough in most areas so that kinetic models would not be needed for aminostratigraphic correlations. However, the conflicts between aministratigraphic correlations related to present temperature gradients (Fig. 19) and those fixed to available U-series data (Fig. 21) are serious reminders that the geochemistry of aminostratigraphy and U-series dating methods need frequent testing and evaluation, particularly by the study of samples from unequivocal stratigraphic sequences. Given that racemization data always conform to local stratigraphic control when the results do not have to be interpreted beyond a single outcrop, these conflicts are all the more enigmatic. Region III (North Carolina) is pivotal for the resolution of these issues.

Simple, qualitative aminostratigraphic correlations within limited geographic regions reveal at least three (usually four) aminozones along the Atlantic coast of the U.S. Our regional aminozone approach results in the "lumping" rather than "splitting" of data, because the precision within a local aminozone is rarely good enough to resolve age differences smaller than about 20 to 25 percent of the age estimate, unless clear lithostratigraphic evidence supports more subtle age distinctions based on amino acids. Nevertheless, in a few cases, early- and late-stage 5 sites may have been resolved by aminostratigraphic data. Although the same approximate number of aminozones is seen in different regions, aminostratigraphic position alone cannot be used to correlate aminozones between regions (Table 5). The short chronology that results from the extended linear model or the empirical approach places most of the post-Yorktown–Waccamaw–James City aminozones in the last 250,000 yr. The long chronology that results from the non-linear kinetic model places these same aminozones from ca. 100,000 to about 900,000 yr B.P.

The results from the Charleston, South Carolina, area are particularly troublesome. Our preference for the aminostratigraphic age assignments in this region (and for the localities in the Myrtle Beach area with identical D/L values) indicates that little or no stage 5 material has been sampled. This is in obvious conflict with the U-Th dates and also contradicts the assumption that stage 5 shorelines would be most commonly preserved. However, geomorphic evidence (reviewed in Oaks and DuBar, 1974) suggests that the late Pleistocene is more deeply embayed (therefore removed?) in this area than in northeastern North Carolina and Virginia. In addition, the coastal plain U-series dates are unusual in that so few cluster around the widely recognized "125,000" high sea level, because of either an unusual geologic history or some unknown geochemical effect on the U-series isotopic system in coral fragments preserved in silicate sedimentary material. Refuting the large body of stage 5 U-series dates for the Charleston-area sites is obviously difficult, but these dates require thermal histories for the coastal plain that would not be easily predictable from the present gradient of temperature along the coast. The relative sea-level history required by the collective interpretation of the high-resolution, U-series-calibrated, aminostratigraphic model is unusual in comparison with isotopic models or coastal reef-terrace records of sea levels for the past 200,000 yr. An unavoidable conclusion to be derived from the different approaches to coastal plain aminostratigraphy is that some of the U-series coral dates must be rejected, so the widely held notion that coral U-series dates are accurate is not necessarily correct for the solitary corals of the Atlantic Coastal Plain.

The complexity of Atlantic Coastal Plain aminostratigraphy requires that a serious effort be made to coordinate the multidisciplinary collection efforts of various workers studying these deposits. Intergeneric comparisons and the lithostratigraphic context of analyzed mollusk and coral samples need to be established more rigorously than at present if current conflicts and ambiguities are to be resolved.

ACKNOWLEDGMENTS

The research described here was funded by grants from the U.S. Geological Survey to the University of Delaware (Grant Nos. 14-08-0001-G592 and G680) and by grants from the National Science Foundation to both the University of Maine (NSF-EAR-8216279) and the University of Delaware (NSF-EAR-8407024). G. H. Miller and G. L. Kennedy have provided many helpful comments on the manuscript, and R. Morris, J. P. Owens, L. McCartan, T. Cronin, J. Demarest, and E. Keenan have been frequent participants in discussions of field or laboratory aspects of this research.

APPENDIX 1

Comparison of isoleucine and leucine data

Leucine and isoleucine have usually received the greatest attention in aminostratigraphic studies of mollusks. Because of analytical facilities available at different laboratories, for a single sample extract to be analyzed for both the alloisoleucine/isoleucine value and the D/L leucine value has been rare. Figure A-1 presents the next best comparison: one in which leucine and isoleucine data have been obtained on the same *Mercenaria* shells (at different times, usually by different laboratories). The data shown in Figure A-1 are from McCartan and others (1982) except for samples from higher latitudes, which have been analyzed for alloisoleucine/isoleucine values in the University of Delaware laboratory (see results in Fig. 20). Wehmiller (1984b) has summarized additional results that permit a comparison of leucine and isoleucine data on the same samples.

No simple conversion factor can be used to equate D/L leucine and alloisoleucine/isoleucine values in *Mercenaria* (Fig. A-1). How much of the scatter in Figure A-1 is due to sampling artifacts is not clear, because most of the isoleucine analyses reported in McCartan and others (1982) were obtained on growth-edge fragments, while the analyses performed in our laboratory usually employ fragments cut from the hinge portion of the shell. In addition, growing evidence exists (B. S. Boutin, unpublished) that some of the scatter in results such as those in Figure A-1 is due to diagenetic leaching that has the effect of moving the allo/iso data off the typical trend line before the leucine D/L data are affected. This effect, if real, could be explained by the fact that D-alloisoleucine and L-isoleucine have different chemical properties (making them separable by ion-exchange chromatography), while D-leucine and L-leucine do not.

Boutin (unpublished) has found that the following equation can be used for comparison of leucine and isoleucine data obtained in our laboratory:

$$A/I = 1.2712 \, [D/L]^{1.4625} \quad r^2 = 0.99$$

This equation is valid for analyses on the same sample extract or on separate, but adjacent, fragments cut from the same shell. A significantly

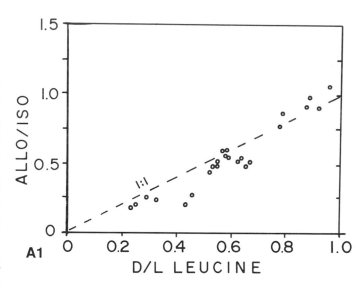

A1

different equation is obtained from the pairs of isoleucine and leucine values given in Table 1:

$$A/I = 1.0328 \, [D/L]^{1.277} \quad r^2 = 0.94$$

The difference between the two equations results from the different shells and different shell positions analyzed for the latter comparison.

Note that the allo/iso (A/I) values determined in our laboratory are on the low side of the range of results reported by all laboratories, as seen in Appendix IV.

APPENDIX II

Discussion of kinetic model curves for leucine racemization

Under the subhead "Aminostratigraphy, Kinetic Model Dates, and Philosophy of Geochronology," we presented equations to describe the

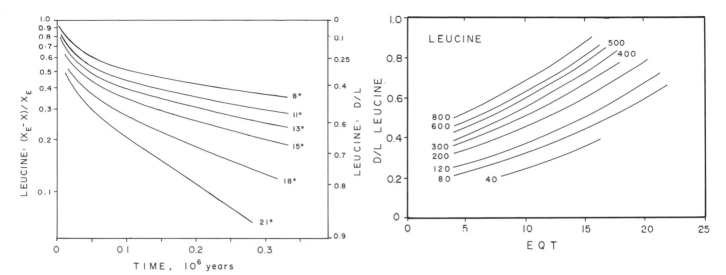

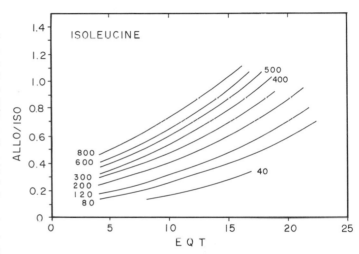

non-linear racemization pathway for leucine. Figure A-2 shows the curves derived from the equations for several effective temperatures in the range of interest. These curves are all quite similar to those presented in Wehmiller and Belknap (1982), with one important exception. In the region below an (Xe-X)/Xe value of about 0.1 (corresponding to a D/L leucine value of 0.8 or greater), the curves steepen, rather than retaining their slope as has been implied in our previous work. The steepening of these curves is a direct consequence of the aforementioned equations. The steepening is not merely a mathematical artifact, however, because it is also a result of the physically reasonable assumption that racemization must reach "true" equilibrium (D/L = 1.0) at different times, depending on temperature. Simple linear extrapolation (on a log plot) of the intermediate portions of the curves in Figure A-2 would never reach true equilibrium.

The consequences of this steepening of the curves are primarily theoretical, because of the widely accepted idea that the age-resolution capabilities of aminostratigraphy are greatly reduced when dealing with such extensively racemized samples. However, the curves shown in Figure A-2, when plotted in the format of Figure 19, show an important difference when compared with our earlier (Wehmiller and Belknap, 1982; Fig. 8) presentation of this model. In Figure 19, the isichrons of D/L leucine values are projected upward to the equilibrium value; in our earlier work (Wehmiller, 1982; Wehmiller and Belknap, 1982) we implied that these isochrons would become less steep and converge toward the equilibrium value at lower latitudes. Our earlier approach would have led to age estimates for extensively racemized samples (e.g., aminozones IIId, IVd) about 25 percent greater than those inferred from Fig. 19.

The implications of this kinetic model, and of the conversion factors for leucine and isoleucine, are discussed more fully in Appendix III.

APPENDIX III

Leucine and isoleucine latitude correlation isochrons compared

These figures permit comparison of leucine and isoleucine kinetic model isochrons, based on the leucine–isoleucine data presented in Appendix I and the leucine kinetic model in Appendix II. The D/L or A/I values are plotted versus effective temperature (EQT). Figure A-3 is

essentially the same as Figure 19. Figure A-4 is the same plot for isoleucine versus EQT, with A/I values derived from the leucine curves of Figure A-3 using the conversion factor equation given in Appendix II. Figure A-5 shows various isochrons for leucine and isoleucine plotted together.

APPENDIX IV

Summary of data obtained in the University of Delaware laboratory on the interlaboratory comparison samples

Wehmiller (1984b) reported the enantiomeric ratio results of analyses obtained by several laboratories on three interlaboratory comparison samples (ILC A, ILC B, and ILC C). We report here a summary of alloisoleucine/isoleucine vlaues obtained on these powder samples since mid-1985, because our laboratory has been using this measurement much more frequently than in the past and because the allo/iso values are most frequently reported by other laboratories. Listed below are the mean allo/iso values determined by laboratories known to us to be using analytical procedures* similar to ours. Most of these values were reported in 1984 (Wehmiller, 1984b). Most of the differences between laboratories seems to be related to instrumental rather than preparative effects (Wehmiller, 1984b).

ILC A	ILC B	ILC C
0.16	0.45	1.1
0.14	0.50	1.0
0.193	0.485	1.12
0.176	0.559	0.986
0.14	0.48	0.87
0.167	0.439	0.78

Miller and Mangerud (1985; Table 2) have summarized additional measurements in their laboratories on these samples. Our results for these samples are listed below:

Mean A/I ratio by:	ILC A	ILC B	ILC C
Peak area	0.126 (14)	0.436 (49)	0.970 (17)
Peak height	0.148 (14)	0.492 (49)	1.092 (17)

The numbers in parenthesis are the number of separate analyses (of different powder hydrolyzates) that have been made. Each analysis involves at least two chromatograms of the sample hydrolyzate. Coefficients of variation are less than 5 percent for all samples, but the data for ILC B generally show greater variability than for the other two powder samples. Our peak area ratios are a bit lower than most of those reported above, but we continue to use these values as tests of internal laboratory consistency. We note, however, that our peak height ratios are quite comparable with those reported by Miller and Mangerud (1985), who favor the use of peak height ratios. We have used our peak area ratios in all calculations of comparisons between allo/iso and D/L leucine values.

During the period in which these allo/iso values were determined on the ILC samples in our laboratory, we have duplicated the D/L values reported earlier (Wehmiller, 1984b) for other amino acids, as determined by gas chromatographic methods.

APPENDIX V

Intergeneric comparisons of Mercenaria, Mulinia, *and* Anadara

Graphs of D/L data for coexisting pairs of samples (mean values of each genus for those situations where multiple samples exist at a site) from all sites (or groups of localities) that we have studied. Locality identifications as in Table 1.

For the *Mulinia* versus *Mercenaria* graph (Fig. A-6), all results are shown in terms of equivalent alloisoleucine/isoleucine (A/I) values. The open circles, representing single shells of each genus at a particular site, represent leucine D/L values (Belknap, 1979 or unpublished) from selected sites, converted into A/I values using the conversion in Appendix I.

For the *Anadara* versus *Mercenaria* graph (Fig. A-7), either leucine (D/L) or allo/iso (A/I) values for the generic pairs are shown. Only the MCK (Moyock) samples fall far from the 1:1 line.

_____

*Our analyses have been performed using a 15 cm × 2.2 mm, HPLC, ion-exchange column. Some laboratories are using a 25-cm column. We are not aware of any systematic difference in allo/iso values obtained using the longer columns.

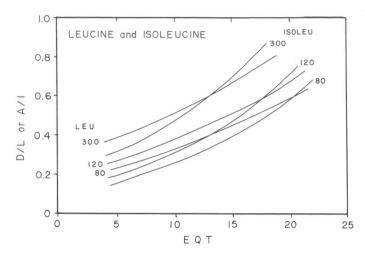

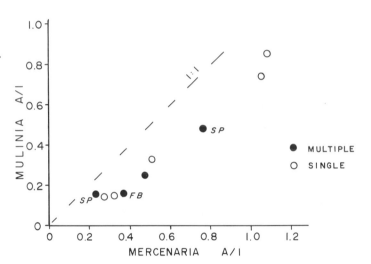

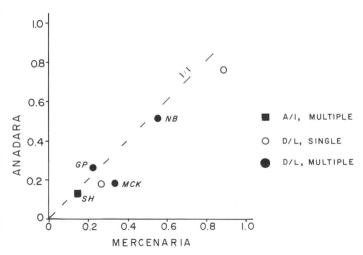

REFERENCES CITED

Atwater, B. F., Ross, B. E., and Wehmiller, J. F., 1981, Stratigraphy of late Quaternary estuarine deposits and amino acid sterochemistry of oyster shells beneath San Francisco Bay: Quaternary Research, v. 16, p. 181–200.

Bada, J. L., 1985, Amino acid racemization dating of fossil bones: Annual Reviews of Earth and Planetary Science, v. 13, p. 214–268.

Bada, J. L., and Schroeder, R. A., 1972, Racemization of isoleucine in calcareous marine sediments; Kinetics and mechanism: Earth and Planetary Science Letters, v. 15, p. 1–11.

Balknap, D. F., 1979, Application of amino acid geochronology to stratigraphy of the late Cenozoic marine units of the Atlantic Coastal Plain [Ph.D. thesis]: Newark, University of Delaware, Department of Geology, 550 p.

—— , 1984, Amino acid racemization in "mid-Wisconsin" C-14 dated formations on the Atlantic Coastal Plain: Geological Society of America Abstracts with Programs, v. 16, p. 2–3.

Belknap, D. F., and Wehmiller, J. F., 1980, Amino acid racemization in Quaternary mollusks; Examples from Delaware, Maryland, and Virginia, in Hare, P. E., Hoering, T. C., and King, K., Jr., Biogeochemistry of amino acids: New York, John Wiley and Sons, p. 401–414.

Boulton, G. S., and others, 1982, A glacio-isostatic facies model and amino acid stratigraphy for late Quaternary events in Spitsbergen and the Arctic: Nature, v. 298, p. 437–441.

Bowen, D. Q., Sykes, G. A., Reeves, A., Miller, G. H., Andrews, J. T., Brew, J. W., and Hare, P. E., 1985, Amino acid geochronology of raised beaches in southwest Britain: Quaternary Science Reviews, v. 4, p. 279–318.

Carew, J. S., Wehmiller, J. F., Mylroie, J. E., and Lively, R. S., 1984, Estimates of late Pleistocene sea level high stands from San Salvador, Bahamas, in Teeter, J. W., ed., Proceedings of the second symposium on the geology of the Bahamas: San Salvador, Bahamas, College Center of the Finger Lakes, p. 153–175.

Carey (Corrado), J. C., 1981, Pleistocene aminostratigraphy of the South Carolina coastal plain near Charleston and Myrtle Beach [M.S. thesis]: Blacksburg, Virginia Polytechnic Institute and State University, 122 p.

Chappell, J., and Shackleton, N. J., 1986, Oxygen isotopes and sea level: Nature, v. 324, p. 137–140.

Colquhoun, D. S., 1974, Cyclic surficial stratigraphic units of the middle and lower coastal plains, central South Carolina, in Oaks, R. Q., Jr., and DuBar, J. R., eds., Post Miocene stratigraphy central and southern Atlantic Coastal Plain: Logan, Utah State University Press, p. 179–190.

Corrado, J., and Hare, P. E., 1981, Aminostratigraphy of the South Carolina coastal plain by monospecific fossil analyses (*Mulinia lateralis* Say): Carnegie Institution of Washington Yearbook, v. 80, p. 387–389.

Corrado, J. C., Weems, R. E., Hare, P. E., and Bambach, R. K., 1986, Capabilities and limitations of applied aminostratigraphy as illustrated by analyses of *Mulinia lateralis* from the late Cenozoic marine beds near Charleston, South Carolina: South Carolina Geology, v. 30, no. 1, p. 19–46.

Cronin, T. M., Szabo, B. J., Ager, T. A., Hazel, J. E., and Owens, J. P., 1981, Quaternary climates and sea levels of the Atlantic Coastal Plain: Science, v. 211, p. 233–240.

Cronin, T. M., Bybell, L. M., Poore, R. Z., Blackwelder, B. W., Liddicoat, J. C., and Hazel, J. E., 1984, Age and correlation of emerged Pliocene and Pleistocene deposits, U.S. Atlantic Coastal Plain: Palaeogeography, Palaeoclimatology, Palaeoecology, v. 47, p. 21–51.

Delcourt, P. A., and Delcourt, H. R., 1984, Late Quaternary paleoclimates and biotic responses in eastern North America and the western North Atlantic Ocean: Palaeogeography, Palaeoclimatology, Palaeoecology, v. 48, p. 263–284.

Dodge, R. E., Fairbanks, R. G., Benninger, L. K., and Maurrasse, F., 1983, Pleistocene sea levels from raised coral reefs of Haiti: Science, v. 219, p. 1423–1425.

DuBar, J. R., Dubar, S. S., Ward, L. W., Blackwelder, B. W., Abbott, W. H., and Huddleston, P. F., 1980, Cenozoic biostratigraphy of the Carolina outer coastal plain, in Frey, R. W., ed., Excursions in southeastern geology: Geological Society of America 1980 Annual Meeting Field Trip Guidebook,

v. 1, p. 179–236.

Frank, H., Nicholson, G. J., and Bayer, E., 1977, Rapid gas chromatographic separation of amino acid enantiomers with a novel chiral stationary phase: Journal of Chromatographic Science, v. 15, p. 174–176.

Grant, D. R., 1980, Quaternary stratigraphy of southwestern Nova Scotia; Glacial events and sea level changes: Halifax, Geological Association of Canada Field Trip Guidebook, May 1980 Meeting, Field Trip no. 9, p. 1–63.

Gustavson, T. C., 1976, Paleotemperature analysis of the marine Pleistocene of Long Island, New York, and Nantucket Island, Massachusetts: Geological Society of America Bulletin, v. 87, p. 1–18.

Hearty, P. J., and Hollin, J. T., 1986, Isotope stage 5 marine transgressions in the Carolinas: Geological Society of America Abstracts with Programs, v. 18, p. 633.

Hearty, P. J., Miller, G. H., Stearns, C. E., and Szabo, B. J., 1986, Aminostratigraphy of Quaternary shorelines in the Mediterranean Basin: Geological Society of America Bulletin, v. 97, p. 850–858.

Imbrie, J., McIntyre, A., and Moore, T. C., Jr., 1983, The ocean around North America at the last glacial maximum, in Porter, S., ed., Late Quaternary environments of the United States; Vol. 1, The Late Pleistocene: Minneapolis, University of Minnesota Press, p. 230–236.

Jacobs, J. M., 1980, Stratigraphy and lithology of Quaternary landforms on the eastern coast of Chesapeake Bay [M.S. thesis]: Newark, University of Delaware, College of Marine Studies, 84 p.

Johnson, G. H., and Peebles, P. C., 1986, Quaternary geologic map of the Hatteras 4° × 6° Quadrangle, United States: U.S. Geological Survey Miscellaneous Investigations Series, Map I-1420 (NI-18), scale 1:1,000,000.

Keenan, E. M., 1983, Amino acid racemization dating; Theoretical considerations and practical applications [Ph.D. thesis]: Newark, University of Delaware, Department of Geology, 351 p.

Kennedy, G. L., Lajoie, K. R., and Wehmiller, J. F., 1982, Aminostratigraphy and faunal correlations of late Quaternary marine terraces, Pacific coast U.S.A.: Nature, v. 299, p. 545–547.

King, K., Jr., 1980, Applications of amino acid biogeochemistry for marine sediments, in Hare, P. E., Hoering, T. C., and King, K., Jr., eds., Biogeochemistry of amino acids: New York, John Wiley and Sons, p. 377–392.

Kvenvolden, K., Blunt, D. J., and Clifton, H. E., 1979, Amino acid racemization in Quaternary shell deposits at Willapa Bay, Washington: Geochimica et Cosmochimica Acta, v. 43, p. 1505–1520.

Lajoie, K. R., Kern, J. P., Wehmiller, J. F., Kennedy, G. L., Matthieson, S. A., Sarna-Wojcicki, A. M., Yerkes, R. F., and McCrory, P. A., 1979, Quaternary shorelines and crustal deformation, San Diego to Santa Barbara, California, in Abbott, P. L., ed., Geological excursions in the southern California area: San Diego, California, San Diego State University, Department of Geology, v. 3–15.

Lajoie, K. R., Wehmiller, J. F., and Kennedy, G. L., 1980, Inter- and intrageneric trends in apparent racemization kinetics of amino acids in Quaternary mollusks, in Hare, P. E., Hoering, T. C., and King, K., Jr., eds., Biogeochemistry of amino acids: New York, John Wiley and Sons, p. 305–340.

McCartan, L., Owens, J. P., Blackwelder, B. W., Szabo, B. J., Belknap, D. F., Kriausakul, N., Mitterer, R. M., and Wehmiller, J. F., 1982, Comparison of amino acid racemization geochronometry with lithostratigraphy, biostratigraphy, uranium-series coral dating, and magnetostratigraphy in the Atlantic Coastal Plain of the southeastern United States: Quaternary Research, v. 18, p. 337–359.

McCartan, L., Lemon, E. M., Jr., and Weems, R. E., 1984, Geologic map of the area between Charleston and Orangeburg, South Carolina: U.S. Geological Survey Miscellaneous Investigations Series, Map I-1472, scale 1:250,000.

Miller, G. H., 1985, Aminostratigraphy of Baffin Island shell-bearing deposits, in Andrews, J. T., ed., Quaternary environments; Eastern Canadian Arctic, Baffin Bay, and western Greenland: London, Allen and Unwin, p. 394–427.

Miller, G. H., and Hare, P. E., 1980, Amino acid geochronology; Integrity of the carbonate matrix and potential of molluscan fossils, in Hare, P. E., Hoering, T. C., and King, K., Jr., eds., Biogeochemistry of amino acids: New York, John Wiley and Sons, p. 415–444.

Miller, G. H., and Mangerud, J., 1985, Aminostratigraphy of European marine interglacial deposits: Quaternary Science Reviews, v. 4, p. 215–278.

Miller, G. H., Sejrup, H. P., Mangerud, J., and Anderson, B. G., 1983, Amino acid ratios in Quaternary mollusks and foraminifera from western Norway; Correlation, geochronology, and paleotemperature estimates: Boreas, v. 12, p. 107–124.

Miller, W., III, 1982, The paleoecologic history of late Pleistocene estuarine and marine fossil deposits in Dare County, North Carolina: Southeastern Geology, v. 23, p. 1–13.

——, 1985, The Flanner Beach Formation (Middle Pleistocene) in eastern North Carolina: Tulane Studies in Geology and Paleontology, v. 18, no. 3, p. 93–125.

Mirecki, J. E., 1985, Amino acid racemization dating of some coastal plain sites, southeastern Virginia and northeastern North Carolina [M.S. thesis]: Newark, University of Delaware, 118 p.

Mitterer, R. M., 1974, Pleistocene stratigraphy in southern Florida based upon amino acid diagenesis of fossil *Mercenaria:* Geology, v. 2, p. 425–428.

——, 1975, Ages and diagenetic temperatures of Pleistocene deposits of Florida based upon isoleucine epimerization in *Mercenaria:* Earth and Planetary Science Letters, v. 28, p. 275–282.

Mixon, R. B., 1985, Stratigraphic and geomorphic framework of uppermost Cenozoic deposits in the southern Delmarva Peninsula, Virginia and Maryland: U.S. Geological Survey Professional Paper 1067-G, 53 p.

Mixon, R. B., Szabo, B. J., and Owens, J. P., 1982, Uranium-series dating of mollusks and corals, and age of Pleistocene deposits, Chesapeake Bay area, Virginia and Maryland: U.S. Geological Survey Professional Paper 1067-E, 18 p.

Neumann, A. C., and Moore, W. S., 1975, Sea level events and Pleistocene coral ages in the northern Bahamas: Quaternary Research, v. 5, p. 215–224.

Oaks, R. Q., Jr., and DuBar, J. R., eds., 1974, Post-Miocene stratigraphy central and southern Atlantic Coastal Plain: Logan, Utah State University Press, 275 p.

Oaks, R. Q., Jr., Coch, N. K., Sanders, J. E., and Flint, R. F., 1974, Post-Miocene shorelines and sea levels, southeastern Virginia, *in* Oaks, R. Q., Jr., and DuBar, J. R., eds., Post-Miocene stratigraphy, central and southern Atlantic Coastal Plain: Logan, Utah State University Press, p. 53–87.

Oldale, R. N., Cronin, T. M., Valentine, P. C., Spiker, E. C., Blackwelder, B. W., Belknap, D. F., Wehmiller, J. F., and Szabo, B. J., 1982, The stratigraphy, structure, absolute age, and paleontology of the upper Pleistocene deposits at Sankaty Head, Nantucket Island, Massachusetts: Geology, v. 10, p. 246–252.

Peebles, P. C., 1984, Late Cenozoic landforms, stratigraphy, and history of sea level oscillations in southeastern Virginia and northeastern North Carolina [Ph.D. thesis]: Williamsburg, Virginia, College of William and Mary, School of Marine Science, 225 p.

Peebles, P. C., Johnson, G. H., and Berquist, C. R., 1984, The middle and late Pleistocene stratigraphy of the outer coastal plain, southeastern Virginia: Virginia Minerals, v. 30, no. 2, p. 14–22.

Rahaim, S. D., 1986, The aminostratigraphy of the Sankaty Sand, Nantucket Island, Massachusetts [M.S. thesis]: Newark, University of Delaware, 121 p.

Rampino, M. R., and Sanders, J. E., 1981, Upper Quaternary stratigraphy of southern Long Island, New York: Northeastern Geology, v. 3, p. 116–128.

Ricketts, H. C., 1986, Examination of boreholes of the subsurface Gardiners Clay (Pleistocene) in the context of the Cretaceous–Quaternary section of Great Neck, Long Island, New York: Northeastern Geology, v. 8, p. 13–22.

Sirkin, L., and Stuckenrath, R., 1980, The Portwashingtonian warm interval in the northern Atlantic Coastal Plain: Geological Society of America Bulletin, v. 91, p. 332–336.

Snyder, S. W., Belknap, D. F., Hine, A. C., and Steele, G. A., 1982, Seismic stratigraphy, lithostratigraphy, and amino acid racemization of the Diamond City Formation; Reinterpretation of a reported "mid-Wisconsin high" sea-level indicator from the North Carolina coastal plain: Geological Society of America Abstracts with Programs, v. 14, p. 84.

Stone, B. D., and Borns, H., Jr., 1986, Pleistocene glacial and interglacial stratigraphy of New England, Long Island, and adjacent Georges Bank and Gulf of Maine: Quaternary Science Reviews, v. 5, p. 39–52.

Stuiver, M., and Borns, H. W., Jr., 1975, Late Quaternary marine invasion in Maine; Its chronology and associated crustal movement: Geological Society of America Bulletin, v. 86, p. 99–104.

Susman, K. R., and Heron, S. D., Jr., 1979, Evolution of a barrier island, Shackleford Banks, Carteret County, North Carolina: Geological Society of America Bulletin, v. 90, p. 205–215.

Szabo, B. J., 1985, Uranium-series dating of fossil corals from marine sediments of the United States Atlantic Coastal Plain: Geological Society of America Bulletin, v. 96, p. 398–406.

Ward, W. T., 1985, Correlation of east Australian Pleistocene shorelines with deep-sea core stages; A basis for coastal chronology: Geological Society of America Bulletin, v. 96, p. 1156–1166.

Wehmiller, J. F., 1980, Intergeneric differences in apparent racemization kinetics in mollusks and foraminifera; Implications for models of diagenetic racemization, *in* Hare, P. E., and others, eds., Biogeochemistry of amino acids: New York, John Wiley and Sons, p. 341–355.

——, 1981, Kinetic model options for interpretation of amino acid enantiomeric ratios in Quaternary mollusks; Comments on a paper by Kvenvolden and others, 1979: Geochimica et Cosmochimica Acta, v. 45, p. 261–264.

——, 1982, A review of amino acid racemization studies in Quaternary mollusks; Stratigraphic and chronologic applications in coastal and interglacial sites, Pacific and Atlantic coasts, United States, United Kingdom, Baffin Island, and tropical islands: Quaternary Science Reviews, v. 1, p. 83–120.

——, 1984a, Relative and absolute dating of Quaternary mollusks with amino acid racemization; Evaluation, application, questions, *in* Mahaney, W. C., ed., Quaternary dating methods: Amsterdam, Elsevier, p. 171–193.

——, 1984b, Interlaboratory comparison of aminoacid enantiomeric ratios in fossil Pleistocene mollusks: Quaternary Research, v. 11, p. 109–121.

——, 1986, Amino acid racemization geochronology, *in* Hurford, A. J., Jager, E., and Ten Cate, J.A.M., eds., Dating young sediments: Bangkok, United Nations, CCOP Technical Publication 16, p. 139–158.

Wehmiller, J. F., and Belknap, D. F., 1978, Alternative kinetic models for the interpretation of amino acid enantiomeric ratios in Pleistocene mollusks; Examples from California, Washington, and Florida: Quaternary Research, v. 9, p. 330–348.

——, 1982, Amino acid age estimates, Quaternary Atlantic Coastal Plain; Comparison with U-series dates, biostratigraphy, and paleomagnetic control: Quaternary Research, v. 18, p. 311–336.

Wehmiller, J. F., and Emerson, W. K., 1980, Calibration of amino acid racemization in late Pleistocene mollusks; Results from Magdalena Bay, Baja California Sur, Mexico, with dating applications and paleoclimatic implications: The Nautilus, v. 94, p. 31–36.

Wehmiller, J. F., Lajoie, K. R., Kvenvolden, K. A., Peterson, E. A., Belknap, D. F., Kennedy, G. L., Addicott, W. O., Vedder, J. G., and Wright, R. W., 1977, Correlation and chronology of Pacific coast marine terraces of continental United States by amino acid stereochemistry; Technique evaluation, relative ages, kinetic model ages, and geologic implications: U.S. Geological Survey Open-File Report 77-680, 196 p.

Wehmiller, J. F., Belknap, D. F., and York, L. L., 1987, Comments on Corrado et al. (1986); Capabilities and limitations of applied aminostratigraphy as illustrated by analyses of *Mulinia lateralis* from the late Cenozoic marine beds near Charleston, South Carolina: South Carolina Geology, v. 31, p. 109–118.

York, L. L., 1984, Aminostratigraphy of Stetson Pit and Ponzer areas of North Carolina by Pleistocene mollusk analysis [M.S. thesis]: Newark, University of Delaware, 188 p.

York, L. L., and Wehmiller, J. F., 1987, Aminostratigraphy of subsurface Quaternary marine sediments, Dare County, North Carolina: Geological Society of America Abstracts with Programs, v. 19, p. 138.

York, L. L., Wehmiller, J. F., Cronin, T. M., and Ager, T. A., 1988, Stetson Pit, Dare County, North Carolina; An integrated chronologic, faunal, and floral record of subsurface coastal sediments *in* Kukla, G., and Adam, D., eds., 1987 INQUA Proceedings Volume, Long Continental Records: Palaeogeography, Palaeoclimatology, Palaeoecology (in press).

MANUSCRIPT ACCEPTED BY THE SOCIETY FEBRUARY 16, 1987

Geological Society of America
Special Paper 227
1988

Paleomagnetism of Quaternary deposits

Don J. Easterbrook
Department of Geology, Western Washington University, Bellingham, Washington 98225

ABSTRACT

Measurement of the paleomagnetism of Quaternary sediments does not yield a numerical age as do isotopic dating methods. In order to convert paleomagnetic data into an age, it must be correlated to known conditions of the geomagnetic field that have been dated by some other method. Paleomagnetic data useful for this purpose include magnetic polarity (normal or reversed), field declination and inclination, secular variation, and magnetic susceptibility.

Diamictons, such as till, glaciomarine drift, and mudflows, may carry a stable magnetic remanence if they contain enough silt and clay in the matrix of the deposit. Glaciomarine drifts provide good examples of diamictons which, although poorly sorted, retain stable and reliable magnetism.

Remanence and anisotropy of magnetic susceptibility in tills may have distinctly different orientations, indicating that remanence is not significantly affected by the preferred orientation of larger grains during shearing. Clay/silt-rich tills in Nebraska, Iowa, and Minnesota give reliable normal and reversed DRMs that record the earth's magnetic field at the time of deposition, despite anisotropy of susceptibility measurements that show a microfabric from glacial shearing. Thus, some, but not necessarily all, tills may be suitable for reliable paleomagnetic measurements if they have enough fine-grained matrix.

Small magnetic grains in some silt/clay-rich tills orient themselves parallel to the earth's magnetic field within water-filled pore spaces. Magnetic grains have sufficient freedom to rotate into alignment because hydrostatic pore pressure carries part of the glacial load and the pore fluid does not transmit shear stresses that might otherwise result in mechanical grain rotation (Easterbrook, 1983). The DRM is fixed in till when enough pore water is expelled to restrict grain rotation.

Supporting evidence is necessary to use polarity changes in sediment for age determination. Boundaries between specific magnetic polarity changes are not identifiable without supporting evidence, because: (1) the change in magnetic polarity may represent any one of several possible reversal boundaries, (2) the change in polarity may belong to one of the many excursions of the magnetic field, and (3) significant erosion of the lower polarity epoch may be followed by deposition of sediment much younger than the beginning of the polarity change.

The degree of magnetic foliation in a till may be determined by the principal axes of a susceptibility ellipsoid, which is significantly more spherical for glaciomarine drift than till. Lodgement till deposited in a water-saturated high-shear environment at the base of a glacier may contain a well-defined magnetic fabric that accurately reflects its petrofabric. Massive till-like glaciomarine drift, consisting of clastic particles dropped from floating ice in marine waters, contains elongate particles that are more randomly oriented than those of subglacial till because it lacks the pervasive shear associated with subglacial till.

An example of the use of combined paleomagnetic measurements and fission-track dating of Pleistocene deposits in the Central Plains indicates that: (1) "Nebraskan" till is not the oldest drift in the region, (2) "Nebraskan" tills at various classic sections are not the same age, (3) the oldest till is older than 2 Ma, and (4) the early Pleistocene glacial sequence in the region is considerably more complex than previously thought.

INTRODUCTION

As a tool for dating of Quaternary deposits, paleomagnetism is somewhat different than other methods because the results of remanence measurements do not give an age, only the nature of the earth's magnetic field at the time of deposition. Paleomagnetic data usually include such information as normal or reverse polarization, declination and inclination of the geomagnetic field, intensity of magnetization, position of the virtual geomagnetic field, and variation of declination and inclination with time. In order to convert this information into an age, it must be correlated to known conditions of the geomagnetic field that have been dated by some other method. Thus, paleomagnetism offers techniques for correlation of Quaternary sediments and lavas by comparison of various magnetic parameters. Where these can be tied to similar magnetic features previously dated by other methods, the age of the unit may be determined. Some common methods of comparison include magnetic polarity (normal or reversed), field declination and inclination, secular variation, and magnetic susceptibility.

A number of reviews of paleomagnetic dating of Pleistocene deposits (e.g., Tarling, 1971; Barendregt, 1981, 1984; Stupavsky and Gravenor, 1984) provide good insights into the general aspects of the methodology, and more detailed technical accounts may be found in the vast literature on paleomagnetism. The objective of this paper is to provide a general overview for geologists, geomorphologists, stratigraphers, palynologists, and other earth scientists who need to know the age of sediments and thus want a general knowledge of paleomagnetism of Quaternary deposits.

GENERAL OVERVIEW

Paleomagnetic data is particularly useful for distinguishing sediments deposited during periods of different magnetic polarity, especially for correlating sediments spanning the Brunhes/Matuyama Polarity Chron. The history of polarity changes in the earth's magnetic field has been fairly well documented and dated, and a paleomagnetic time scale developed (Cox, 1969; Mankinen and Dalrymple, 1979). Times of dominant normal or reversed polarity spanning intervals of the order of about a million years or more are known as polarity epochs. Shorter intervals of polarity change within epochs are called events, and still shorter variations in the earth's magnetic field are known as excursions. Secular variation—small-scale fluctuations of declination and inclination of the magnetic field with time—can sometimes be used for chronologic comparison within a given polarity episode. Reversals of the earth's dipole field are worldwide, so that epochs and events may be recognized on a global scale. Excursions and secular variations have the advantage of much greater age precision than dipole reversals, but are likely to be limited to local or regional extent and are more difficult to find and document.

ACQUISITION OF REMANENT MAGNETISM IN GLACIAL SEDIMENTS

Detrital remanent magnetization (DRM), the acquisition of remanent magnetism in sediments, has been well studied for certain types of sediments (e.g., McNish and Johnson, 1938; Griffiths and others, 1960; Collinson, 1965; Irving and Major, 1964; Tucker, 1983).

Magnetite, the dominant magnetic mineral in most sediments, is usually the principal source of remanence, although other minerals may also contribute. The main remanence in sediments comes from single and pseudo-single domain grains up to 17 microns in diameter (Evans and others, 1968). Grains larger than that probably behave as multi-domain particles.

If enough single or pseudo-single domain grains are present in a sediment, a high-coercivity remanence is generated. Thus, fine-grained sediments are best for measuring remanence. Brownian movement in small particles randomizes DRM, so grains less than about 0.1 micron in diameter are not significant contributors to the remanence of a sediment (King and Rees, 1966; Stacey, 1972). For the geologist, what this means is that silt and silty clay are best for magnetization measurements, whereas sand and gravel are not suitable.

The important factors in acquisition of DRM in sediments are: (1) the earth's ambient magnetic field, (2) the nature and amount of magnetic minerals, (3) particle size distribution, and (4) effect of hydrodynamic and shear forces. The strength (intensity) of magnetization of a sample is determined by the strength of the earth's magnetic field at the time of deposition, the nature of the minerals carrying the DRM and, to a lesser degree, the nature of the depositional process.

Waterlaid sediments

The magnetic orientation of particles deposited in water may occur as particles settle out of the water column or after the particles have been deposited, but while the unconsolidated sediments are still fully saturated with pore water and able to rotate in conjunction with the earth's magnetic field.

The physical environments of deposition of waterlaid sediments vary considerably. tal grains that fall through a column of quiet water have ample time to orient themselves parallel to the earth's magnetic field, whereas grains deposited in flowing water or under the influence of currents may be affected by hydrodynamic forces that exert mechanical torques strong enough to overcome the ambient magnetic field and cause grain rotation (Benedict, 1943; Nagata, 1962; King and Rees, 1966; Verosub, 1977; Henshaw and Merrill, 1979). The effect of hydrodynamic forces on grains is most important on larger particles.

Particle inclination may be affected by grain shape. In regions of relatively high latitude (and thus high magnetic inclination), if one end of an elongate grain touches the bottom before the other end, the grain may rotate into a shallower inclination than that of the ambient magnetic field. Thus, sediments sometimes exhibit shallower inclinations than expected.

Although hydrodynamic and settling rotation of larger grains may produce errors in DRM, small magnetic particles orient essentially parallel to the magnetic field (Rees, 1965; King and Rees, 1966).

The critical particle size beyond which viscous flow may significantly alter the DRM is about 10 to 180 microns. The critical limit for silt-sized particles is about 40 to 50 microns (Keen, 1963). Thus, silt/clay grains are more likely to exhibit a reliable DRM than coarser material.

Windblown sediments

Fine-grained windblown sediments allow ample opportunity for magnetic grains to orient themselves with the earth's magnetic field. Two main types of windblown sediments offer possibilities for retaining a record of paleomagnetic fields: detrital particles and tephra.

Silt and clay in loess usually yield good paleomagnetic measurements. Extensive paleomagnetic studies of loess have been made in Europe (e.g., Koci, 1974; Bucha and others, 1975, 1978; Hus and others, 1976; Koci and Sibrava, 1976; Fink and others, 1978; Fink and Kukla, 1977; Kukla and Koci, 1972; Kukla, 1975; Koci and Sibrava, 1976). Studies of loess have also been made in North America (e.g., Easterbrook and Boellstorff, 1981, 1984; Foley, 1982; Packer, 1979; Kukla and Opdyke, 1980).

Paleomagnetic investigations of tephra are relatively rare. The lack of cohesion in volcanic ash often makes sampling without disturbing the material a real problem. Ash from the 1981 eruption of Mount St. Helens was studied by Steele (1981) soon after deposition and found to record the magnetic field well.

Diamictons

Poorly sorted sediments, such as till, glaciomarine drift, and mudflows, may carry a stable magnetic remanence if they contain enough silt and clay in the matrix of the deposit. Magnetic particles within critical size ranges acquire a remanent magnetism, whereas larger grains produce an essentially random component of remanence, which seems to cancel itself out without affecting the remanence in the smaller particles (Easterbrook, 1983; Easterbrook and Othberg, 1976).

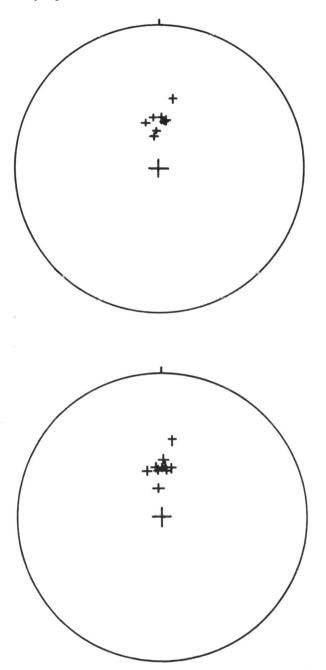

Figure 1. Equal-area plots of 10 samples from a single horizon of glaciomarine drift in the Puget Lowland, Washington. 1a is for natural remanent magnetism, and 1b is after a.f. demagnetization of 200 oe.

Glaciomarine drifts provide good examples of diamictons, which although poorly sorted, retain stable and reliable remanence. Figure 1 shows equal-area plots of multiple samples from a single horizon of glaciomarine drift in the Puget Lowland of Washington. The glaciomarine drift was deposited by release of debris from ice floating in marine water and settling of particles to the sea floor beneath. As seen from the plots in Figure 1, the glaciomarine drift gives reproducible results for both NRM and 200 oe of 10 samples from a single horizon.

The statistical consistency of 6-spin magnetometer measurements can be calculated using a Fisherian model for three-dimensional Gaussian distributions in which the probability density (P) is:

$$P = \frac{K}{4\text{pi sinh K}} \exp (K \cos \theta)$$

where θ is the angle between measured directions and the true mean direction, and K is the precision parameter. The reliability of the mean measured direction can be defined by measuring the radius of a circle on the surface of a sphere, the cone of confidence. These two precision parameters, K and α-95, represent measures of reliability, the greatest reliability being for highest K and lower α-95 values. Alpha-95 values for remanence directions of glaciomarine drift (Fig. 1) are 5.2° for NRM and 5.4° after a.f. demagnetization of 200 oe.

Another measure of data scatter is angular standard deviation (ASD). Values less than 10 are considered good. ASDs for the glaciomarine data in Figure 1 range from 1 to 5 for each specimen, showing good reliability. Figure 2 shows data for 20 samples of Kulshan glaciomarine drift from multiple localities.

The effect of shearing stress at the base of a glacier during deposition of till has long been considered an impediment to using till for remanence measurement. Deposition of lodgement till by moving ice at the base of a glacier produces mechanical shearing and preferential alignment of elongate grains in the till. Initially, magnetic remanence of tills was thought to be related to the direction of shear stress produced by ice movement, rather than to the earth's magnetic field. Early investigations of paleomagnetism in till included studies of remanence (Gravenor and others, 1973, 1979; Gravenor and Stupavsky, 1976; Stupavsky and Gravenor, 1975; Stupavsky and others, 1979; Easterbrook, 1977, 1978, 1983; Easterbrook and Boellstorff, 1981; Barendregt and others, 1977); and magnetic susceptibility (Barendregt and others, 1976; Stupavsky and others, 1974a,b).

Studies have shown that remanence and anisotropy of magnetic susceptibility in tills have distinctly different orientations (Gravenor and Stupavsky, 1974; Crandall, 1979; Easterbrook, 1983), indicating that remanence is not seriously affected by the preferred orientation of larger grains during shearing.

Because of the rather large variability of physical properties among tills of different origin, not all tills are suitable for paleomagnetic measurements. Obviously, coarse-grained tills are not suitable because of the lack of fine-grained material in which the remanence is carried. In addition to lodgement till, flow till may be formed by flow of material (a kind of glacial mudflow) in which elongate grains may be preferentially oriented by viscous motion. Ablation till is deposited as ice melts and drops debris, giving random particle orientations. Stupavsky and others (1979) found poor correlation of remanence from place to place in a till near Lake Erie and attributed it to variability in till deposition. On the other hand, measurements of more than 500 samples of

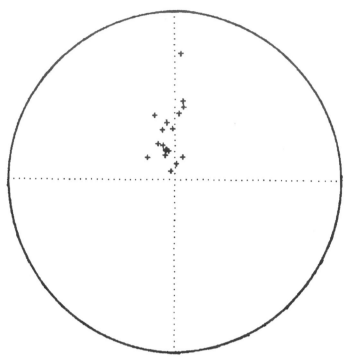

Figure 2. Equal-area plot of 20 samples of Kulshan glaciomarine drift from multiple localities. Average delineation is 354°, and average inclination is 71°, with an α95 of 9.5°.

till by Easterbrook (1977, 1983) between 1971 and 1983 led to the conclusion that many clay/silt-rich tills in Nebraska, Iowa, and Minnesota yielded reliable normal and reversed DRMs that record the earth's magnetic field at the time of deposition. Anisotropy of susceptibility measurements on some of these tills showed a microfabric, presumably from glacial shearing. Considering the many possible variations in till deposition, not to mention potential sediment disruption during sampling of compact till, some, but not all, tills are suitable for reliable paleomagnetic measurements. The validity of remanence data of tills must ultimately be evaluated through detailed alternating field demagnetization studies on a sample-to-sample basis.

The mechanism of acquisition of remanence in lodgement till has been interpreted by several investigators. Gravenor and others (1973) studied "basal tills" around Lake Erie and Lake Ontario and suggested that remanence might be acquired as magnetic minerals fell through a layer of water or slurry at the ice-sediment interface. Subsequent study of these deposits led to the conclusion that they were subaqueous flows of glacial sediments, rather than lodgement tills (Evenson and others, 1977; Eyles and Eyles, 1983; Eyles and others, 1982, 1983, 1985; Stupavsky and Gravenor, 1974), so the origin of their DRM must be reconsidered in light of the new interpretation of the mechanism of sedimentation.

Easterbrook (1977, 1981, 1983) proposed that small magnetic grains orient themselves parallel to the earth's magnetic field essentially in situ within water-saturated pore spaces in till. The magnetic grains have sufficient freedom to rotate into magnetic alignment because the hydrostatic pore pressure carries part of the glacial load. The pore fluid does not transmit shear stresses, which might otherwise result in mechanical grain rotation (Easterbrook, 1983). The DRM is fixed in till when enough pore water is expelled to restrict any further grain rotation. This mechanism has been studied for other types of sediment by Blow and Hamilton (1978), Denham and Chave (1982), Otofuji and Sasajima (1981), Tucker (1980), and Tarling (1974). The main factors in this mechanism are: (1) enough silt/clay particles in the till matrix, (2) saturation of pore spaces to allow grain rotation, and (3) subsequent dewatering of the till to prevent further grain movement (Day and Eyles, 1984; Easterbrook, 1964, 1982; Verosub and others, 1979).

MEASUREMENT OF PALEOMAGNETISM IN GLACIAL SEDIMENTS

Field sampling

Reliable measurement of remanent magnetization requires careful collection of samples in the field. Sample holders must be nonmagnetic (usually plastic), and care must be taken to avoid drying of moist samples in the holder, which can result in rotation of the specimen inside the holder. This can be accomplished easily by taping the openings of the holder in the field and later sealing them with paraffin or rubber cement. Orientation of the sample in the field is measured with a magnetic or solar compass prior to extraction of the specimen and marked on the sample holder (or on the material itself if the sediment is carved).

In order to assess the consistency of remanence measurements, two or more parallel profiles are collected at each site. The degree of similarity between multiple samples from the same horizon is a measure of "noise" in the recorded geomagnetic signal. Variations may be due to post-depositional disturbance of the sediment, sample distortion during sampling, or poor preservation of the geomagnetic signal.

The effect of distortion of samples during field collection is often difficult to evaluate. Laboratory experiments have shown that remanence may be reset by mechanical shock (Stupavsky and others, 1979; Symons and others, 1980; Gravenor and others, 1979). Thus, heavy pounding of sample holders into a sediment may induce partial or total resetting of the remanence. However, the effect on remanence imparted by varying degrees of percussion is not yet clearly defined. Differences in sediment compaction and moisture content require varying application of force to drive a sample holder into the sediment. For moist, loosely compacted silt or clay, a sample holder may often be easily pushed by hand into the sediment, but for drier, more compacted deposits, pounding of the sample holder may be necessary to penetrate the sediment. Fortunately, those sediments most vulnerable to shock resetting, (saturated, loosely compacted silt and clay) usually require the least force to sample. Gravenor and others, (1979)

measured variations in remanence after samples had been subjected to varying degrees of shock and concluded that pounding of sample holders into sediment should be avoided. The alternative to pushing or pounding sample holders is carving of the sediment by hand and manually inserting the specimen into the holder. This procedure avoids the possible problem of shock-induced resetting of the remanence, but is very time consuming and introduces other potential sources of error. Over a period of 15 years, Easterbrook (1983) compared remanence measurements of replicate samples obtained by carving and by percussion and found no significant difference between the two methods for moist, loosely compacted deposits. Plastic cylinders are not very strong, so heavy pounding usually shatters them if sediments offer much resistance to penetration. Where sample carving is impractical, pushing or tapping of sample holders into soft sediments does not seem to produce spurious remanence data. However, the remanence of carved samples should be routinely compared with driven samples as a check, especially for deposits that offer penetration resistance.

A somewhat different problem may result from frictional distortion of sediments along the edge of a sample holder. This can usually be visually detected in clear plastic sample holders and can sometimes be alleviated by wetting the plastic prior to sampling.

Demagnetization

The natural remanent magnetism (NRM) of a sediment may include a variable amount of viscous remanent magnetism (VRM) that has affected the sediment remanence since its deposition. Thus, measurements of NRM alone can easily lead to erroneous conclusions. However, VRM can be removed by stepwise alternating field (a.f.) demagnetization. For example, a sediment deposited in a reversed geomagnetic field and then subjected to a normal geomagnetic field for hundreds of thousands of years may yield normal NRM measurements, and the reversed nature of the remanence may appear only after demagnetization. Therefore, measurement of only NRM without demagnetization is often meaningless, and results based on NRMs must be considered highly tentative.

Demagnetization involves randomly rotating a specimen in a magnetic field of a certain intensity. A portion of the remanent magnetization will be randomized in a magnetic field of that particular strength, leaving the more stable (higher coercivity) remanent fraction. Demagnetization of the specimen at successively higher levels of alternating field intensity progressively removes the viscous remanent magnetism. Orthogonal projections of magnetic vectors (Zijderveld, 1967) may be used to show the point at which low coercivity overprints are removed by demagnetization.

Sample reliability

The degree to which measurement of the magnetic remanence of a sample portrays the geomagnetic field at the time of deposition depends on evaluation of multiple factors. (1) Are the measurements from parallel profiles consistent? (2) Are data from multiple samples of a single profile reproducible? (3) Can a stable remanence be isolated by demagnetization? (4) Have sampling techniques distorted the remanence? (5) Have post-depositional changes disturbed the magnetism? (6) Have laboratory techniques adequately measured the DRM?

Assessing the validity of magnetic data involves more than just deciding if results are good or bad, because the effect of each of these factors is superimposed upon the sample. Some of the factors may be evaluated statistically. For example, the magnetic remanence of a sample is usually measured in the laboratory with a spinner magnetometer. The sample is spun in a magnetic field and measurements are recorded for each of six orientations of the sample. When these measurements are processed by computer, the degree of variability of data can be calculated as angular standard deviation (ASD). Thus, the within-specimen remanence variation can be quickly evaluated during the spin process and measurements may be repeated before going to the next level of demagnetization. Specimens with high ASDs may then be screened out of a batch of samples as unsuitable or given a lower degree of credibility.

Plotting of remanence measurements of multiple samples from a given stratigraphic horizon is useful for evaluating between-specimen remanence variability at different levels of demagnetization. Fisher statistics may be used to show the degree of variability of remanence values. The radius of the cone of 95 percent confidence of declination and inclination values plotted on the surface of a sphere (alpha-95) is a measure of the degree of similarity of remanence measurements. Low values of alpha-95 indicate a high degree of similarity of remanent magnetization values for multiple specimens.

USE OF PALEOMAGNETISM IN INTERPRETING GLACIAL SEDIMENTS

Reversal boundaries

One of the most-used tools for correlation of sedimentary sequences is change of magnetic polarity from normal to reverse or reverse to normal. Interpretation of normally polarized sediments underlain by reversely polarized deposits as the boundary between the Brunhes Normal Polarity Chron and the Matuyama Reversed Polarity Chron (dated at about 730,000 yr, Mankinen and Dalrymple, 1979) is common. However, gaps in the sedimentary record may lead to significant error in such an interpretation for several reasons:

1. Without supporting evidence, the change in magnetic polarity may represent any one of several reversal boundaries.

2. The change in polarity may belong to one of the many excursions of the magnetic field.

3. The polarity change may represent a stratigraphic unconformity wherein the lower polarity unit may have been significantly eroded, then covered with sediment much younger than the lower part of the polarity epoch.

Therefore, the mere finding of normal-polarity over reversed-polarity sediments does not in itself demonstrate the presence of the Brunhes/Matuyama boundary. In almost all cases, supporting evidence is necessary to use polarity changes in sediment for age control.

Secular variation

Geomagnetic secular variation, the fluctuation of the geomagnetic field with time, can sometimes provide a valuable means of correlating the remanence of sediments from place to place and, when tied to curves dated by isotope methods, may be used to correlate profiles over short periods of geologic time. For a detailed discussion of secular variation, see Verosub (this volume).

Anisotropy of magnetic susceptibility

The preferred orientation of inequant grains in a sample may be estimated by measuring anisotropy of magnetic susceptibility (Fuller, 1964; Hrouda, 1982). The anisotropy of magnetic susceptibility is a measure of the preferred orientation of a large number of multi-domain magnetic minerals, most commonly rod-shaped magnetite. The orientation of these grains (the magnetic fabric) reflects any pervasive petrofabric elements within the sediment (Fuller, 1964; Gravenor and others, 1973). For example, lodgement till is deposited at the base of a glacier where ice motion exerts a mechanical shear stress on the underlying sediment and produces alignment of particles, as elongate grains of sand size and larger are rotated into the plane of maximum shearing stress. Thus, the orientation of elongate grains may be used as a means of distinguishing lodgement till from diamictons deposited in the absence of shearing stress.

Magnetic fabric has been shown to be largely independent of the detrital remanent magnetism present in a specimen. Anisotropy of magnetic susceptibility is measured in an alternating external magnetic field. Detrital remanent magnetization is measured in a constant magnetic field (usually zero). This distinction permits separation of the magnetic effects of remanent and induced magnetization. Multidomain magnetite crystals acquire an induced magnetic moment in low magnetic fields, whereas the smaller grains, which carry the remanent magnetization, for the most part, do not. Thus the torque on a multi-domain grain suspended in an alternating field is always in the same direction, whereas the torque on the smaller grains changes with each alternation of the field. The upper size limit of pseudo-single-domain magnetite crystals is approximately 17 microns in length (Evans and others, 1968). Magnetite of this size or smaller is largely inert to anisotropy of magnetic susceptibility measurements. These grains are responsible for the "hard" detrital remanent magnetization of a sample and do not affect investigations into the magnetic fabric of a sample. Multi-domain magnetite crystals up to 180 microns in length are affected by the geomagnetic couple (Rees and Woodhall, 1975), in the absence of external shear. Thus, magnetite grains in the size range from 180 to 17 microns are affected by the geomagnetic couple. Enough coarse-grained magnetite is typically present in the sediments to justify considering their magnetic fabric as primarily a function of shear during deposition.

Lodgement till is deposited in a water-saturated high-shear environment at the base of a glacier (Boulton, 1970, 1971, 1975, 1978, 1982; Boulton and others, 1974) and has been shown to contain a well-defined, highly variable magnetic fabric that accurately reflects its petrofabric (Fuller, 1964; Gravenor and Stupavsky, 1975; Lawson, 1979). Massive glaciomarine drift resembling till consists of clastic particles dropped from floating ice in marine waters (Armstrong and Brown, 1954; Easterbrook, 1963). The constituent elongate particles of glaciomarine sediments should therefore be more randomly oriented than those of subglacial till because the environment of deposition of the glaciomarine sediments lacks a pervasive element of shear of the magnitude associated with subglacial till. To test this hypothesis and to determine if glacial tills contain magnetic signatures that can be used to distinguish them from glaciomarine drift, Crandall (1979) measured the anisotropy of three sediment types: glaciomarine drift and till from the Puget Lowland of Washington, and till from Nebraska. Samples were measured with a SSM-1A Schoensted magnetometer, and data were reduced using the method of successive approximation (Nye, 1960).

The most practical way to describe the anisotropy of magnetic susceptibility of a specimen is by calculating a best-fitting second-rank tensor or by characterizing the anisotropy of the specimen as an ellipsoid. Extraction of the eigenvectors from the susceptibility tensor can be performed by standard mathematical techniques. These eigenvectors should represent the axes of the best-fit ellipsoid. However, for tills, the method of fitting an ellipsoid by successive approximation (Nye, 1960) is preferred because, as noted by Gravenor and Stupavsky (1975), "it does not break down for, (1) samples with small magnetic susceptibility anisotropy; and (2) magnetic susceptibility that cannot be adequately represented by an ellipsoid because of experimental measurement errors, or within-specimen inhomogeneity caused by short wave length variation in till fabric."

Two types of fabrics are commonly found in sediments: (1) a planar fabric (bedding) produced by the gravitational settling of particles, frequently the best-developed grain-fabric element present in sediments; and (2) a linear fabric due to the effect of forces tangent to the bed (Ellwood, 1980; Hamilton and Rees, 1970; Rees, 1961). The magnetic fabric most commonly observed in undeformed sediments has elements that have been described as magnetic foliation and lineation (Rees, 1965). The magnetic foliation is parallel to the bedding, and the magnetic lineation lies within the magnetic foliation plane (Rees and others, 1968). The parameter used in the study by Crandell (1979) to estimate the degree of magnetic foliation is Kmax–Kmin/Kint, where Kmax, Kint, and Kmin are the principal axes of a susceptibility ellipsoid (parameters are zero for a sphere) (Rees, 1965). These indices are nearly independent of the bulk susceptibility of the magnetic

TABLE 1. VALUES OF PRINCIPAL AXES FOR SAMPLES WHOSE MAGNETIC SUSCEPTIBILITY COULD BE CHARACTERIZED BY AN ELLIPSOID*

(Max-Min)/Int	Populations
Computed t value	test statistic - 99.5% level of confidence
Glaciomarine drift vs. Vashon till 9.12	$t_{(47)}99.5 = 2.576$ ∴ means are significantly different
Glaciomarine drift vs. Till from Nebraska 15.74	$t_{(49)}99.5 = 2.576$ ∴ means are significantly different
Till from Nebraska vs. Vashon till 10.8	$t_{(52)}99.5 = 2.576$ ∴ means are significantly different

(Max-Int)/Int	Populations
Computed t value	test statistic - 99.5% level of confidence
Glaciomarine drift vs. Vashon till 38.0	$t_{(47)}99.5 = 2.576$ ∴ means are significantly different
Till from Nebraska vs. Glaciomarine drift 6.54	$t_{(49)}99.5 = 2.576$ ∴ means are significantly different
Vashon till vs. Till from Nebraska	$t_{(52)}99.5 = 2.576$ ∴ means are not significantly different for this level of confidence

*Data from Crandall (1979).

material and of its concentration in the sediment, and so may be used to compare different sediments.

Table 1 shows the values of the principal axes for each sample whose anisotropy of magnetic susceptibility could be characterized by an ellipsoid. Figures 3 and 4 are graphical pres-

Kmax - Kmin / Kint

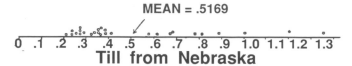

MEAN = .5169

Till from Nebraska

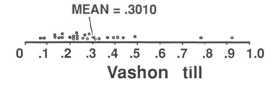

MEAN = .3010

Vashon till

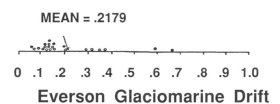

MEAN = .2179

Everson Glaciomarine Drift

Figure 3. Graphical presentation of values for (Kmax-Kmin)/Kint magnetic fabric parameter (from Crandall, 1979).

entations of the fabric parameters. Mean values for each population were compared using a student's test, modified for this situation. The results of this comparison concur with the hypothetical model, i.e., the ellipsoid of susceptibility of glaciomarine drift has a significantly more spherical shape than either of the tills.

A tendency for glaciomarine sediments to have a relatively poorly defined plane of magnetic foliation relative to till is clear, in accord with the typically massive nature of glaciomarine sediments. The lack of bedding in glaciomarine sediments indicates that both the nonmagnetic and magnetic components of the sediment are not constrained by the dynamics of this depositional environment to preferentially come to rest in nearly horizontal orientations (Domack and Lawson, 1985; Easterbrook, 1964; Glen and others, 1957; Gravenor, 1985; Powell 1984). The lack of a pervasive element of shear in the marine environment is thought to be responsible for the poorly developed magnetic lineation in glaciomarine drift relative to till.

The ellipsoid of susceptibility of a sediment is dependent upon grain size, mineralogy, and conditions controlling the amount of shear in the environment of deposition (Rahman and others, 1975; Rees and Woodhall, 1975). Significantly different values of magnetic foliation between the tills is a reflection of variations in these parameters. Divergence in this fabric element may be caused by: (1) differences in higher pore pressures at the glacier base, (2) frozen ground under one glacier and not under another, (3) dissimilar particle size distributions, (4) differences in mineralogy, and (5) differences in the flow velocities of glaciers. The values of magnetic lineation, however, are not significantly different. Because factors controlling the magnetic signature of a sediment are complex and difficult to isolate, differences in magnetic signatures in themselves may not be adequate to distinguish between subglacial tills.

An example of the use of paleomagnetism in solving stratigraphic problems

The status of middle to early Pleistocene and older sediments (0.5 to 2.2 Ma) in Nebraska, Iowa, and South Dakota is presently confused as a result of re-evaluation of the chronology and correlation of pre-Illinoian stratigraphy in the Central Plains region by Boellstorff (1978). "Nebraskan" has long been used for the oldest glaciation in America, "Aftonian" for the first interglacial cycle, and "Kansan" for the second-oldest glaciation; but this usage is now obsolete because tills at localities long considered "Nebraskan" or "Kansan" have recently been shown to be non-correlative and widely separated in age, on the basis of fission-track dating of volcanic ash, paleomagnetism, and lithology of tills (Boellstorff, 1978; Easterbrook and Boellstorff, 1984). Measurements of the remanent magnetism of till combined with fission-track dating of tephra provided a basis for direct comparison of tills over wide areas.

"Nebraskan" till, whose type locality in Nebraska was defined by Shimek (1909), lies beneath a thick section of loess containing Pearlette 0 volcanic ash from Yellowstone National Park, dated at 0.6 Ma from K-Ar dating of the Lava Creek flow in the source area and confirmed with fission-track dating of the ash. Composition of the till places it within a group of tills known as type A (Boellstorff, 1978), somewhat younger than Pearlette S ash, fission-track dated at 1.2 Ma. The till is reversely magnetized, which, in conjunction with the ash chronology, places it in the late Matuyama Reversed Epoch (0.73 to .90 Ma or >0.97 Ma) (Easterbrook and Boellstorff, 1984).

A stratigraphic section near Afton, Iowa, described by Kay and Apfel (1929), consists (by definition) of "Afton" soil underlain by "Nebraskan" till and overlain by "Kansan" till. Beneath this sequence, coring by Boellstorff (1978) revealed sediments containing volcanic ash underlain by till lying on bedrock. The ash and underlying till are pre-"Nebraskan" because they lie beneath the Nebraskan-Afton-Kansas sequence. Glass from the ash yielded a fission-track age of about 2.2 Ma, so the underlying till (which is lithologically similar to Elk Creek till) pre-dates type "Nebraskan" till be at least 1 m.y. (Boellstorff, 1978). The pre–2.2 Ma till is mostly reversely magnetized (Easterbrook and Boellstorff, 1984), and since the earth's magnetic field was reversed during the lower Matuyama Epoch, an age between 2.2 and 2.4 Ma for the till is consistent with the paleomagnetic data.

Two tills overlain by volcanic ash at Hartford, South Dakota, have long been considered "Nebraskan" and "Kansan,"

Kmax - Kint / Kint

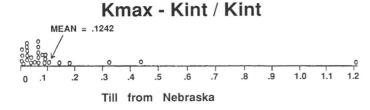

Till from Nebraska

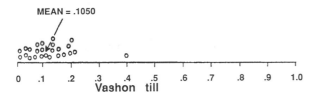

Vashon till

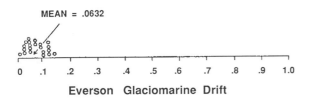

Everson Glaciomarine Drift

Figure 4. Graphical presentation of values for (Kmax-Kint)/Kint magnetic fabric parameter (from Crandall, 1979).

respectively (Flint, 1955; Steece and others, 1960), on the basis that the ash was the late-Kansan Pearlette and, therefore, the two underlying tills must correspond to advances during the Nebraskan and Kansan Glaciations. The ash at Hartford was fission-track dated at 0.76 and ±.09 Ma (Boellstorff, 1978). The upper ("Kansan") till has normal polarity with an average inclination of 41.5°, and the lower ("Nebraskan") till is also normal with an average inclination of 52.6° (Easterbrook and Boellstorff, 1984). Since the type Nebraskan of Shimek (1909) is reversely magnetized, neither the "Nebraskan" nor the "Kansan" tills at Hartford can correlate with it (Easterbrook and Boellstorff, 1984). Because these two tills predate the 0.76-Ma ash, postdate the 1.2-Ma ash, and postdate the Brunhes-Matuyama boundary, they must have been deposited either during the lowermost part of the Brunhes Normal Epoch or, alternatively, during the Jaramillo normal event (0.90 to 0.97 Ma) of the Matuyama Epoch.

An 18-m-thick type B till in a deep core at David City, Nebraska, is reversely magnetized, with an average inclination of −30° (Easterbrook and Boellstorff, 1984), and overlies Pearlette S ash, fission-track dated at 1.2 Ma. A 9-m-thick type A till higher in the section is also reversely magnetized, but with very shallow inclinations. Twenty-four m of normally magnetized loess tops the sequence. The type B till is younger than 1.2 Ma, and its reversed magnetism places it in the latest part of the Matuyama Reversed Epoch.

Four type A tills, all similar in composition, have been recognized in the Central Plains (Boellstorff, 1978). Three of these tills are normally magnetized and one is reversely magnetized (Easterbrook and Boellstorff, 1984). The three oldest tills lie stratigraphically between the Pearlette S ash (1.2 Ma) and the Pearlette O ash (0.6 Ma). The youngest type A till postdates the Pearlette O ash. Among the type A tills are the Nickerson, Cedar Bluffs, Santee, and Clarkson tills.

Twenty-three m of Nickerson till (the lowermost of two tills cored at the type locality in Nebraska) are normally polarized (Easterbrook and Boellstorff, 1984). Although no volcanic ash is present, physical stratigraphy and ash chronology elsewhere suggests that the Nickerson till was deposited between 0.7 and 1.2 Ma, i.e., during the lowermost part of the Brunhes Normal Epoch.

At its type locality, Cedar Bluffs till about 8 m thick overlies a few meters of clay and 14 m of an older till. The Cedar Bluffs till is normally magnetized, with an average inclination of 47 degrees (Easterbrook and Boellstorff, 1984). The underlying till is also normal, with an average inclination of 50 degrees. The type Santee till consists of about 8 m of normally magnetized till overlain by a few meters of loess and underlain by about 18 m of silt. Loess on top of the till also has normal polarity. The upper few meters of the underlying Fullerton silt is normally magnetized; samples below are reversed.

Based on fission-track and paleomagnetic dating of early and middle Pleistocene deposits in the Central Plains: (1) "Nebraskan" till is not the oldest drift in the region, (2) "Nebraskan" tills at various classic sections are not the same age, (3) the oldest till is older than 2 Ma, (4) the early Pleistocene glacial sequence in the region is considerably more complex than previously thought, (5) at least one type A till, and one type B till, are reversely magnetized and older than 0.73 Ma, and (6) at least three type A tills are normally magnetized and two are older than Pearlette O ash (0.6 Ma).

CONCLUSIONS

Paleomagnetism offers several techniques for correlation of Quaternary sediments, but in order to convert paleomagnetic data into an age, it must be correlated to known conditions of the geomagnetic field that have been dated by some other method. Some useful paleomagnetic parameters include magnetic polarity (normal or reversed), field declination and inclination, secular variation, and magnetic susceptibility.

Poorly sorted sediments, such as till, glaciomarine drift, and mudflows, may carry a stable magnetic remanence if they contain a sufficient amount of silt and clay in the matrix of the deposit. Glaciomarine drifts provide good examples of diamictons which, although poorly sorted, retain stable and reliable remanence.

Remanence and anisotropy of magnetic susceptibility in tills have distinctly different orientations indicating that remanence is not seriously affected by the preferred orientation of larger grains during shearing. Clay/silt-rich tills in Nebraska, Iowa, and Minnesota give reliable normal and reversed DRMs that record the earth's magnetic field at the time of deposition. Anisotropy of susceptibility measurements on some of these tills show a microfabric from glacial shearing. Some, but not all, tills are suitable for reliable paleomagnetic measurements. The validity of remanence data of tills depends on having abundant silt/clay in the matrix and detailed a.f. demagnetization studuies.

Small magnetic grains in some silt/clay-rich tills orient themselves parllel to the earth's magnetic field within water-saturated pore spaces. The magnetic grains have enough freedom to rotate into magnetic alignment because the hydrostatic pore pressure carries part of the glacial load and the pore fluid does not transmit shear stresses that might otherwise result in mechanical grain rotation (Easterbrook, 1983). The DRM is fixed in till when enough pore water is expelled to restrict any further grain rotation.

Supporting evidence is necessary to use polarity changes in sediment for age control. Establishing ages of sedimentary sequences by changes in magnetic polarity from normal to reverse or reverse to normal is complicated because, without supporting evidence: (1) the change in magnetic polarity may represent more than one of several possible reversal boundaries, (2) the change in polarity may belong to one of the many excursions of the magnetic field, and (3) significant erosion of the lower polarity epoch may be followed by deposition of sediment much younger than the lower part of that polarity epoch.

The degree of magnetic foliation in a till may be determined by Kmax-Kmin/Kint, where Kmax, Kint, and Kmin are the principal axes of a susceptibility ellipsoid. The ellipsoid of susceptibility of glaciomarine drift is significantly more spherical than till. Lodgement till deposited in a water-saturated high-shear environment at the base of a glacier may contain a well-defined magnetic fabric that accurately reflects its petrofabric. Massive till-like glaciomarine drift, consisting of clastic particles dropped from floating ice in marine waters, contains elongate particles that are more randomly oriented than those of subglacial till because the environment of deposition of the glaciomarine sediments lacks the pervasive shear associated with subglacial till.

Fission-track and paleomagnetic dating of Pleistocene deposits in the Central Plains indicates that: (1) "Nebraskan" till is not the oldest drift in the region, (2) "Nebraskan" tills at various classic sections are not the same age, (3) the oldest till is older than 2 Ma, and (4) the early Pleistocene glacial sequence in the region is considerably more complex than previously thought.

REFERENCES CITED

Armstrong, J. E., and Brown, W. L., 1954, Late Wisconsin marine drift and associated sediments of the lower Fraser Valley, British Columbia, Canada: Geological Society of America Bulletin, v. 65, p. 349–364.

Barendregt, R. W., 1981, Dating methods of Pleistocene deposits and their problems; VI, Paleomagnetism: Geoscience Canada, v. 8, p. 56–63.

——, 1984, Using paleomagnetic remanence and magnetic susceptibility data for the differentiation, relative correlation, and absolute dating of Quaternary sediments, in Mahaney, W. C., ed., Quaternary dating methods: Amsterdam, Elsevier, p. 101–122.

Barendregt, R. W., Stalker, A. M., and Foster, J. H., 1976, Differentiation of tills in the Pakowki–Pinhorn area of southern Alberta on the basis of their magnetic susceptibility: Geological Survey Canada Paper 76–1C, p. 189–190.

Barendregt, R. W., Foster, J. H., and Stalker, A. M., 1977, Paleomagnetic remanence characteristics of surface tills found in Pakowki–Pinhorn area of southern Alberta: Geological Survey Canada Paper 77–1B, p. 271–272.

Benedict, E. T., 1943, A method of determination of the direction of the magnetic field of the earth in geological epochs: American Journal of Science, v. 241, p. 124–129.

Blow, R. A., and Hamilton, N., 1978, Effect of compaction on the acquisition of a detrital remanent magnetization in fine-grained sediments: Geophysical Journal, v. 52, p. 13–23.

Boellstorff, J., 1978, Chronology of some late Cenozoic deposits from the central U.S. and the Ice Ages: Nebraska Academy of Sciences Transactions, v. VI, p. 35–49.

Boulton, G. S., 1970, On the deposition of subglacial and melt-out tills at the margins of certain Svalbard glaciers: Journal of Glaciology, v. 9, p. 231–245.

——, 1971, Till genesis and fabric in Svalbard, Spitsbergen, in Goldthwait, R. P., ed., Till; A symposium: Columbus, University of Ohio Press, p. 41–72.

——, 1975, Processes and patterns of subglacial sedimentation; A theoretical approach, in Wright, A. E., and Mosley, F., eds., Ice ages; Ancient and modern: Liverpool, England, Seal House Press, p. 7–42.

——, 1978, Boulder shapes and grain-size distributions of debris as indicators of transport paths through a glacier and till genesis: Sedimentology, v. 25, p. 773–799.

——, 1982, Subglacial processes and the development of glacial bedforms, in Davidson-Arnott, R., Nickling, W., and Fahey, B. D., eds., Research in glacial, glaciofluvial, and glaciolacustrine systems: Norwich, England, Geo Books, p. 1–31.

Boulton, G. S., Dent, D. L., and Morris, E. M., 1974, Subglacial shearing and crushing and the role of water pressure in tills from southeast Iceland: Geografiska Annaler, v. 56A, p. 135–145.

Bucha, V., Sibrava, V., and Lozek, V., 1975, Paleomagnetic correlations of Pleistocene sediments in central Europe, in Quaternary glaciations in the Northern Hemisphere: International Geological Correlation Program Project 24 Report no. 2, p. 9–36.

Bucha, V., Koci, A., and Sibrava, V., 1978, Paleomagnetic correlation of Quaternary sequences in glaciated and nonglaciated areas, in Quaternary glaciations in the Northern Hemisphere: International Geological Correlation Program Project 24 Report no. 5, p. 63–70.

Collinson, D. W., 1965, Depositional remanent magnetization in sediments: Journal of Geophysical Research, v. 70, p. 4663–4668.

Cox, A., 1969, Geomagnetic reversals: Science, v. 163, p. 237–245.

Crandall, R., 1979, Diatoms and magnetic anisotropy as means of distinguishing glacial till from glaciomarine drift [M.S. thesis]: Bellingham, Western Washington University, 62 p.

Day, T. E. and Eyles, N., 1984, Genetic influences on the remanent magnetization characteristics of glacial diamics: Geological Society of America Abstracts with Program, v. 16, p. 484.

Denham, C. R., and Chave, A. D., 1982, Detrital remanent magnetization; Viscosity theory of the lock-in zone: Journal of Geophysical Research, v. 87, p. 7126–7130.

Domack, E. W. and Lawson, D. E., 1985, Pebble fabric in an ice-rafted diamicton: Journal of Geology, v. 93, p. 577–592.

Easterbrook, D. J., 1963, Late Pleistocene events and relative sea level changes in the northern Puget Lowland, Washington: Geological Society of America

Bulletin, v. 80, p. 1465–1484.

——, 1964, Void ratios and bulk densities as means of identifying Pleistocene tills: Geological Society of America Bulletin, v. 75, p. 745–750.

——, 1977, Paleomagnetic chronology and correlation of Pleistocene deposits: Geological Society of America Abstracts with Program, v. 9, p. 961–962.

——, 1978, Paleomagnetism of glacial tills, *in* Symposium on genesis of glacial deposits: International Quaternary Association Commission, Abstract, Zurich, Switzerland.

——, 1981, Paleomagnetic chronology of "Nebraskan-Kansan" tills in the midwestern U.S.: International Geological Correlation Program Project 24 Report no. 6, p. 72–82.

——, 1982, Physical criteria for distinguishing glacial deposits: American Association Petroleum Geologists Memoir 31, p. 1–10.

——, 1983, Remanent magnetism in glacial tills and related diamictons, *in* Evenson, E. B., Schluchter, C., and Rabassa, J., eds., Tills and related deposits: Rotterdam, The Netherlands, Balkema, p. 303–313.

Easterbrook, D. J., and Boellstorff, J., 1981, Age and correlation of early Pleistocene glaciations based on paleomagnetic and fission-track dating in North America: International Geological Correlation Program Project 24 Report no. 6, p. 189–207.

——, 1984, Paleomagnetism and chronology of early Pleistocene tills in the central United States, *in* Mahaney, W. C., ed., Correlation of Quaternary chronologies: Norwich, England, Geo Books, p. 73–90.

Easterbrook, D. J., and Othberg, K. L., 1976, Paleomagnetism of Pleistocene sediments in the Puget Lowland, Washington, *in* Quaternary glaciations of the Northern Hemisphere: International Geological Correlation Program Project 24 Report no. 3, p. 189–207.

Ellwood, B. B., 1980, Application of the anisotropy of magnetic susceptibility method as an indicator of bottom-water flow direction: Marine Geology, v. 34, p. M83–M90.

Evans, M. E., McElhinny, M. W., and Gifford, A. C., 1968, Single domain magnetite and high coercivities in a gabbroic intrusion: Earth and Planetary Letters, v. 4, p. 142–146.

Evenson, E. G., Dreimanis, A., and Newsome, T. W., 1977, Subaquatic flow tills; A new interpretation for the genesis of some laminated till deposits: Boreas, v. 6, p. 115–133.

Eyles, C. H., and Eyles, N., 1983, Sedimentation in a large lake; A reinterpretation of the late Pleistocene stratigraphy at Scarborough Bluffs, Ontario, Canada: Geology, v. 11, p. 146–152.

Eyles, N., Sladen, J. A., and Gilroy, S., 1982, A depositional model for stratigraphic complexes and facies superimposition in lodgement tills: Boreas, v. 11, p. 317–333.

Eyles, N., Eyles, C. H., and Day, T. E., 1983, Sedimentologic and palaeomagnetic characteristics of glaciolacustrine diamict assemblages at Scarborough bluffs, Ontario, Canada, *in* Evenson, E., Schlucter, C., and Rabassa, J., eds., Tills and related deposits: Rotterdam, The Netherlands, Balkema, p. 23–45.

Eyles, C. H., Eyles, N., and Miall, A. D., 1985, Models of glaciomarine sedimentation and their application to the interpretation of ancient glacial sequences: Palaeogeography, Palaeoclimatology, Palaeocology, v. 51, p. 15–84.

Fink, J., and Kukla, G. J., 1977, Pleistocene climates in central Europe at least 17 interglacials after the Olduvai event: Quaternary Research, v. 7, p. 363–371.

Fink, J., Koci, A., Kohl, H., and Pevzner, M. A., 1978, Paleomagnetic research in the northern foothills of the Alps and the question of correlation of terraces in the upper reach of the Danube, *in* Quaternary glaciations in the Northern Hemisphere: International Geologic Correlation Program Project 24 Report no. 5, p. 108–116.

Flint, R. F., 1955, Pleistocene geology of eastern South Dakota: U.S. Geological Survey Professional Paper 262, 17 p.

Foley, L. L., 1982, Quaternary chronology of the Palouse Loess near Washtucna, eastern Washington [M.S. thesis]: Bellingham, Western Washington University, 137 p.

Fuller, M. D., 1964, A magnetic fabric in till: Geological Magazine, v. 99, p. 233–237.

Glen, J. W., Donner, J. J., and West, R. G., 1957, On the mechanism by which stones in till become oriented: American Journal of Science, v. 255, p. 194–205.

Gravenor, C. P., 1985, Magnetic and pebble fabrics of glaciomarine diamictons in the Champlain Sea, Ontario, Canada: Canadian Journal of Earth Sciences, v. 22, p. 422–434.

Gravenor, C. P. and Stupavsky, M., 1974, Magnetic susceptibility of the surface tills of southern Ontario: Canadian Journal of Earth Sciences, v. 11, p. 658–663.

——, 1975, Convention for reporting magnetic anisotropy of till: Canadian Journal of Earth Sciences, v. 12, p. 1063–1069.

——, 1976, Magnetic, physical and lithologic properties and age of till exposed along the east coast of Lake Huron, Ontario: Canadian Journal of Earth Sciences, v. 13, p. 1655–1666.

Gravenor, C. P., Stupavsky, M., and Symonds, D. T., 1973, Paleomagnetism and its relationship to till deposition: Canadian Journal of Earth Sciences, v. 10, p. 1068–1078.

Gravenor, C. P., and others, 1979, DRM errors in Pleistocene tills including the Seminary till, Scarborough, Ontario: EOS Transactions of the American Geophysical Union, v. 60, no. 18, p. 246.

Griffiths, D. H., King, R. F., Rees, A. I., and Wright, A. E., 1960, The remanent magnetism of some recent varved sediments: Proceedings of the Royal Society of London, v. 59, p. 359–383.

Hamilton, N., and Rees, A. I., 1970, The use of magnetic fabric in paleocurrent estimation, *in* Runcorn, S. K., ed., Paleogeophysics: New York, Academic Press, p. 445–464.

Henshaw, P. C., and Merrill, R. T., 1979, Characteristics of drying remanent magnetization in sediment: Earth and Planetary Science Letters, v. 43, p. 315–320.

Hrouda, F., 1982, Magnetic anisotropy of rocks and its application in geology and geophysics: Geophysical Surveys, v. 5, p. 37–82.

Hus, J., Paepe, R., Geeraerts, R., Somme, J., and Vanhoorne, R., 1976, Paleomagnetic investigations of sediments, *in* Quaternary glaciations in the Northern Hemisphere: International Geological Correlation Program Project 24 Report no. 3, p. 99–128.

Irving, E., and Major, A., 1964, Post-depositional detrital remanent magnetization in a synthetic sediment: Sedimentology, v. 3, p. 135–143.

Kay, G. F., and Apfel, E. T., 1929, The pre-Illinoian Pleistocene geology of Iowa: 34th Annual Report, Iowa Geological Survey, 304 p.

Keen, M. J., 1963, The magnetization of sediment cores from the eastern basin of the north Atlantic Ocean: Deep-Sea Research, v. 10, p. 607–622.

King, R. F., and Rees, A. I., 1966, Detrital magnetism in sediments; An examination of some theoretical models: Journal of Geophysical Research, v. 71, p. 561–571.

Koci, A., 1974, Paleomagnetic investigation of sediments, *in* Quaternary glaciations in the Northern Hemisphere: International Geological Correlation Program Project 24 Report no. 1, p. 110–122.

Koci, A., and Sibrava, V., 1976, The Brunhes-Matuyama boundary at central European localities, *in* Quaternary glaciations in the Northern Hemisphere: International Geological Correlation Program Project 24 Report no. 3, p. 135–141.

Kukla, G. J., 1975, Loess stratigraphy of central Europe, *in* Butzer, D. W., and Isaac, G. L., eds., After the Australopithecines: The Hague, Moulton, p. 99–188.

Kukla, G. J., and Koci, A., 1972, End of the last interglacial in the loess record: Quaternary Research, v. 2, p. 374–383.

Kukla, G. T., and Opdyke, N. D., 1980, Matuyama loess at Columbia Plateau, Washington: American Quaternary Association Abstracts with Programs, no. 6, p. 122.

Lawson, D. E., 1979, A comparison of the pebble fabrics in ice and the deposits of the Matanuska Glacier, Alaska: Journal of Geology, v. 87, p. 629–645.

Mankinen, E. A., and Dalrymple, G. B., 1979, Revised geomagnetic polarity time scale for the interval 0–5 m.y. B.P.: Journal of Geophysical Research, v. 84, no. B2, p. 615–626.

McNish, A. G., and Johnson, E. A., 1938, Magnetization of unmetamorphosed varves and marine sediments: Terrestrial Magazine, v. 43, p. 401–407.

Nagata, T., 1962, Notes on detrital remanent magnetization of sediments: Journal of Geomagnetism and Geoelectricity, v. 14, p. 99–106.

Nye, J. F., 1960, Physical properties of crystals: Oxford, Clarendon Press, 322 p.

Otofuji, Y., and Sasajima, S., 1981, A magnetization process of sediments; Laboratory experiments on post-depositional remanent magnetization: Geophysical Journal, v. 66, p. 241–259.

Packer, D. R., 1979, Paleomagnetism and age dating of the Ringold Formation and loess deposits in the state of Washington: Oregon Geology, v. 41, no. 8, p. 119–132.

Powell, R. D., 1984, Glaciomarine processes and inductive lithofacies modelling of ice shelf and tidewater glacier sediments based on Quaternary examples: Marine Geology, v. 57, p. 1–52.

Rahman, A. U., Gough, D. I., and Evans, M. E., 1975, Anisotropy of magnetic susceptibility of the Martin Formation, Saskatchewan, and its sedimentological implications: Canadian Journal of Earth Sciences, v. 12, p. 1465–1473.

Rees, A. I., 1961, The effect of water currents on the magnetic remanence and anisotropy of susceptibility of some sediments: Geophysical Journal, v. 5, p. 235–251.

—— , 1965, The use of anisotropy of susceptibility in the estimation of sedimentary fabric: Sedimentology, v. 4, p. 257–271.

Rees, A. I., and Woodhall, W. A., 1975, The magnetic fabric of some laboratory deposited sediments: Earth and Planetary Science Letters, v. 25, p. 121–130.

Rees, A. I., Von Rod, U., and Shepard, F. P., 1968, Magnetic fabric of sediments from the La Jolla submarine canyon and fan, California: Marine Geology, v. 6, p. 145–178.

Shimek, B., 1909, Aftonian sands and gravels in western Iowa: Geological Society of America Bulletin, v. 20, p. 399–408.

Stacey, F. D., 1972, Role of Brownian motion in the control of detrital remanent magnetization of sediments: Pure and Applied Geophysics, v. 98, p. 139–145.

Steece, F. V., Tipton, M. J., and Agnew, A. F., 1960, Guidebook, 11th Annual Field Conference, Midwestern Friends of the Pleistocene: South Dakota Geological Survey, 21 p.

Steele, W. K., 1981, Remanent magnetization of ash from the 18 May 1980 eruption of Mount St. Helens: Geophysical Research Letters, v. 8, p. 213–216.

Stupavsky, M., and Gravenor, C. P., 1974, Water release from the base of active glaciers: Geological Society of America Bulletin, v. 85, p. 433–436.

—— , 1975, Magnetic fabric around boulders in till: Geological Society America Bulletin, v. 86, p. 1534–1536.

—— , 1984, Paleomagnetic dating of Quaternary sediments; A review, *in* Mahaney, W. C., ed., Quaternary dating methods: Amsterdam, Elsevier Science Publishers, p. 123–140.

Stupavsky, M., Symonds, D.T.A., Gravenor, C. P., 1974a, Paleomagnetism and magnetic fabric of the Leaside and Sunnybrook tills near Toronto, Ontario: Geological Society of America Bulletin, v. 85, p. 1233–1236.

—— , 1974b, Paleomagnetism of the Port Stanley till, Ontario: Geological Society of America Bulletin, v. 85, p. 141–144.

Stupavsky, M., Gravenor, C. P., Symons, D.T.A., 1979, Paleomagnetic stratigraphy of the Meadowcliffe till, Scarborough Bluffs, Ontario; A Late Pleistocene excursion?: Geophysical Research Letters, v. 6, p. 269–272.

Symons, D.T.A., Stupavsky, M., and Gravenor, C. P., 1980, Remanence resetting by shock-induced thixotropy in the Seminary till, Scarborough, Ontario, Canada: Geological Society America Bulletin, v. 91, p. 593–598.

Tarling, D. H., 1971, Principles and applications of paleomagnetism: London, Chapman and Hall, 164 p.

—— , 1974, A paleomagnetic study of Eocambrian tillites in Scotland: Journal of the Geological Society of London, v. 130, p. 163–177.

Tucker, P., 1980, A grain mobility model of post-depositional realignment: Geophysical Journal, v. 63, p. 149–163.

—— , 1983, Magnetization of unconsolidated sediments and theories of DRM, *in* Creer, K. M., Tuchulka, P., and Barton, C. E., eds., Geomagnetism of baked clays and recent sediments: New York, Elsevier, p. 9–19.

Verosub, K. L., 1977, Depositional and post-depositional processes in the magnetization of sediments: Review of Geophysics and Space Physics, v. 15, p. 129–143.

Verosub, K. L., Ensley, R. R., and Ulrick, J. S., 1979, The role of water content in the magnetization of sediments: Geophysical Research Letters, v. 6, p. 226–232.

Zijderveld, J.D.A., 1967, Demagnetization of rocks; Analysis of results, *in* Collinson, D. W., Creer, K. M., and Runcorn, S. K., eds., Methods in paleomagnetic methods: Newcastle upon Tyne, England.

MANUSCRIPT ACCEPTED BY THE SOCIETY FEBRUARY 16, 1988

Geological Society of America
Special Paper 227
1988

Geomagnetic secular variation
and the dating of Quaternary sediments

Kenneth L. Verosub
Department of Geology, University of California at Davis, Davis, California 95616

ABSTRACT

Long-term changes in the inclination and declination of the magnetic field at a site
are manifestations of geomagnetic secular variation. If a master curve of secular variation is available, then correlation of the paleomagnetic record of undated Quaternary
sediments with the master curve can lead to determination of the age of the sediments.
Because secular variation is coherent only over distances on the order of a few thousand
kilometers, separate master curves must be developed for each region. Historical records, lava flows, and archaeological sites can all provide information about secular
variation, but only rapidly deposited sediments can provide the continuous record
needed to construct a master curve. The quality of the sedimentary secular variation
record depends, however, on the processes by which the sediments acquire their magnetization. These processes create inherent limitations on the agreement in space and
time between records. So-called "second-generation" paleomagnetic studies of lacustrine sequences are now yielding credible master curves. These studies are characterized
by careful attention to coring procedures, good stratigraphic control, a firm chronologic
framework, replicate paleomagnetic sampling, and auxiliary rock magnetic studies. Sediments from lakes in Oregon and Minnesota have provided master curves for western
and east-central North America, respectively. Analysis of these master curves shows that
dating of sedimentary sequences by geomagnetic secular variation is feasible and that it
can provide new opportunities for high-resolution studies of climatologic and sedimentologic processes.

INTRODUCTION

A compass needle does not, in general, point to true north;
an adjustment must be made for the declination of the geomagnetic field. With time, the declination at a particular site changes,
as part of a phenomenon known as geomagnetic secular variation. If chronologic records of the secular variation at a given site
were available, the age of an undated nearby site could be determined by comparing the paleomagnetic directions associated
with the second site to the chronologic record of the first site.
However, this method of dating materials has seldom been used,
primarily because published curves of secular variation from adjacent sites showed very little similarity. Within the past five
years, paleomagnetists have been able to obtain records of secular
variation with the quality and reproducibility needed for dating.
Concurrently, new insights into the nature of the geomagnetic
field and the processes involved in the acquisition of a remanent

magnetism have provided a better understanding of the potential
applications and limitations of the method.

THE NATURE OF THE EARTH'S
MAGNETIC FIELD

The deviation of a compass needle from true north is due to
the complex nature of the earth's magnetic field. If the field were
mathematically equivalent to a geocentric axial dipole, that is, a
bar magnet located at the center of the earth and aligned with its
rotation axis, the magnetic declination would everywhere be zero,
and all compass needles would point to true north. The local
magnetic declination arises because the dipole portion of the
earth's magnetic field is inclined about 11° with respect to the
rotation axis, and the mathematical representation of the field

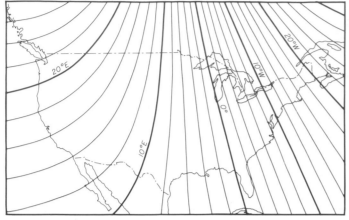

Figure 1. Magnetic declination over North America in 1975. (Redrawn from Defense Mapping Agency Hydrographic Center, Chart 42, 7th edition, June 1975.)

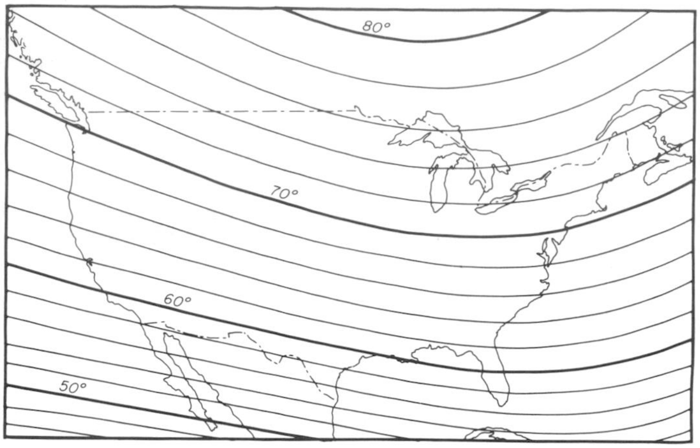

Figure 2. Magnetic inclination over North America in 1975. (Redrawn from Defense Mapping Agency Hydrographic Center, Chart 30, 7th edition, June 1975.)

requires non-dipole as well as dipole components. These factors account for the local magnetic declination, which ranges from 21°W to 24°E in the continental United States (Fig. 1).

The earth's magnetic field also has an inclination, which is the angle in a vertical plane between the local horizontal and the field direction. By definition, a positive inclination points down into the earth; a negative inclination points up out of the earth. If the earth's magnetic field were a geocentric axial dipole, there would be a unique relationship between the colatitude of a site, p, and its magnetic inclination, I, which is given by:

$$\text{Tan } I = 2 \text{ Cot } p \qquad (1)$$

At northern mid-latitudes, the field is quite steep. For example, in the center of North America the inclination is about 70°. Thus, the strength of the vertical component, which is the component of the magnetic field not measured by a compass, is almost three times that of the horizontal component, which is the component that affects a compass. Again, as a result of the tilt of the dipole component and the presence of the non-dipole components, the local magnetic inclination deviates from the values given in equation 1. In the continental U.S., the inclination ranges from about 56° to 76° (Fig. 2).

For the most part, the magnetic field is generated by fluid motions within the outer core of the earth. As these fluid motions change with time, they manifest themselves as changes in the earth's magnetic field at the surface. The inclination and declination measured at a particular site on the surface of the earth change with time, and these changes are known as secular variation. The ranges and amplitudes of secular variation as measured in Paris and London over the past 400 yr are shown in Figure 3.

An alternate way of presenting the data in Figure 3 is shown in Figure 4. Here the secular variation observed in Paris and London has been transformed into virtual geomagnetic pole (VGP) paths. The VGP transformation assumes that each geomagnetic direction is produced by a geocentric (but not necessarily axial) dipole source. The VGP represents the point where the axis of the dipole source intersects the surface of the earth. The distance of the VGP from the observation site is determined from the equation given above using the observed inclination. The VGP is located this distance away along a great circle that passes through the observation site in the direction of the observed declination. Clearly, the idea that a geocentric dipole source produces an obesrved geomagnetic direction is not consistent with the explanation of secular variation presented above. However, the representation of field directions as VGPs has proven to be a useful means of comparing data from different sites, especially those located at different latitudes, and the technique is very common even though it has no physical reality.

The overall features of the magnetic field were previously believed to persist over long periods of time, and the gross structure of the field was thought to drift slowly westward across the surface of the earth (Bullard and others, 1950; Yukutake, 1967) due to differential rotation of the core with respect to the mantle. These conclusions were based mainly on analyses of the field dating back to 1835 when Gauss compiled the first worldwide systematic survey of the magnetic field. In recent years, however, researchers have carefully combed through old ship's logs and other records to develop surveys of the field going back as far as about 1600 (Thompson and Barraclough, 1982; Bloxham and Gubbins, 1985). Analysis of the global magnetic field over this longer interval of time clearly shows that the field structure is quite complex, with non-dipole field components that grow, evolve, and decay on a time scale that may be as short as 100 yr

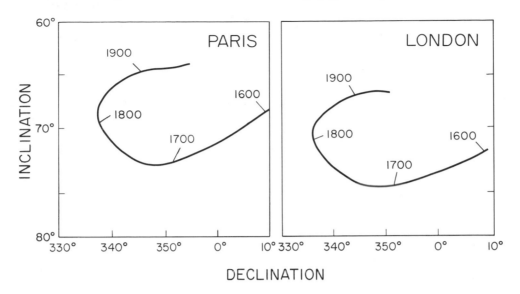

Figure 3. Secular variation of the geomagnetic field in Paris and London during the last 400 years (from Thellier, 1981). The steeper inclinations observed in London are due to the fact that it is farther north than Paris.

(Bloxham and Gubbins, 1985). Furthermore, the non-dipole features of the field move north and south as well as east and west, although there may be a net westward motion of the total field (Thompson and Barraclough, 1982).

These new findings lead to the conclusion that the pattern of secular variation seen at different sites at a given latitude will not necessarily be correlated over very large distances. With regard to the use of secular variation in the dating of Quaternary sediments, separate curves of secular variation, known as master curves, must be established for each different region, and these curves can only be used over the area in which the variations of the field can be expected to have been coherent. Unfortunately, specifying how large such an area would be is not yet possible, but the available data suggest that a given feature should be coherent over distances on the order of 1,000 to 2,000 km (Thompson and others, 1985). For the United States, four separate curves may be necessary, covering for example, the western, north-central, south-central, and eastern states. A corollary of the fact that the pattern of secular variation may not be correlated from region to region is that the master curve for each region must be determined from continuous records within the region. Discontinuous curves cannot be used because we cannot rely on data from outside the region to fill the gaps which may exist. Thus, consideration of the sources of paleomagnetic records of secular variation is appropriate.

SOURCES OF RECORDS OF SECULAR VARIATION

Data concerning secular variation can be obtained from four basic sources: historical records, pyroclastic flows, archeological features, and rapidly deposited sediments. As noted above, good

historical records of magnetic field directions exist only for the last 150 yr; with difficulty, records covering the last 400 yr have been recovered. Other types of historical records also provide information about the geomagnetic field before A.D. 1600. For example, records of auroral observations are useful because a given aurora will be seen over a circular region that is centered on the geomagnetic rather than the rotational pole (Siscoe and Verosub, 1983). However, these sources are unlikely to provide information about the details of secular variation.

In certain areas, however, volcanic eruptions occur quite frequently, so that a well-dated sequence of pyroclastic flows could provide a record of secular variation. In fact, the magnetization of a flow is set as the flow cools from about 700°C to about 600°C, and the recorded direction usually provides an accurate record of the magnetic field at the time of the eruption. The problem is that any sequence of flows represents a discontinuous series in time, and the time missing between successive pairs of flows can differ greatly from one pair to the next. In order to construct a secular variation record, one must also be fortunate enough to find material suitable for radiocarbon dating of each individual flow. Moreover, if the record is assembled from pyroclastic flows at different sites that do not form a stratigraphic sequence, determining whether the dating of one particular site contains a systematic age offset due to local geochemical conditions is difficult. However, as is shown below, the paleomagnetic direction recorded in a pyroclastic flow that can be associated with a tephra layer found in lake sediments can provide a very valuable datum for a lake sediment record.

Archaeomagnetic studies are another source of data about secular variation of the geomagnetic field, and indeed, some of the first detailed records of secular variation were based on archeomagnetic data (Watanabe and DuBois, 1965; Burlatskaya and others, 1965; Kawai and others, 1965). The most common archaeological features used in archaeomagnetic studies are hearths. Here, the magnetization is set as a result of the last intense heating of the clay or stones that line the hearth. Although a given hearth can yield only a single magnetic field direction, construction of a chronological sequence of sites is sometimes possible, using dendrochronology or artifacts. From these, one can obtain a record of secular variation that is perhaps several hundred or even 1,000 yr long (Sternberg, 1983). Care must be taken in interpreting such a record because a dendrochronological date usually reflects the earliest occupation of a site, the artifact may reflect some intermediate time of occupation, and the heating of a hearth usually reflects the latest occupation of a site. Thus, for sites that have been occupied over long periods of time, discrepancies may exist between the accepted date of the site and the correct date of the archaeomagnetic direction. Moreover, archaeologists are primarily interested in archaeomagnetism as an auxiliary means of dating "problem" sites rather than as the primary source of the secular variation curves. As a result, suites of well-dated hearths that could have been used in the construction of a secular variation curve have not been called to the attention of archaeomagnetists.

The fourth source of secular variation data is the paleomagnetic record from rapidly deposited sediments, such as those found in lakes, caves, and the nearshore marine environment. In point of fact, marine sediments have seldom been used to obtain records of secular variation. However, recent work (Karlin, 1983) has shown that rapidly deposited marine sediments do faithfully record secular variation, and the number of such studies can be expected to increase in the near future.

In contrast, work on cave sediments has a long history. In what might be considered one of the first examples of successful secular variation dating, Ellwood (1971) showed that the paleomagnetic record from cave deposits in Georgia could be matched with the archaeomagnetic record from southwestern North America. Later, Creer and Kopper (1976) published secular variation records from caves in the Mediterranean. Recently, several researchers have been examining the paleomagnetic record of stalagmites and stalagtites (Latham and others, 1982; Morinaga and others, 1986). Evidence is growing that these structures, which are collectively known as speleothems, preserve a fairly detailed and continuous record of secular variation (Verosub and others, 1986; Morinaga and others, 1986).

At first glance, lake sediments would appear to be ideal as sources of secular variation data, and indeed, some of the first paleomagnetic studies were designed to determine secular variation from lake sediments (Ising, 1942; Johnson and others, 1948). Slow deposition in quiet waters was thought to produce an accurate and continuous record of the geomagnetic field. Furthermore, because the time scale for changes in the field was on the order of a few hundred years, and typical paleomagnetic samples included about 2.5 cm of sediment, adequate resolution of 50 yr

or less per sample could be achieved with a sedimentation rate of 0.5 mm/yr or more. Such rates are not unusual for lacustrine environments.

Unfortunately, many years passed before reliable records of secular variation could be obtained from lake sediments. One of the first such records came from Lake Windermere in England (Mackereth, 1971; Creer and others, 1972). Other lakes in Great Britain soon yielded paleomagnetic records that showed the same general features as the Lake Windermere record (Thompson, 1975). However, in general, a master curve of secular variation clearly could not be developed from a single core or section having only a few radiocarbon dates. In many cases, published curves from adjacent lakes, or even the same lake, showed almost no similarities (Creer and others, 1976; Vitorello and Van der Voo, 1977). In retrospect, the use of a single core from a given lake was clearly an inadequate approach because it provided no means of assessing the variability of the paleomagnetic recording process. In addition, the gravity corers used in the early studies often twisted as they penetrated the sediment, which imposed a continuous monotonic offset on the declination record, or they bent as they penetrated, which distorted the inclination record. Other problems arose from the lack of azimuthal orientation, from shearing of the sediment within the core barrel, and from gaps in the record where casing was used and a hole was reentered. In addition, in order to obtain a chronology for the entire core from a handful of radiocarbon dates, a uniform rate of sedimentation had to be assumed. More recent work has shown that local climatic fluctuations can produce significant changes in the rate of sedimentation and that uniform rates are the exception rather than the rule.

In the past five years, paleomagnetists have overcome many of the problems that plagued earlier studies, and consistent, believable records of secular variation are becoming available for many areas of the world. Several factors have contributed to the success of these "second-generation" secular variation studies, not the least of which has been more careful attention to the coring technique. Development of the Mackereth piston corer (Mackereth, 1969) and of techniques for slowly driving core barrels into the sediment from small moored platforms (Cushing and Wright, 1965) have made feasible the collecting of several cores up to 10 cm in diameter and up to 6 m long from a small area in a given lake. In some cases, determination of the azimuthal orientation of the cores has been possible. Where reentry of a hole was deemed desirable, the use of casing to determine accurately the depth of each core segment has been possible. In a neighboring hole, the depth control could then be used to obtain a sequence of cores that overlapped those of the first hole.

The chronological framework of the studies has also been significantly improved. Anywhere from 10 to 40 radiocarbon dates may be used to resolve changes in the rate of sedimentation. Ancillary geochemical studies provide a firmer basis for correction to the radiocarbon dates if they are needed. In some cases, identifiable tephra layers or changes in the pollen spectra can be used to provide additional chronologic control.

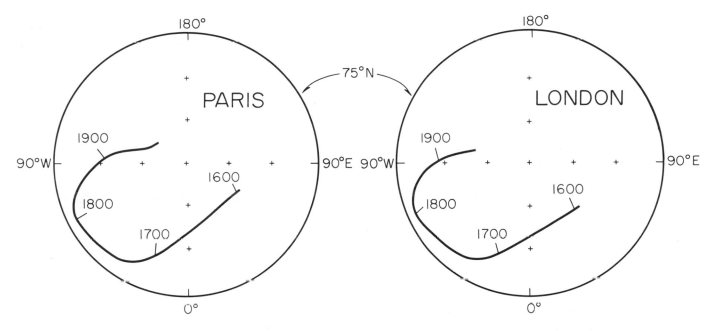

Figure 4. Stereographic projections of the virtual geomagnetic pole (VGP) paths corresponding to the secular variation in Paris and London during the last 400 yr. The VGP transformation takes into account the difference in latitude between London and Paris and produces curves that are nearly congruent.

More attention has also been paid to determining the nature of the magnetic carriers and the processes by which the sediment has acquired its magnetization. These factors play an important role in understanding the uses and limitations of master curves of secular variation.

THE NATURE OF DETRITAL REMANENT MAGNETISM

The paleomagnetism of sediments is known as detrital remanent magnetism (DRM), and it arises from the collective alignment of small magnetic carriers with the earth's magnetic field. In general, the magnetic carriers that contribute most to the paleomagnetic signal are those with diameters smaller than 10 microns. Except where the sediment itself is derived from material containing hematite grains, the primary magnetic mineral is fine-grained magnetite. In a typical situation, the magnetite weathers out of parent rock material and is carried by water into a relatively quiet sedimentary environment where it is deposited in a sediment. As a magnetic carrier settles through the water column, the magnetic moment of the carrier tends to rotate to become parallel to the earth's magnetic field. Thus, when a carrier reaches the sediment/water interface, its orientation has been influenced by the direction of the existing magnetic field. Although not every carrier is actually aligned with the magnetic field, the net magnetization of all of the carriers is parallel to the external field. This magnetization is known as a depositional DRM (Verosub, 1977), and it would represent an accurate record of the earth's magnetic field if no other factors affected the magnetic carriers near the sediment/water interface. In fact, however, if a clearly defined substrate exists at the sediment/water interface, then spherical magnetic carriers may roll as they reach the substrate, and elongated magnetic carriers may rotate as their long axes become horizontal. If these effects are averaged statistically, net shallowing of the inclination, which can amount to as much as 10° (King and Rees, 1966), can be shown to occur. Other factors which can affect the accuracy of a depositional DRM are water currents (Rees, 1961) and deposition on a sloping substrate (Hamilton and King, 1964).

In most cases, however, the sediment/water interface does not represent a clearly defined substrate but is instead a transition zone, which at the top is a very dilute dispersion of sediment with a very high water content and at the bottom is a fully consolidated and dewatered sediment. Recent studies have shown that if the sediment has a high sand content, the magnetic carriers are initially free to rotate in the spaces between the larger sediment grains (Payne and Verosub, 1982). When the water content falls below a critical value, the magnetic carriers become immobile, and the magnetization becomes locked (Tucker, 1980; Denham and Chave, 1982). If the sediment has a high clay content, the magnetic carriers are apparently immobilized by their interaction with the clay floccules (Ellwood, 1979). However, certain types of disturbances, such as those produced by bioturbation, can remobilize the magnetic carriers and produce a realignment with the earth's magnetic field (Ellwood, 1984). In fact, many deep-sea sediments are thoroughly bioturbated, and yet they possess a satisfactory record of the earth's magnetic field. Although the mechanisms by which bioturbation leads to the remagnetization of the sediment are probably quite complex, localized increases in the water content associated with thixotropy are believed to be important (Payne and Verosub, 1982).

Any magnetization that is produced after the magnetic carriers reach the sediment/water interface is called post-depositional DRM (Verosub, 1977). Although such a magnetization is less likely to possess the alignment errors that may affect a depositional DRM, post-depositional DRM is not acquired at the sediment/water interface but at some deeper level in the sediment. Thus, if the magnetization is primarily reset by bioturbation, the magnetization will not become locked until enough new sediment has accumulated that the magnetic carriers are beyond the range of the bioturbation. If no bioturbation occurs or if the water content is still high below the zone of bioturbation, the rate at which the sediments dewater will determine the point at which the magnetization becomes fixed. The offset between the sediment/water interface and the point at which the magnetization becomes locked was recently demonstrated in a study that compared variations in the concentration of beryllium-10 with variations in the paleomagnetic intensity of sediments (Raisbeck and others, 1985). Both parameters are believed to be directly related to variations in the intensity of the earth's magnetic field. In the study, the same pattern of variation was found, but the paleomagnetic changes were displaced 10 to 15 cm with respect to the changes in beryllium-10 concentration.

Where dewatering is the most important process, the magnetization may actually be acquired over a finite interval because different magnetic carriers may have different critical water contents. From the point of view of linear filter theory, the observed paleomagnetic record represents a convolution of the variations of the geomagnetic field with a filter function that reflects the interval over which the magnetization is required (Hyodo, 1984). If the shape of the filter function can be inferred, the data can be deconvolved to recover the original behavior of the field. An additional problem with the dewatering process is that of compaction. Although the evidence is not conclusive, some indications exist that the compaction associated with dewatering can lead to the development of a post-depositional inclination error (Anson and Kodama, 1987).

All of these considerations are relevant to attempts to construct master curves of secular variation and to use these curves for the dating of sediments. For example, consider a thin layer of volcanic ash (tephra) that is deposited as a marker horizon in a sediment. If a depositional DRM is present, the sediment at the marker horizon will represent the field recorded at the time of deposition of the tephra layer, but the direction of magnetization might contain an inclination error. On the other hand, if a post-depositional DRM is present, the field at the time of the deposition of the tephra layer will be recorded by sediment somewhat below the tephra layer. In addition, the paleomagnetic signal may be smoothed by the acquisition process, and it may contain a compaction-related inclination error. In general, these factors are

not serious enough to prevent the use of master curves of secular variation in the dating of Quaternary sediments, but they do underscore the need to understand two important factors: (1) the processes by which sediments have acquired their magnetization, and (2) the inherent limitations on the agreement in both space and time that can be expected from two paleomagnetic records.

MASTER CURVES OF SECULAR VARIATION

As noted above, second-generation paleomagnetic studies of lacustrine sediments are now yielding believable curves of secular variation. One such study was done on cores from Fish Lake, a small lake in Steens Mountain, Harney County, Oregon (Verosub and others, 1986). In this study, 11 10-cm-diameter cores were obtained from five separate holes distributed over an area of less than 50 m² on the lake bottom The cores were obtained from a moored raft using a chain-hoist–operated piston corer, and the interval sampled by each core was controlled to provide multiple overlap of the entire section. Individual core segments were 1 to 3 m long. A total of 20.9 m of sediment was recovered, representing complete recovery of the entire 8.9-m-thick section of postglacial lake sediments.

The cores contained six distinct tephra layers, as well as numerous thin, distinctively colored bands (Fig. 5). Because of the presence of the tephra layers and the bands, overlapping segments of different cores could be aligned with an uncertainty of less than 0.3 cm. The thickest tephra layer, Tephra V, was identified as the Mazama tephra associated with the collapse of Mt. Mazama and the formation of the Crater Lake (Oregon) caldera about 6,800 yr B.P. Age control is based on 18 radiocarbon dates from Fish Lake as well as 19 radiocarbon dates from two nearby lakes, which contained the same six tephra layers. The sediments span the interval from 13,500 yr B.P. to the present, although the sedimentation rate was low from 13,500 yr B.P. until about 10,000 yr B.P.

A total of 455 paleomagnetic samples were collected in small plastic boxes. The samples encompassed 16 m of sediment and represent continuous sampling with overlap of the entire section. Each box was fully oriented with respect to the axis of its core segment. Rock magnetic studies indicated that the magnetic carrier was pseudo-single-domain magnetite, which is not surprising because most of the detritus coming into Fish Lake is derived from Miocene basalt. The paleomagnetic directions were very stable during alternating field demagnetization, and the intensities had median destructive fields of 20 mT.

The declination records from two overlapping cores from Fish Lake are shown in Figure 6. Unlike many first-generation paleomagnetic studies, the Fish Lake records have very low scatter and a high degree of serial correlation within each core as well as excellent agreement between corresponding sections of overlapping cores. Although the cores were not azimuthally oriented, the large overlaps and precise stratigraphic correlation between the cores made the alignment of the cores possible, to obtain a single, unoriented, composite record. The absolute orientation of

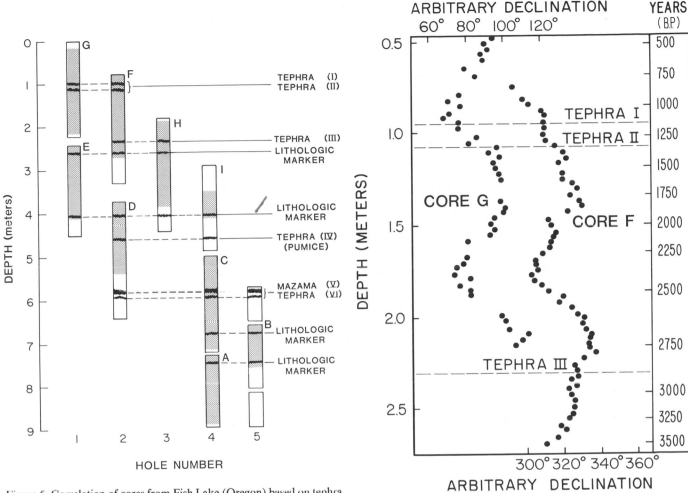

Figure 5. Correlation of cores from Fish Lake (Oregon) based on tephra layers and lithologic markers. The five holes were distributed over an area of less than 50 m² on the lake bottom. The shaded areas correspond to portions of the cores sampled for paleomagnetic study. Tephra V has been associated with the cataclysmic eruption of Mt. Mazama (Crater Lake) in Oregon. Tephra VI represents a precursory eruption from Mt. Mazama.

Figure 6. Declination record of two cores from Fish Lake (Oregon). The samples have been demagnetized at a level of 20 mT. The declinations are arbitrary because the cores were not azimuthally oriented. The non-linear relation between radiocarbon years and depth is caused by variations in sedimentation rate.

this composite record was then recovered by matching the declination at the level of the Mazama tephra in the core with that of the Mazama pyroclastic deposits sampled in outcrop in Oregon. Various consistency checks showed that the cores had penetrated vertically and had not twisted during penetration (Verosub and others, 1986).

Comparison of the inclination at the Mazama tephra in the core with that measured from outcrop samples of Mazama pyroclastic deposits showed that an inclination error was present. A similar conclusion was reached by comparing inclinations at the top of the core with historical records. The age of the samples at the top of the core was determined by locating the decrease in Gramineae (grass) pollen in the core a result of European settlement and cultivation in the area beginning in 1877. After correc-

tion for the inclination error, the data were smoothed using a Gaussian weighting function corresponding to an 80-yr moving window (Verosub and others, 1986). The resulting record of secular variation for the last 10,000 yr is shown in Figure 7. The same data represented as VGPs is shown in Figure 8.

Recently Hanna and others (1986, 1987) examined the secular variation record from two other lakes in the Pacific Northwest. A 3-m composite core from Blue Lake, a small lake in southwestern Idaho, yielded a secular variation record spanning the past 3,500 yr. Comparison of the record from Blue Lake with that from Fish Lake showed that, although there was good morphological agreement between the two records, the ages of major features showed significant disagreement. However, by matching the inclination extremum of the two records, the Blue Lake rec-

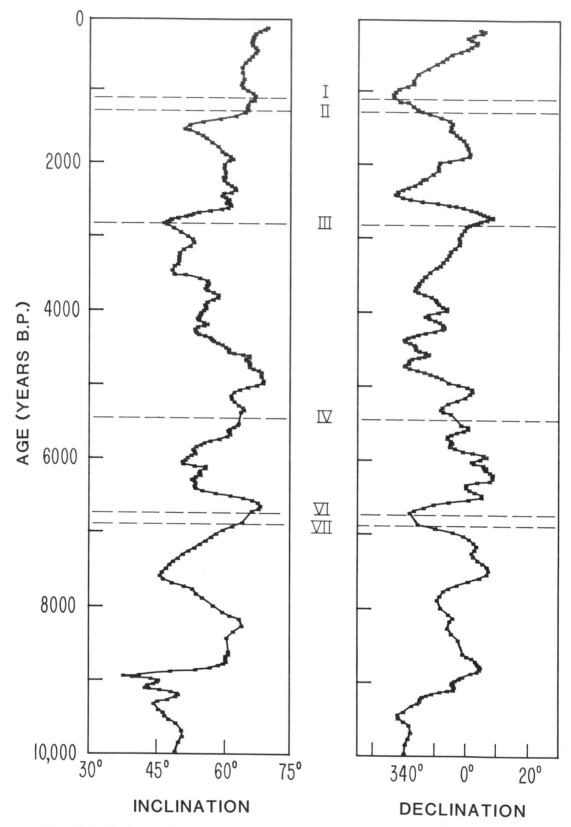

Figure 7. Declination and inclination records from Fish Lake (Oregon). The data have been corrected for an inclination error and have been smoothed using a 7-point Gaussian weighting function. The location of the six tephra layers is shown by Roman numerals. This record is proposed as a master curve for Holocene secular variation in western North America.

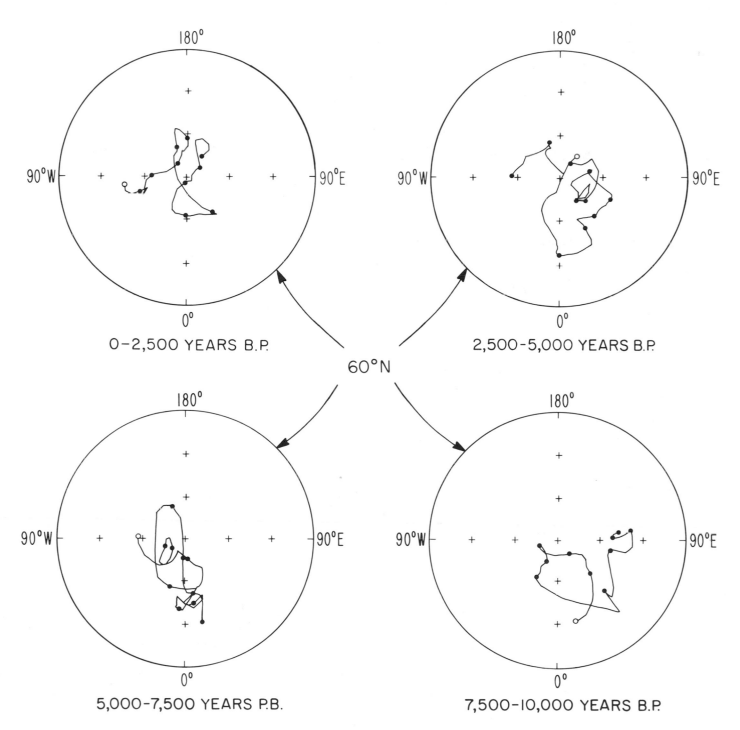

Figure 8. Stereographic projections of the virtual geomagnetic pole paths corresponding to 2,500-yr segments of Figure 7. The closed circles mark 250-yr intervals. The open circles mark the youngest point of each record.

ord could be dated relative to the Fish Lake chronology. The resulting agreement between the declination records validated the morphological correlation and demonstrated the existence of possible systematic errors associated with the radiocarbon dating (Hanna and others, 1986).

On the other hand, a 1.4-m core from Wildhorse Lake, located in the same area as Fish Lake, yielded a 3,200-yr record spanning the interval from about 3,000 to 6,200 yr B.P. Both inclination and declination showed excellent correspondence between the features in the record from Wildhorse Lake and those from Fish Lake (Hanna and others, 1987). On the basis of the agreement of the secular variation records from all three lakes, the Fish Lake record is proposed as a master curve for secular variation in western North America during the past 10,000 yr.

Verosub and Mehringer (1984) have compared a portion of the record from Fish Lake with an archaeomagnetic record compiled by Sternberg (1983) from sites in the southwestern United States. The archaeomagnetic record spans the interval from A.D. 750 to 1450 and is based on 73 independently dated archaeological features, mainly hearths. The VGP path, which is shown in the lower portion of Figure 9, can be compared to the appropriate portion of the Fish Lake record, which is shown in the upper portion of the same figure. Because the dating of the archaeological features was based on dendrochronology, the dates are given in calendar years, and the radiocarbon ages of the Fish Lake record must be converted to calendar ages. In the upper part of Figure 9, the arrows represent intervals of 100 radiocarbon yr. The calendar age ranges from the radiocarbon calibration curves of Klein and others (1982) are given for four points. For the remaining points, only the midpoint of the range is shown.

The agreement between the two records is as good as, or better than, that between any other published pair of records, leaving little doubt that the two curves record the passage of the same non-dipole feature of the geomagnetic field (Verosub and Mehringer, 1984). However, although the curves have similar shapes and the same sense of movement, the archaeomagnetic curve is translated toward the south and rotated slightly counterclockwise with respect to the Fish Lake curve. Furthermore, in only three out of four cases does the age range of a point shown in the upper part of Figure 9 include the age of the corresponding archaeomagnetic point. The discrepancies between the two curves could be caused by a variety of factors, including possible dating problems in both records, questions about the relationship between dates and the time of magnetization, and the 900-km distance between the two sites. Whatever the causes, the fact remains that in this case the best that can be done is to find a correspondence between the overall features of the curves and not between individual points. The implications of these observations for the dating of sediments is discussed below.

Another example of a second-generation paleomagnetic study is that of Lund and Banerjee (1985) on sediments from two Minnesota lakes. Lake St. Croix, located on the Minnesota-Wisconsin border, formed by post-glacial blockage of the St. Croix River, and Kylen Lake developed in a drumlin field in

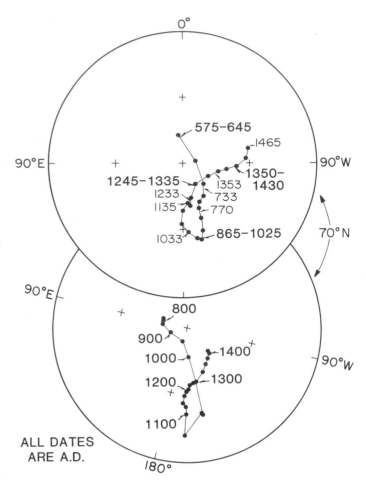

Figure 9. Stereographic projections of a portion of the paleomagnetic record from Fish Lake (Oregon) (upper) and an archaeomagnetic record from the southwestern U.S. (lower). The chronology of the paleomagnetic record is based on radiocarbon dates that have been converted to calendar dates. The chronology of the archaeomagnetic record is based on tree-ring dates that correspond directly to calendar dates. The lower projection has been rotated clockwise 15°.

northeastern Minnesota. Both lakes were cored from their frozen surfaces using a modified piston corer. The diameter of the cores was 5 cm at Lake St. Croix, and 10 cm at Kylen Lake. Two sets of staggered cores were recovered from each lake. The cores from Lake St. Croix were 1 m long and provided replicate sampling of the complete 19-m section of post-glacial sediment. The cores from Kylen Lake were 3 m long, and a 10-m section of post-glacial sediment was recovered. Azimuthal orientation was achieved by aligning the coring device with magnetic north and then transferring this alignment to each core segment. The Kylen Lake cores contained four distinct lithologic units. The boundaries between these units, as well as a distinct twin layer of plant detritus, provided four stratigraphic marker horizons. No such marker horizons were available in the homogeneous sediments from Lake St. Croix.

Age control for Lake St. Croix was based on nine radiocarbon dates. Two independent age determinations, one based on pollen changes associated with cultural influences and one based on stratigraphic evidence relating to the formation of the lake, were used to infer a systematic 980-yr offset in the radiocarbon dates. With this correction, the sediments span the interval from 9,600 yr B.P. to the present. Fifteen radiocarbon dates were used to establish the chronologic framework for Kylen Lake. The recovered sediments span the interval from 13,900 to 4,300 yr B.P.

Paleomagnetic sampling horizons were spaced every 3 to 10 cm, and two to four samples were collected from each horizon. A small secondary magnetization was present in some samples, but alternating field demagnetization at 10 mT was sufficient to remove it. Rock magnetic studies showed that magnetite was the only magnetic phase present in the sediment. After demagnetization, the cleaned paleomagnetic directions of all samples from a given horizon were averaged.

Although an attempt had been made to collect azimuthally oriented cores, analysis of the declination data indicated that the procedure was not completely successful. Therefore, the cores were reoriented either by matching endpoints and trends at the tops and bottoms of adjacent cores from the same set, or by cross-correlating the records from overlapping segments of the replicate sets. As at Fish Lake, various consistency tests showed that the cores had penetrated vertically and had not twisted during penetration.

The uppermost portion of the record from Lake St. Croix showed very good agreement with secular variation in the area, as computed from a spherical harmonic analysis of global historical data. This agreement indicates that an inclination error is not present in the paleomagnetic record and is consistent with previous experiments, which had shown that upon redeposition the sediments of Lake St. Croix accurately record the external magnetic field direction (Levi and Banerjee, 1975). (The reason for an inclination error in some sedimentary records but not in others is not completely understood.) Based on the agreement between the historical data and the paleomagnetic data and on the similarities of the intensity variations at Kylen Lake and Lake St. Croix, Lund and Banerjee concluded that if there were a lock-in zone, it would be less than 10 cm deep at both sites.

The data from each lake were smoothed using the cross-validation technique of Clark and Thompson (1978). The resulting curves of inclination and declination are shown in Figure 10. The curves show good agreement for the age range in which they overlap, although the declination features in the record from Lake St. Croix have a lower amplitude than those in the Kylen Lake record. The differences between the ages of the same features in the two records has been attributed to errors in the radiocarbon chronology of Kylen Lake (Lund and Banerjee, 1985).

Other second-generation paleomagnetic studies have been published for sites in Great Britain (Turner and Thompson, 1981), Iceland (Thompson and Turner, 1985), Greece (Creer and others, 1981), Israel (Thompson and others, 1985), Argentina (Creer and others, 1983), and Australia (Barton and McElhinny,

1981; Constable and McElhinny, 1985). Several of the second-generation European curves have been combined to produce master curves for that region (Turner and Thompson, 1981; Creer and Tucholka, 1982a). Creer and Tucholka (1982b) have also produced a master curve for east-central North America. However, that curve contains data from both first- and second-generation paleomagnetic studies, and the lack of tight chronological control on the former may have compromised the reliability of the first (older) half of the record.

APPLICATIONS

Now that reliable master curves of secular variation are becoming available, we are in a better position to evaluate how they can be used for the dating of Quaternary materials. In doing so we must bear in mind that the presence or absence of an inclination error or a compaction error, as well as the possibility of an offset in time arising from the existence of a post-depositional DRM, leads to an inherent uncertainty in the absolute age of a point on the master curve of as much as 100 to 200 yr.

Probably the oldest proposed use of secular variation curves is for the dating of chronologically isolated sites, such as a thin sedimentary layer, a single lava flow, or an archaeologic feature. In principal, the dating would be done by determining the paleomagnetic direction for the undated material and comparing its VGP with the appropriate master curve. As shown in Figure 8, if the only constraint is that the sample is Holocene in age, finding an unambiguous correlation to the master curve is highly unlikely. Even if the age can be confined to a 1,000- or 2,000-yr interval, the situation may not improve significantly. A case in point is provided by the correlation of the archaeomagnetic curve with the paleomagnetic curve in Figure 9. As discussed above, the correlation requires a 15° rotation of the archaeomagnetic curve, and even then some significant age differences remain. In fact, if the actual archaeomagnetic direction of any point on the archaeomagnetic curve is compared to the paleomagnetic record, it will either yield an erroneous age or miss the paleomagnetic curve completely.

The above example may be an extreme case because of the distance between the sites and the fact that different materials are involved. If the sites are close together and if alignment errors are not present in either the master curve or the isolated direction, the situation may be more favorable for the correlation of the isolated direction (Fig. 11). Of course, the isolated direction might still fall near a crossing point of the VGP curve or in a region where the VGP is moving slowly. On the other hand, it may correspond to a point where the master curve makes a tight turn, or the motion of the VGP is very rapid. In this case, paleomagnetic dating might be quite useful if proper consideration were given to the uncertainties in the single paleomagnetic direction and the uncertainties associated with the master curve.

Another favorable situation involves the use of secular variation curves to discriminate and date tephra layers with similar

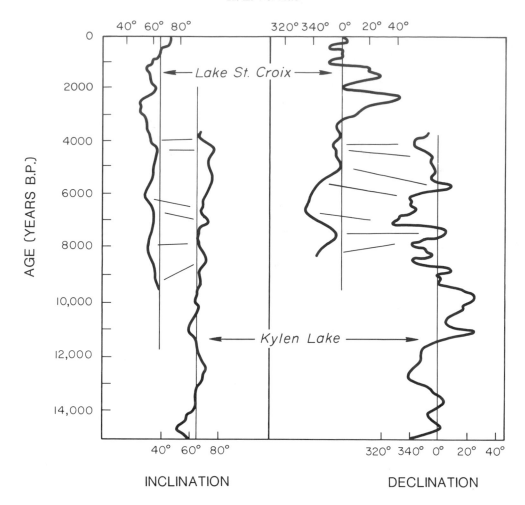

Figure 10. Declination and inclination records from Lake St. Croix and Kylen Lake in Minnesota (from Lund and Banerjee, 1985).

chemical composition. Tephrostratigraphy depends on the existence of discernible chemical differences between tephra layers from different sources and between tephra layers erupted at different times from the same source. Two tephra layers erupted from the same source within a few hundred years of each other may not be distinguishable chemically, and indeed the existence of two distinct eruptions may only be known because the two tephra layers have been found in stratigraphic succession at a particular site. If only one tephra layer is found at a second site, standard tephra studies may not be able to distinguish which of the two layers is present.

A case in point involves the tephra layers attributed to eruptions from Glacier Peak in Washington, which occurred between 11,000 and 13,000 yr B.P. (Porter, 1978). At Sheep Mountain Bog, near Missoula, Montana (Mehringer and others, 1984), the chemical signature of a single tephra layer suggests that it corresponds to the oldest layer, which is considered to be older than 12,000 yr B.P. However, radiocarbon dating implies an age of 11,200 yr B.P., which is the age associated with the youngest

layer. Because the paleomagnetic directions associated with each of the tephra layers are known (Westgate and Evans, 1978), a paleomagnetic study of a suitable core from Sheep Mountain Bog should be able to resolve this problem.

Despite the original assumption that master curves of secular variation would be used for dating chronologically isolated samples, the principal utility of such curves will probably be in the dating of other sedimentary sequences. Such sequences could be from a lake that was used in establishing the master curve in the first place or from another lake that is within the area covered by the given master curve. The primary constraint on this approach is that the sedimentary sequence must be long enough to provide distinct secular variation features that can be correlated to the master curve. The time interval involved depends on the morphology of the master curve, but in general, several hundred to 1,000 yr of record are probably necessary. Of course, the dating of the second sedimentary sequence is still limited by the uncertainties involved in the magnetization process, but for the same lake or a nearby lake with the same sedimentologic regime, these

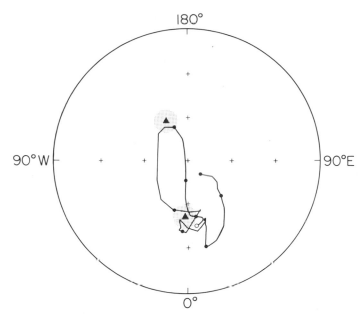

Figure 11. Stereographic projection of VGP path from Fish Lake (Oregon) for the time interval 8,000 to 6,000 yr B.P. The open circle represents the youngest point on the record. Solid circles correspond to 250-yr intervals. The triangles illustrate two possible results of an attempt to correlate a single paleomagnetic direction to the VGP path.

factors will be minimized. Any additional chronological information, such as a radiocarbon date, significantly increases the feasibility of using the method. The presence of a tephra layer (or other distinct horizon) that has already been correlated to the master curve is particularly valuable because the tephra layer provides a direct tie to one particular point on the master curve, while the secular variation of the undated sequence can provide information about the duration of the sequence.

If two high-resolution records can be correlated, and if the two records are from the same lake or from different lakes with similar sedimentologic environments, the corresponding features of the secular variation curves provide a means of obtaining relative dating at a much higher resolution than would be possible from two suites of radiocarbon dates. Consequently, secular variation studies open the way for several new approaches to the detailed study of Quaternary sediments. For example, in determining the precise distribution of a tephra layer, knowing whether or not a thin distal deposit exists at a particular site may be useful. While this can be determined by microscopic examination of the sediment, examination of a long segment of a core is often impractical. However, by matching the secular variation features in the core with those in another core in which the tephra is present, the search could be narrowed to a small segment of the core, and the microscopic examination would become feasible (Fig. 12). Identification of the distal margins of a tephra layer represents one means of determining the dispersal pattern and

total volume of an eruption, and therefore, it contributes to the evaluation of volcanic hazards.

Another possible role for high-resolution relative dating relates to pollen or any other recorder of climate. Here one could analyze the climate parameter at a particular feature of the secular variation curve. On a small scale, this approach could be used to study climatic variation in a series of nearby lakes at different elevations, and on a larger scale, it could be used to study regional differences in climate at the same elevation. Of course, verification that the same magnetic features in different lakes represented a time-stratigraphic horizon would be necessary.

Within a single lake, the magnetization process for fine-grained sediments is likely to be uniform. Therefore, secular variation features can provide a series of time-stratigraphic horizons within the lake. Minor variations in the sedimentology and lithology on the same horizon at different sites will be due to the interaction of the sources of sediment with the circulation patterns in the lake (Tolonen and Oldfield, 1986; Stott, 1986). Larger-scale variations that might be evident by comparing different horizons at the same site in the basin might be caused by changes in the source areas, changes in the drainage basin, or changes in the circulation patterns caused by changes in the morphology of the lake (Oldfield and others, 1979; Dearing and others, 1986; Richardson, 1986). All of these might ultimately be related to climatic or tectonic processes.

Two parameters that can be affected by these processes are the abundance of magnetic carriers and the grain size distribution of the magnetic carriers. For example, an episode of intense weathering in the watershed might lead to an increase in the proportion of small magnetic carriers, while a stream capture in the appropriate source area might lead to an increase in the amount of magnetic material incorporated into the sediment. These changes can be quantified using measurements of the magnetic susceptibility of the sediment and another rock magnetic property, known as the anhysteretic remanent magnetization (ARM). The ARM is the magnetization acquired by a sample that is exposed to a weak direct magnetic field at the same time that it is exposed to a decaying alternating magnetic field. The ratio of the susceptibility to the ARM provides a measure of the size of the magnetic carriers, while the magnitudes of these quantities provide a measure of their abundance (King and others, 1982). If these parameters change with time at different sites in a lake, the secular variation curves can be used to determine whether or not the changes are synchronous. If they are, they probably reflect some general environmental change. If they are not, then the variations probably represent some local change in the lake system. Relative dating using the secular variation curves can then be used to track the penetration of this local change into the lake basin.

CONCLUSION

Second-generation paleomagnetic studies of lacustrine sediments are characterized by tight stratigraphic control, a good

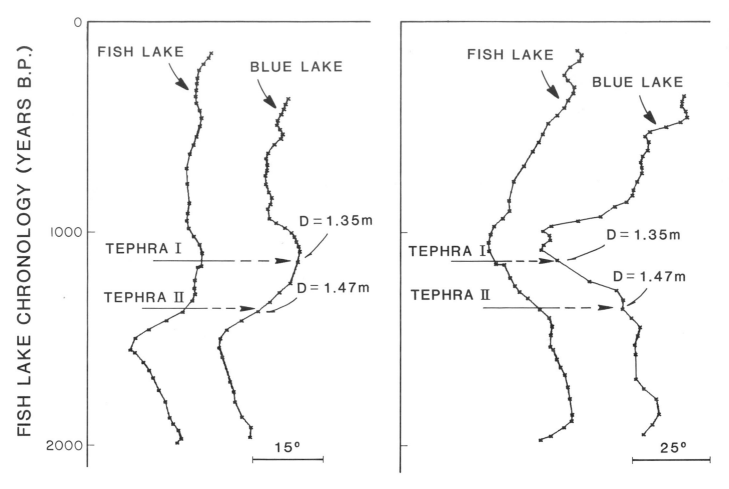

Figure 12. Example of the use of secular variation dating for the location of distal tephra layers. The paleomagnetic correlation implies that Tephras I and II from Fish Lake (Oregon) should be sought at depths of 1.35 and 1.47 m, respectively, in the core from Blue Lake (Idaho).

chronologic framework, replicate sampling, and appropriate rock magnetic investigations. Such studies are capable of providing reproducible high-resolution records, which serve as the basis for the construction of master curves of secular variation. The availability of these curves allows us to evaluate the feasibility of dating of Quaternary materials using geomagnetic secular variation. Due to the inherent errors in the magnetization process, an age determined from a single paleomagnetic direction may be poorly constrained unless special circumstances permit a definitive correlation to a point on the master curve. For sequences of directions spanning several hundred years, an unambiguous correlation is much more likely to exist and may provide age determinations that have greater resolution than is possible from radiocarbon dating. The higher resolution may make it feasible to study climatologic and sedimentologic processes in more detail than was previously possible.

REFERENCES CITED

Anson, G. L. and K. P. Kodama, 1987, Compaction-induced inclination shallowing of the post-depositional remanent magnetization in a synthetic sediment: Geophysical Journal of the Royal Astronomical Society, v. 88, p. 673–692.

Barton, C. E. and M. W. McElhinny, 1981, A 10,000 yr geomagnetic secular variation record from three Australian maars: Geophysical Journal of the Royal Astronomical Society, v. 67, p. 465–485.

Bloxham, J. and Gubbins, D., 1985, The secular variation of Earth's magnetic field: Nature, v. 317, p. 777–781.

Bullard, E. C., Freedman, C., Gellman, H., and Nixon, J., 1950, The westward drift of the earth's magnetic field: Philosophical Transactions of the Royal Society of London, v. A 243, p. 67–92.

Burlatskaya, S. P., Nechaeva, T. B., and Petrova, G. N., 1965, The westward drift of the secular variation of magnetic inclination and variations of the earth's magnetic moment according to "Archeomagnetic" data: Izvestia Earth Physics (English), v. 6, p. 280–285.

Clark, R. M., and Thompson, R., 1978, An objective method for smoothing paleomagnetic data: Geophysical Journal of the Royal Astronomical Society, v. 52, p. 205–213.

Constable, C. G., and McElhinny, M. W., 1985, Holocene geomagnetic secular variation records from northeastern Australian lake sediments: Geophysical Journal of the Royal Astronomical Society, v. 81, p. 103–120.

Creer, K. M., and Kopper, J. S., 1976, Secular oscillations of the geomagnetic field recorded by sediments deposited in caves in the Mediterranean region: Geophysical Journal of the Royal Astronomical Society, v. 45, p. 35–38.

Creer, K. M., and Tucholka, P., 1982a, Secular variation as recorded in lake sediments; A discussion of North American and European results: Philosophical Transactions of the Royal Society of London, v. A 306, p. 87–102.

—— , 1982b, Construction of type curves of geomagnetic secular variation for dating lake sediments from east central North America: Canadian Journal of Earth Sciences, v. 19, p. 1106–1115.

Creer, K. M., Thompson, R., Molyneux, L., and Mackereth, F.J.H., 1972, Geomagnetic secular variation recorded in the stable magnetic remanence of recent sediments: Earth and Planetary Science Letters, v. 14, p. 115–127.

Creer, K. M., Gross, D. L., and Lineback, J. A., 1976, Origin of regional variations recorded by Wisconsinan and Holocene sediments from Lake Michigan, U.S.A., and Lake Windermere, England: Geological Society of America Bulletin, v. 87, p. 531–540.

Creer, K. M., Readman, R. W., and Papamarinopoulos, S., 1981, Geomagnetic secular variations in Greece through the last 6,000 years obtained from lake sediment studies: Geophysical Journal of the Royal Astronomical Society, v. 66, p. 193–219.

Creer, K. M., Valencio, D. A., Sinito, A. M., Tucholka, P., and Vilas, J.F.A., 1983, Geomagnetic secular variations 0–14,000 yr BP as recorded by lake sediments from Argentina: Geophysical Journal of the Royal Astronomical Society, v. 74, p. 223–238.

Cushing, E. J., and Wright, H. E., Jr., 1965, Hand-operated piston corers for lake sediments: Ecology, v. 46, p. 380–384.

Dearing, J. A., Morton, R. I., Price, T. W., and Foster, I.D.L., 1986, Tracing movements of topsoil by magnetic measurements; Two case studies: Physics of the Earth and Planetary Interiors, v. 42, p. 93–104.

Denham, C. R., and Chave, A. D., 1982, Detrital remanent magnetization; Viscosity theory of the lock-in zone: Journal of Geophysical Research, v. 87, p. 7126–7130.

Ellwood, B. B., 1971, An archeomagnetic measurement of the age and sedimentation rate of Climax Cave sediments, southwest Georgia: American Journal of Science, v. 271, p. 304–310.

—— , 1979, Particle flocculation; One possible control on the magnetization of deep-sea sediments: Geophysical Research Letters, v. 6, p. 237–240.

—— , 1984, Bioturbation; Some effects on remanent magnetization acquisition: Geophysical Research Letters, v. 11, p. 653–655.

Hamilton, N., and King, R. F., 1964, Comparison of the bedding errors of artificially and naturally deposited sediments with those predicted from a simple model: Geophysical Journal of the Royal Astronomical Society, v. 8, p. 370–374.

Hanna, R., Verosub, K. L., and Mehringer, P. J., Jr., 1986, Late Holocene secular variation as recorded at Blue Lake, Nez Perce County, Idaho: EOS Transactions of the American Geophysical Union, v. 67, p. 915.

—— , 1987, A secular variation record from Wildhorse Lake, Steens Mountain, Oregon: EOS Transactions of the American Geophysical Union, v. 68, p. 293.

Hyodo, M., 1984, Possibility of reconstruction of the past geomagnetic field from homogeneous sediments: Journal of Geomagnetism and Geoelectricity, v. 36, p. 45–62.

Ising, F., 1942, On the magnetic properties of varved clay: Arkiv for Matematik, Astronomi och Fysik, v. 29, p. 1–37.

Johnson, E. A., Murphy, T., and Torreson, O. W., 1948, Pre-history of the earth's magnetic field: Terrestrial Magnetism and Atmospheric Electricity, v. 53, p. 349–372.

Karlin, R., 1983, Paleomagnetism, rock magnetism, and diagenesis in hemipelagic sediments from the northeast Pacific Ocean and the Gulf of California

[Ph.D. thesis]: Corvallis, Oregon State University, 246 p.

Kawai, N., Kirooka, K., Sasajima, S., Yasukawa, K., Ito, H., and Kume, S., 1965, Archaeomagnetic studies in southwestern Japan: Annals of Geophysics, v. 21, p. 574–578.

King, J. W., Banerjee, S. K., Marvin, J., and Ozdemir, O., 1982, A comparison of different magnetic methods for determining the relative grain size of magnetite in natural materials; Some results from lake sediments: Earth and Planetary Science Letters, v. 59, p. 404–419.

King, R. F., and Rees, A. I., 1966, Detrital magnetism in sediments; An examination of some theoretical models: Journal of Geophysical Research, v. 71, p. 561–571.

Klein, J., Lerman, J. C., Damon, P. E., and Ralph, E. K., 1982, Calibration of radiocarbon dates: Radiocarbon, v. 24, p. 103–150.

Latham, A. G., Schwarcz, H. P., Ford, D. C., and Pearce, G. W., 1982, The paleomagnetism and U-Th dating of three Canadian speleothems; Evidence for the westward drift, 5.4–2.1 ka BP: Canadian Journal of Earth Sciences, v. 19, p. 1985–1995.

Levi, S., and Banerjee, S. K., 1975, Redeposition of and DRM experiments using lake sediments: EOS Transactions of the American Geophysical Union, v. 56, p. 977.

Lund, S. P., and Banerjee, S. K., 1985, Late Quaternary paleomagnetic field secular variation from two Minnesota lakes: Journal of Geophysical Research, v. 90, p. 803–825.

Mackereth, F.J.H., 1969, A short core sampler for subaqueous deposits: Limnology and Oceanography, v. 14, p. 145–151.

—— , 1971, On the variation in direction of the horizontal component of remanent magnetization in lake sediments: Earth and Planetary Science Letters, v. 12, p. 332–338.

Mehringer, P. J., Jr., Sheppard, J. C., and Foit, F. F., Jr., 1984, The age of Glacier Peak tephra in west-central Montana: Quaternary Research, v. 21, p. 36–41.

Morinaga, H., Inokuchi, H., and Yaskawa, K., 1986, Magnetization of a stalagmite in Akiyoshi Plateau as a record of the geomagnetic secular variation in west Japan: Journal of Geomagnetism and Geoelectricity, v. 38, p. 27–44.

Oldfield, F., Rummery, F.T.A., Thompson, R., and Walling, D. E., 1979, Identification of suspended sediment sources by means of magnetic measurements; Some preliminary results: Water Resources Research, v. 15, p. 211–218.

Payne, M. A., and Verosub, K. L., 1982, The acquisition of post-depositional detrital remanent magnetization in a variety of natural sediments: Geophysical Journal of the Royal Astronomical Society, v. 68, p. 625–642.

Porter, S. C., 1978, Glacier Peak tephra in the North Cascade Range, Washington; Stratigraphy, distribution, and relationship to late-glacial events: Quaternary Research, v. 10, p. 30–41.

Raisbeck, G. M., Yiou, F., Bourles, D., and Kent, D. V., 1985, Evidence for an increase in cosmogenic Be-10 during a geomagnetic reversal: Nature, v. 315, p. 315–317.

Rees, A. I., 1961, The effect of water currents on the magnetic remanence and anisotropy of susceptibility of some sediments: Geophysical Journal of the Royal Astronomical Society, v. 5, p. 235–251.

Richardson, N., 1986, The mineral magnetic record in recent ombrotrophic peat synchronized by fine resolution pollen analysis: Physics of the Earth and Planetary Interiors, v. 42, p. 48–56.

Siscoe, G. L. and Verosub, K. L., 1983, High medieval auroral incidence over China and Japan; Implications for the medieval site of the geomagnetic pole: Geophysical Research Letters, v. 10, p. 345–348.

Sternberg, R. S., 1983, Archaeomagnetism in the southwest of North America, *in* Creer, K. M., Tucholka, P., and Barton, C. E., eds., Geomagnetism of baked clays and recent sediments: Amsterdam, Elsevier, p. 159–167.

Stott, A., 1986, Sediment tracing in a reservoir-catchment system using a magnetic mixing model: Physics of the Earth and Planetary Interiors, v. 42, p. 105–112.

Thellier, E., 1981, Sur la direction du champ magnetique terrestre, en France, durant les deux derniers millenaires: Physics of the Earth and Planetary Interiors, v. 24, p. 89–132.

Thompson, R., 1975, Long period European geomagnetic secular variation confirmed: Geophysical Journal of the Royal Astronomical Society, v. 43, p. 847–859.

Thompson, R., and Barraclough, D. R., 1982, Geomagnetic secular variation based on spherical harmonic and cross validation analyses of historical and archaeomagnetic data: Journal of Geomagnetism and Geoelectricity, v. 34, p. 245–263.

Thompson, R., and Turner, G. M., 1985, Icelandic Holocene palaeolimnomagnetism: Physics of the Earth and Planetary Interiors, v. 38, p. 250–261.

Thompson, R., Turner, G. M., Stiller, M., and Kaufman, A., 1985, Near East paleomagnetic secular variation recorded in sediments from the Sea of Galilee (Lake Kinneret): Quaternary Research, v. 23, p. 175–188.

Tolonen, K., and Oldfield, F., 1986, The record of magnetic-mineral and heavy metal deposition at Regent Street Bog, Fredricton, New Brunswick, Canada: Physics of the Earth and Planetary Interiors, v. 42, p. 57–66.

Tucker, P., 1980, A grain mobility model of post-depositional realignment: Geophysical Journal of the Royal Astronomical Society, v. 63, p. 149–163.

Turner, G. M., and Thompson, R., 1981, Lake sediment record of the geomagnetic secular variation in Britain during Holocene times: Geophysical Journal of the Royal Astronomical Society, v. 65, p. 703–725.

Verosub, K. L., 1977, Depositional and postdepositional processes in the magnetization of sediments: Reviews of Geophysics and Space Physics, v. 15, p. 129–143.

Verosub, K. L., and Mehringer, P. J., Jr., 1984, Congruent paleomagnetic and archeomagnetic records from the western United States; A.D. 750 to 1450: Science, v. 224, p. 387–389.

Verosub, K. L., Mehringer, P. J., Jr., and Waterstraat, P., 1986, Holocene secular variation in western North America; The paleomagnetic record from Fish Lake, Harney County, Oregon: Journal of Geophysical Research, v. 91, p. 3609–3623.

Vitorello, I., and Van der Voo, R., 1977, Magnetic stratigraphy of Lake Michigan sediments obtained from cores of lacustrine clay: Quaternary Research, v. 7, p. 398–412.

Watanabe, N., and DuBois, R. L., 1965, Some results of an archaeomagnetic study on the secular variation in the southwest of North America: Journal of Geomagnetism and Geoelectricity, v. 17, p. 395–397.

Westgate, J. C., and Evans, M. E., 1978, Compositional variability of Glacier Peak tephra and its stratigraphic significance: Canadian Journal of Earth Sciences, v. 15, p. 1554–1567.

Yukutake, T., 1967, The westward drift of the Earth's magnetic field in historic times: Journal of Geomagnetism and Geoelectricity, v. 19, p. 103–116.

MANUSCRIPT ACCEPTED BY THE SOCIETY FEBRUARY 16, 1988

Geological Society of America
Special Paper 227
1988

Application of paleomagnetism, fission-track dating, and tephra correlation to Lower Pleistocene sediments in the Puget Lowland, Washington

Don J. Easterbrook
Department of Geology, Western Washington University, Bellingham, Washington 98225
John L. Roland
16263 8th Avenue N.E., Seattle, Washington 98155
Robert J. Carson
Department of Geology, Whitman College, Walla Walla, Washington 99362
Nancy D. Naeser
U.S. Geological Survey, MS 424, Box 25046, Denver Federal Center, Denver, Colorado 80225

ABSTRACT

The early Pleistocene history of the Puget Lowland is marked by repeated advances of the Cordilleran ice sheet into the southern Puget Lowland where deposits of at least three glaciations older than 1.0 Ma are recognized: the Orting (oldest), Stuck, and Salmon Springs, separated by the interglacial Alderton (older) and Puyallup Formations. Until recently, the chronology of these stratigraphic units was unknown and correlations were based entirely on relative age considerations. Paleomagnetic analyses of sediments and petrographic, geochemical, and fission-track analyses of associated tephra were undertaken in order to provide a basis for establishing the chronology of the type sections of these stratigraphic units and to develop a standard for correlations throughout the Puget Lowland. Although paleomagnetic overprinting is common in the sediments sampled, primary components of remanent magnetism were successfully isolated during demagnetization. Previous work (Easterbrook and others, 1981; Westgate and others, 1987) identified the Salmon Springs Drift as reversely magnetized and about 1.0 m.y. old. This investigation establishes the reversed magnetization of the pre–Salmon Springs sediments at or close to their type localities and illustrates the use of the Lake Tapps tephra in regional correlations. The 1.0-Ma fission-track age of the Lake Tapps tephra and the reversed magnetic polarity of the Orting Drift, Alderton Formation, Stuck Drift, Puyallup Formation, and Salmon Springs Drift indicate that all were deposited during the Matuyama Reversed Epoch which began about 2.48 Ma and ended about 0.73 Ma.

INTRODUCTION

The purpose of this investigation was to establish the paleomagnetism of pre–Salmon Springs stratigraphic units at their type localities and to use fission-track and tephra data from Lake Tapps ash to correlate the Salmon Springs Drift and its equivalents in the southern Puget Lowland. The need for such a study was stimulated by the discovery by Easterbrook and others (1981) that the previously accepted age of about 71,000 yr B.P.

(Stuiver and others, 1978) for the Salmon Springs Drift at its type locality was in error. This conclusion followed the discovery of reversely magnetized silt at the type locality (Easterbrook and Othberg, 1976) and a zircon fission-track age of 0.84 + 0.21 Ma from tephra immediately below the silt (Easterbrook and others, 1981). The reversed polarity and fission-track age at the type locality placed the Salmon Springs Drift within the Matuyama Reversed Epoch (Fig. 1). Prior to that investigation, regional correlation or pre-Wisconsin Pleistocene deposits in the central

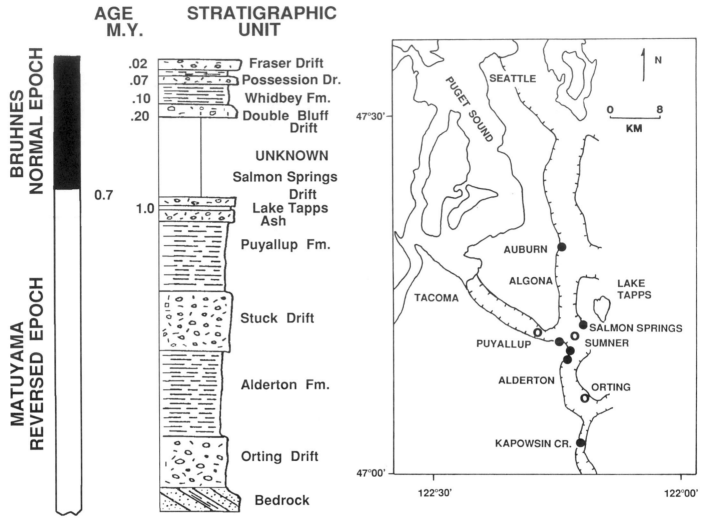

Figure 1. Stratigraphic sequence in the Puget Lowland, Washington.

Figure 2. Map of sample localities in the southeastern Puget Lowland, Washington.

and southern Puget Lowland was based entirely on apparent stratigraphic position. As a result, till beneath Vashon till was often designated Salmon Springs Drift simply because it was the next-oldest till in a section. The establishment of the antiquity of the Salmon Springs Drift immediately cast serious doubt on the validity of most of the correlations of Salmon Springs Drift in the Puget Lowland. However, using paleomagnetic data, fission-track dating, and petrographic and geochemical correlation of tephra, Easterbrook and others (1981) were able to demonstrate correlation of the Salmon Springs Drift with deposits near Auburn, Washington. Later, Westgate and others (1987) extended correlation of the Lake Tapps tephra to other localities in the southeastern and southwestern Puget Lowland.

Among the objectives of this project were: (1) to establish the paleomagnetism of the Puyallup Formation, Stuck Drift, Al-

derton Formation, and Orting Drift at their type localities as a means of verifying their pre–Salmon Springs stratigraphic position as defined by Crandell and others (1958); (2) to measure the fission-track age of tephra at several localities; and (3) to characterize the tephras petrographically and geochemically as potential stratigraphic marker beds. Paleomagnetic and tephra samples were collected during this study to provide data for determination of the age and correlation of pre–Salmon Springs sediments.

This investigation was carried out as part of NSF research grant EAR-8008321. Paleomagnetic analyses were made by Easterbrook and Roland, fission-track dating was done by Naeser and Westgate, and stratigraphic relationships were determined by Easterbrook and Carson. Tephra petrographic and geochemical analyses shown in this paper were made by J. A. Westgate (Westgate and others, 1987).

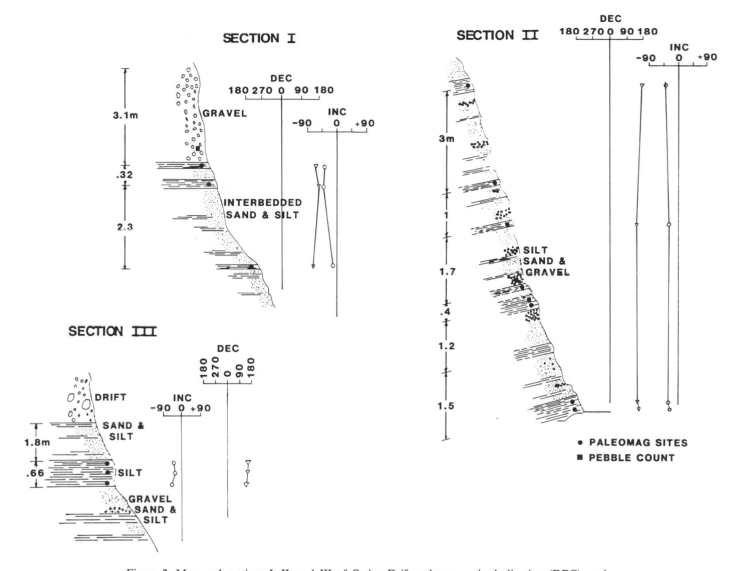

Figure 3. Measured sections I, II, and III of Orting Drift, paleomagnetic declination (DEC), and inclination (INC).

Previous work

Willis (1898) recognized two glaciations in the Puget Lowland, which he named the Vashon and Admiralty, separated by the Puyallup Interglaciation. Additional pre-Vashon glacial sediments were included by Bretz (1913) in the Admiralty Glaciation. Hansen and Mackin (1949) were the first to document more than one pre-Vashon glaciation by identifying two tills, separated by interglacial sediments, beneath Vashon till north of Seattle.

Evidence for four glaciations in the southern Puget Lowland was first recognized by Crandell and others (1958), who defined a stratigraphic framework that was extended throughout the Puget Lowland by relative correlation. This framework was later redefined and extended by Armstrong and others (1965) and Easterbrook and others (1967). It served as the basis for interpre-

tation of Pleistocene stratigraphy and chronology in the Puget Lowland until 1981 when the unexpected discovery of the 0.8 to 0.9 Ma age of tephra in the Salmon Springs Drift required extensive revision of the regional chronology (Easterbrook and others, 1981). The age of the tephra was later revised to 1.0 Ma (Westgate and others, 1987).

Paleomagnetic studies of sediments in the Puget Lowland are reported in Easterbrook (1986), Easterbrook and Briggs (1979), Easterbrook and Othberg (1976), Easterbrook and others (1981, 1985), Naeser and others (1984), and Roland (1984).

Paleomagnetic sampling

Stratigraphic units were sampled for paleomagnetic measurement at their type localities or, where this was not possible

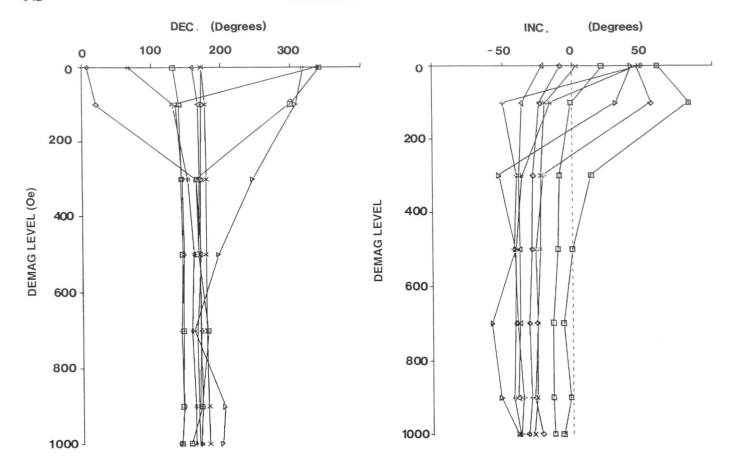

Figure 4. Changes in declination (DEC) and inclination (INC) at varying demagnetization levels for Orting silt.

because of vegetative cover or slumping, from alternate, directly correlated sections nearby. Two or more field-oriented specimens were collected from single stratigraphic horizons to allow replicate measurements from specimens that must have acquired their magnetic remanence contemporaneously. Multiple specimens were collected within 15 cm of each other at each site, and dispersion between paired specimens was used to judge the reliability of the remanent magnetism.

Specimens were collected only from undisturbed silt beds because silt is the most reliable grain size for obtaining consistent detrital remanent magnetism (DRM) directions. Samples were collected by exposing fresh, unoxidized sediment and clearing a surface parallel to bedding. Nonmagnetic plastic cylinders 2.5 cm (1 in) long and 2.5 cm (1 in) in diameter were pushed into the silt by hand or driven perpendicular to bedding by placing a small wood block on the cylinder to absorb the shock of three to ten blows from a hammer. Generally, if the silt was quite cohesive, four to six blows would shatter the plastic cylinder, so samples could not be subjected to severe shock during collection. At selected sites where the sediment was especially compact, samples

were oriented, hand carved, and inserted into the cylinders to minimize mechanical disturbances. Each specimen was oriented with a Brunton compass and magnetic north marked on the cylinder. Magnetic north was later changed to geographic north during computer-assisted directional calculations. The open ends of the cylinder were sealed with masking tape in the field and later sealed with paraffin at both ends to retain moisture and prevent sediment shrinkage, rotation in the cylinder, or acquisition of a drying remanent magnetization (Henshaw and Merrill, 1979).

The remanent magnetism of samples was measured using a Schonstedt analog fluxgate spinner magnetometer (model SSM-1A), interfaced with a Cromenco Z-2D computer system. Resulting magnetic directions, sample intensities, and angular standard deviations were calculated by the computer during measurement. Progressive alternating field (a.f.) demagnetization was carried out on a Schonstedt model GSD-5 tumbling demagnetizer. Based on directional stability, a.f. cleaning steps between 50 and 200 oersteds (oe) were selected according to the magnetic characteristics of the sediment at each site.

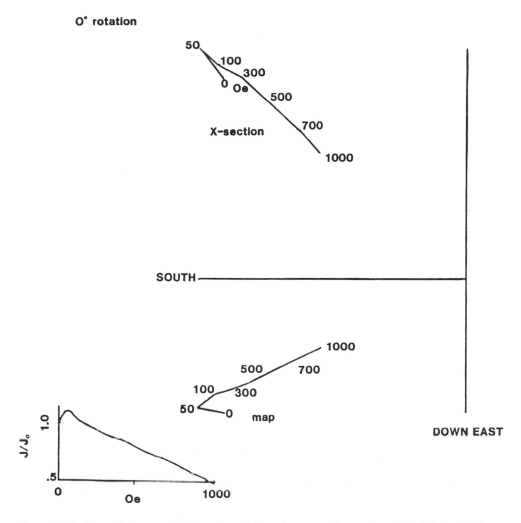

NRM MOMENT- 3.35300E-05

MAXIMUM MOMENT- 3.69900E-05   50 Oe

Figure 5. Orthogonal demagnetization plot and intensity curve for specimen F5 of Orting Drift at measured section III.

Orthogonal plots were used to isolate magnetic components of multicomponnet systems by projecting three-dimensional magnetic vectors onto two-dimensional diagrams (Zijderveld, 1967). Identification of magnetic components overprinted by later magnetic fields was based on the method of converging circles of remagnetization using stereographic projections of directions only (Halls, 1978). Magnetic components were also isolated using vector subtraction, which permits the identification of magnetic vectors removed between selected cleaning steps. All of these methods were used to determine the nature of sample magnetization, and in addition, angular standard deviation (Harrison, 1980) and directional variability were taken into account.

STRATIGRAPHY AND CHRONOLOGY

Introduction

The early Pleistocene sediments in the Puget Lowland that were investigated in this project consist of interbedded glacial and nonglacial deposits. The glacial sediments were deposited by: (1) the Puget lobe of the Cordilleran ice sheet, and (2) alpine glaciers that flowed eastward from the Olympic Mountains. Interglacial sediments were deposited by local surficial processes, mostly fluvial and lacustrine.

The provenance of sediments in the lowland is reflected in

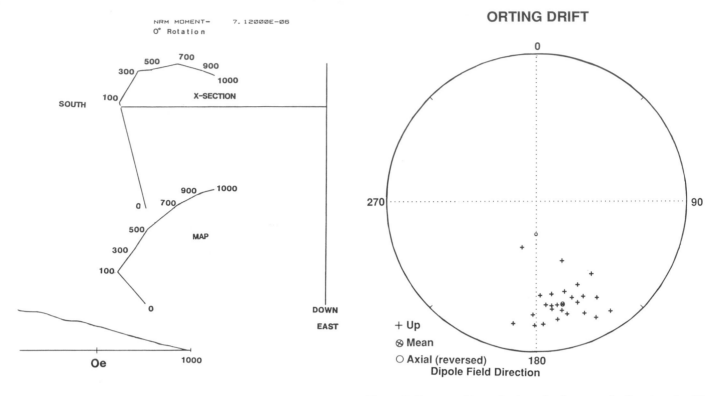

Figure 6. Orthogonal demagnetization plot and intensity curve for sample A1 of the Orting Drift at measured section I.

Figure 7. Stereographic projection of paleomagnetic directions for 25 specimens of Orting silt used to calculate the mean formation direction at sections I, II, and III.

their composition and may be used to distinguish deposits of the Cordilleran ice sheet from sediments that came from the Olympic or Cascade Mountains. Sediments of the Cordilleran ice sheet are characterized by rocks and minerals derived from granitic and regionally metamorphosed terranes of British Columbia. Drift coming from these terranes is composed of granodiorite, gneiss, schist, and quartzite, with distinctive pink garnet in the heavy mineral sand fraction.

The central Cascade Range east of the southern Puget Lowland provides a different source for sediment deposited by streams and alpine glaciers in the lowland. Sediments of this provenance typically contain andesite, basalt, diorite, and granodiorite with much hypersthene and hornblende, and smaller amounts of ilmenite and epidote in the heavy mineral fraction. Sediment, consisting mostly of fluvial sand and volcanic mudflows derived from Mt. Rainier, is characterized by hypersthene-hornblende andesite, which was deposited in the lowland only when the Cordilleran ice sheet was absent from the lowland (Crandell, 1963).

Alpine glacial deposits derived from the Olympic Mountains are dominated by basalt and graywacke and lack granitic components.

Orting Drift

Willis (1898) first applied the name Orting to gravel exposed near the town of Orting in the Puyallup Valley (Fig. 2). Later, Crandell and others (1958) identified till sheets and lenses within the Orting gravel and redefined it as Orting Drift, which they recognized as the oldest Pleistocene glacial deposit in the Puget Lowland on the basis of its intense oxidation. The drift consists of a thick basal gravel of central Cascade provenance overlain by till and gravel of the Puget lobe. The till is generally a compact, unstratified, poorly sorted diamicton containing 10 to 15 percent clasts of northern provenance and 5 to 15 percent garnet in the heavy mineral fraction. Stratified gravel and sand not closely associated with till are usually of central Cascade origin but lack clasts of Mt. Rainier origin, perhaps because the deposits predate Mt. Rainier. Orting gravel is commonly up to 60 m (200 ft) thick and reaches a maximum thickness of 79 m (260 ft). Orting Drift lies on late Tertiary rocks and is overlain unconformably by sediments of the Alderton Formation.

Paleomagnetism of the Orting Drift. Sampling of Orting Drift at its type locality (Crandell and others, 1958) was difficult

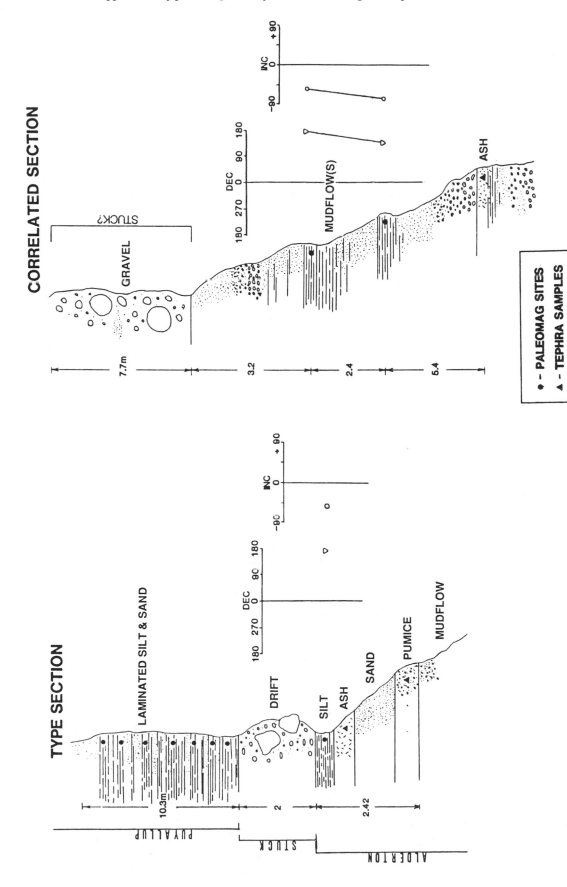

Figure 8. Measured sections and paleomagnetic mean directions for the Alderton type locality and a correlated exposure about 645 m to the south.

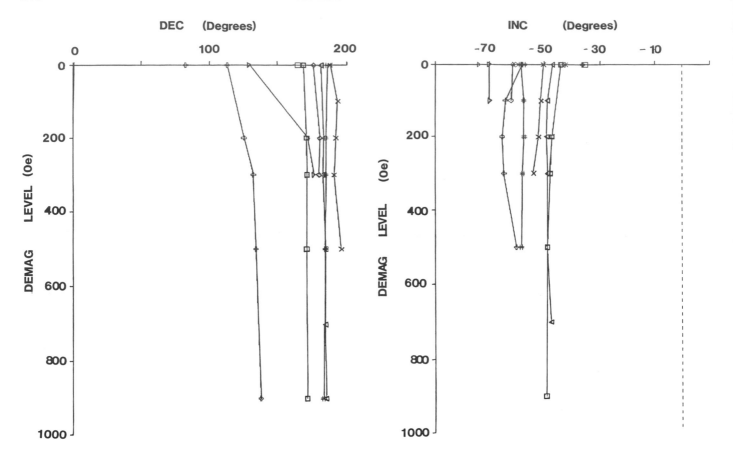

Figure 9. Changes in declination (DEC) and inclination (INC) at varying demagnetization levels for Alderton silt.

because: (1) several outcrops did not contain silt horizons adequate for paleomagnetic measurement, (2) outcrops were poorly exposed, or (3) outcrops were inaccessible. Therefore, samples were collected from alternate localities nearby, all of which had been previously identified as Orting Drift by Crandell (1963). Samples were obtained from glaciofluvial deposits at Kapowsin Creek south of Orting (Fig. 3). Thirty-four specimens were collected and measured from three sections within 67 m (220 ft) of each other. Pebble lithologies and heavy mineral compositions in sand verify the northern provenance (with central Cascades components) of the sediments. Specimens showed similar magnetic behavior at the three Orting sections (Fig. 4). Ideal specimens, such as F5 (Fig. 5), carried a low-coercivity normal overprint with a declination of about 0° and an inclination of 54° that was removed by the 100-oe demagnetization level. Between 300 and 1,000 oe, a stable reversed magnetism was isolated with a southerly declination of 154° and an inclination of –39°.

Orting specimens commonly exhibited southerly declinations with upward inclinations, but with heterogeneous magnetization. Sample A1 (Fig. 6) is a typical example. Following removal of a low-coercivity normal overprint (declination [D] = 48, inclination [I] = 69) between 0 and 100 oe, the declination

swung abruptly southward as step demagnetization continued and the shallow, persistent, normal component deteriorated while a higher-coercivity reversed direction emerged. By the maximum 1,000-oe level, reversed (southerly) magnetization was obvious. Vector subtraction demonstrated that above 700 oe the magnetization was still heterogeneous, yet the vectors removed were southerly and shallow (up). The sediment proved to be reversely magnetized, yet the primary direction remained overprinted through the 1,000-oe level, with both the normal and reversed components being demagnetized together above 700 oe.

Mean directions were calculated for 25 specimens from the three measured sections. The mean formation direction is D = 164°, I = –31° with an alpha-95 of 8° (Fig. 7).

Alderton Formation

The Alderton Formation lies unconformably above the Orting Drift and consists of interbedded mudflows and fluvial sediments, derived almost exclusively from an active ancestral Mt. Rainier. The type section (Crandell and others, 1958), exposed along the west wall of the Puyallup Valley near the town of Alderton (Fig. 2), consists of prominent mudflow deposits of very

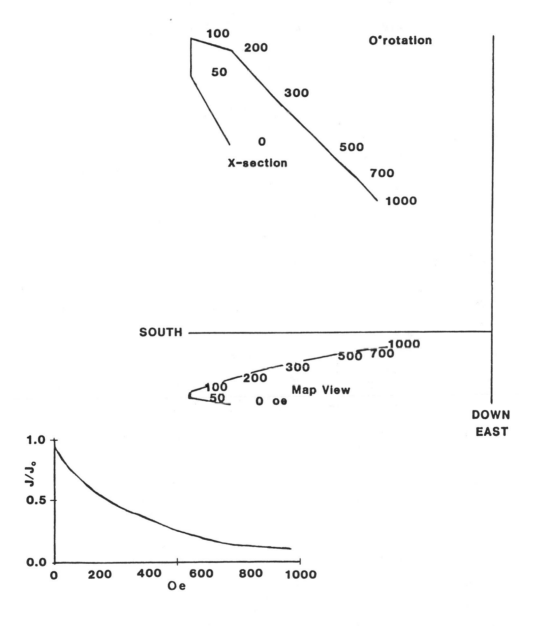

Figure 10. Orthogonal demagnetization plot and intensity curve for sample II of the Alderton Formation at its type locality.

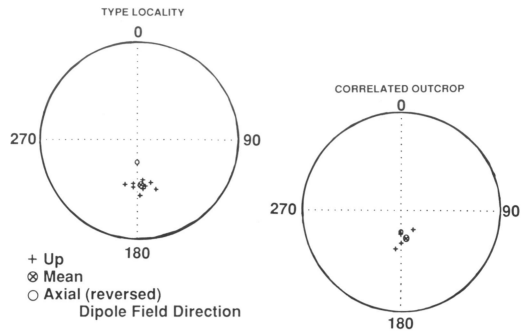

Figure 11. Stereographic projection of directions for specimens from the Alderton type locality (700-oerstedts level) and its correlated exposure (500-oerstedt level).

compact, angular to subrounded pebbles and boulders of hornblende-hypersthene andesite in a sandy-clay matrix. Thicknesses of individual mudflows vary from 1 to 8 m (3 to 25 ft).

Based on the absence of northern-provenance components and the dominance of Mt. Rainier source material, the Alderton sediments are interpreted to be interglacial. Pollen from a peat bed low in the formation indicates a Canadian vegetation zone, changing upward to a humid transition zone similar to the present climate in the area (Crandell and others, 1958).

Paleomagnetism of the Alderton Formation. The Alderton Formation was sampled at its type locality. However, all but 4.5 m (15 ft) of the originally defined section were covered by vegetation and slopewash. Nine samples were collected about 25 cm (10 in) below the base of the overlying Stuck Drift (Fig. 8). Four additional samples were collected from Mt. Rainier mudflows lying beneath Stuck (?) oxidized boulder-cobble gravel at a second exposure, about 645 m (2,110 ft) south of the type section.

The samples from the Alderton type locality exhibited consistently stable magnetic remanence (Figs. 9 and 10). A normal polarity overprint ($D = 10°$, $I = 58°$) was removed between 0 and 50 oe during a.f. demagnetization. By 100 oe, remaining normal components became negligible with respect to the higher coercivity reversed components. Continued demagnetization from 200 to 700 oe revealed a persistent reversed overprint. By the 700-oe level, a reversed component was isolated that was slightly steeper and more southerly than the previous reversed magnetism. The mean direction calculated for the Alderton type locality at the 700-oe level was $D = 176°$, $I = -50°$ (Fig. 11).

Additional paleomagnetic measurements of the samples at the second exposure 645 m from the type section yielded supportive results. The specimens at this site were also stable by 500 to 700 oe demagnetization and gave an average direction of $D = 170°$, $I = -65°$ at the 500-oe level.

Stuck Drift

Stuck Drift lies above the Alderton Formation and represents the second oldest glaciation of the Puget Lowland. The type locality (Crandell and others, 1958) southwest of the town of Alderton consists of oxidized till containing 10 to 15 percent northern-provenance clasts, overlain and underlain by oxidized sand and gravel.

Most Stuck outcrops in the Puyallup Valley contain till. Lacustrine sediments at the top of Stuck sections may belong to the uppermost Stuck or the overlying nonglacial Puyallup Formation. Exposures of Stuck Drift are limited to the walls of the Puyallup Valley, so the southern extent of Stuck Glaciation is not known and correlation elsewhere in the Puget Lowland has not been made.

Paleomagnetism of the Stuck Drift. Because sediment at the Stuck type locality consists of till, sand, and gravel unsuitable for paleomagnetic sampling, an alternate exposure south of the town of Sumner (Fig. 2) was sampled (Fig. 12). Stuck Drift exposed in this vicinity consists of glaciolacustrine sediments overlain by sand and silt of the Puyallup Formation (Crandell, 1963). The outcrop sampled is composed of interbedded sand and silt, which rests directly on oxidized fluvial gravel interpreted

STUCK DRIFT OUTCROP

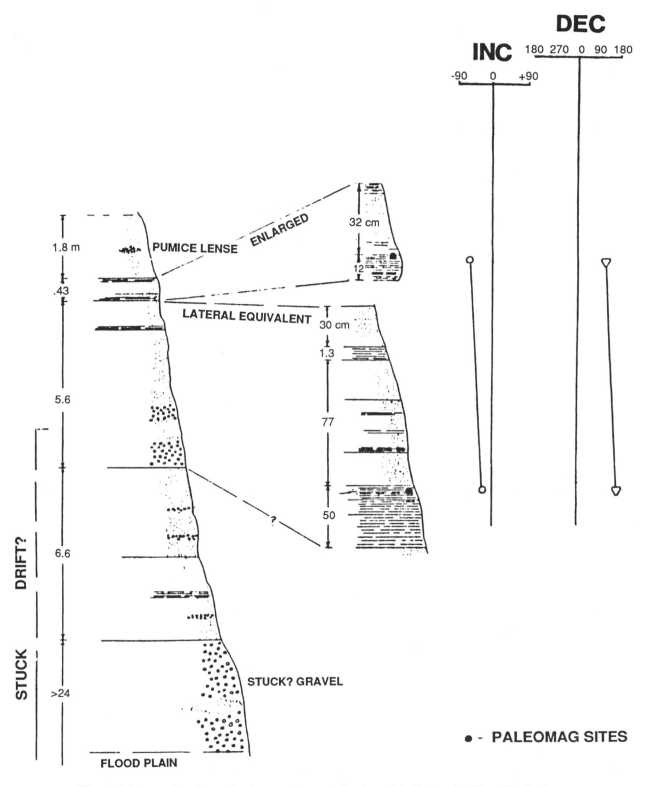

Figure 12. Measured sections and paleomagnetic mean directions of declination (DEC) and inclination (INC) for Stuck Drift about 1.6 km south of Sumner.

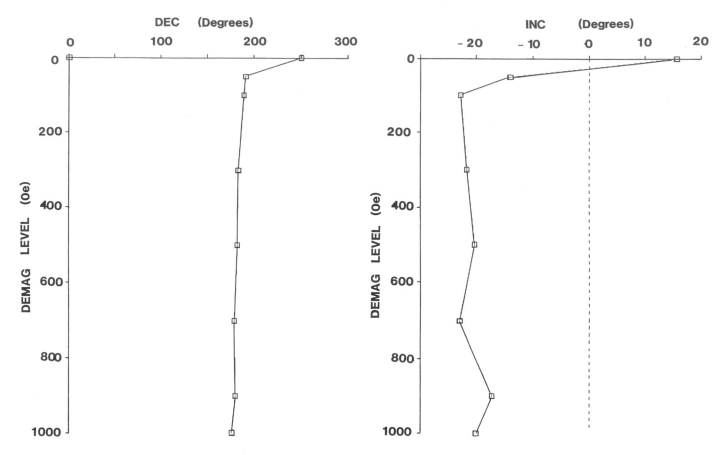

Figure 13. Changes in declination (DEC) and inclination (INC) at varying demagnetization levels for Stuck Drift.

as Stuck Drift. The sand and silt appear to be fluvial and are thought to represent the transition between the upper Stuck Drift and basal Puyallup Formation. Eleven specimens were collected from two units within this transition zone. The lowermost sample site in the transition sediments, closest to the Stuck gravel, may also belong to the Stuck. Sediment clearly of Mt. Rainier provenance is not recognized in the outcrop until just below the upper sample site. The magnetic behavior of samples from the lower site is typified by sample C9. During a.f. demagnetization, low-coercivity normal components were removed by the 100-oe level (Figs. 13 and 14). Remaining steps progressively reduced or removed high-coercivity reversed vectors. The mean magnetic direction calculated at the 700-oe level for two specimens from the lower site is D = 183°, I = –19°. Silt at the upper (Puyallup?) site is also reversely magnetized (D = 126°, I = –51°).

Puyallup Formation

Stuck Drift is overlain by the nonglacial Puyallup Formation, which includes, but is not restricted to, the Puyallup sand of Willis (1898). At its type locality (Crandell and others, 1958) in

the Puyallup Valley, the Puyallup Formation consists of about 41 m (135 ft) of lacustrine silt and sand, alluvial sand and gravel, mudflows, tephra, and peat. The mudflows contain up to 95 percent Rainier provenance clasts, indicating that they were derived from Mt. Rainier. Most gray unoxidized sand beds in the unit are also of Rainier provenance, whereas most oxidized fluvial sand and gravel are typically of mixed Rainier and central Cascade provenance. Depositional patterns represent volcanic-source sedimentation along an ancestral Puyallup River, analogous to the Alderton Formation.

Pollen analyses of peat beds in the Puyallup Formation suggest forest growth in the upper Canadian vegetation zone (probably early postglacial), warming to a middle Canadian vegetation zone during the warmest phase (Crandell, 1963). Additional pollen samples from nearby peat beds suggest a climate similar to the conditions found in the area today. Erosion and weathering at the top of the formation offer evidence suggesting that only a part of this nonglacial interval between the Stuck and Salmon Springs glaciations is present in the exposures of the Puyallup Valley.

Only a portion of the original type locality described by

STUCK

C9

NRM 6.41E-5

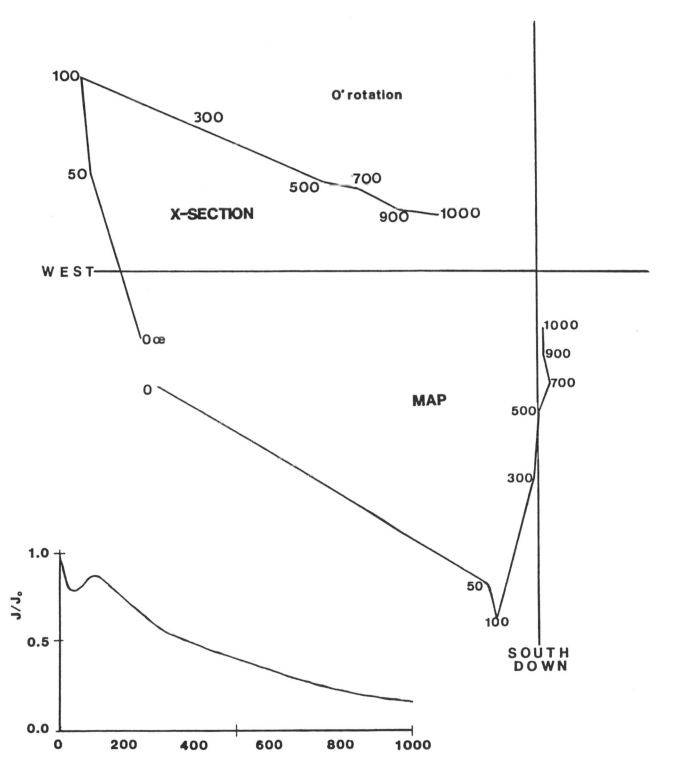

Figure 14. Orthogonal demagnetization plot and intensity curve for Stuck Drift.

PUYALLUP FORMATION

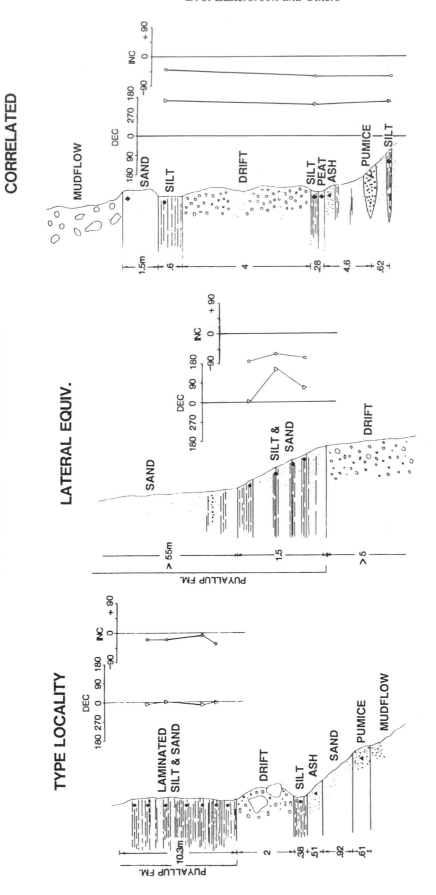

Figure 15. Measured sections and paleomagnetic mean directions for the Puyallup Formation type locality, its lateral equivalent, and exposures at the Corless and Sons gravel pit near Sumner.

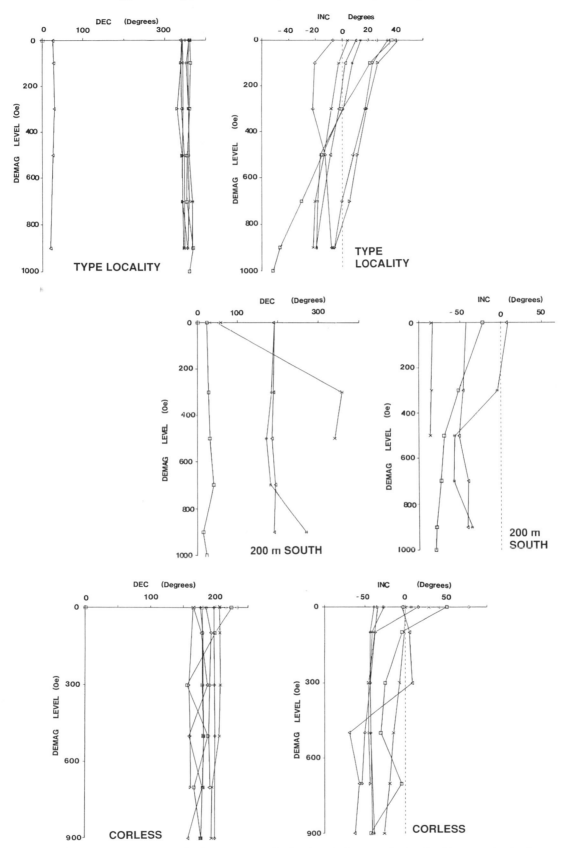

Figure 16. Changes in declination (DEC) and inclination (INC) at varying levels of demagnetization for the Puyallup Formation.

Crandell and others (1958) remains exposed today. The Puyallup Formation here is represented by about 3 m (10 ft) of horizontally laminated, compact, impermeable, pink-gray silt and fine sand overlying the Stuck Drift and Alderton Formation (Fig. 15).

Paleomagnetism of the Puyallup Formation. Twenty-four paleomagnetic samples were collected from the type locality at roughly 25 cm (10 in) vertical intervals where sediment texture permitted. Seven of the specimens were hand carved. In addition to samples collected from the type locality, nine specimens were collected from four sedimentary units at a somewhat sandier, lateral equivalent less than 200 m (655 ft) to the south (Fig. 15). Thirteen specimens from sediments of probable Puyallup age were also collected from an exposure in the Corless and Sons gravel pit near Sumner where silt, sand, peat, ash, and pumice lie beneath oxidized sand and gravel of central Cascade provenance. Silt, sand, and mudflows of Mt. Rainier provenance overlie the gravel and the section is capped by Vashon Drift.

Lacustrine silt immediately above the Stuck Drift at the type locality of the Puyallup Formation carries a complex remanent magnetization. Magnetic behavior of this sediment is character-

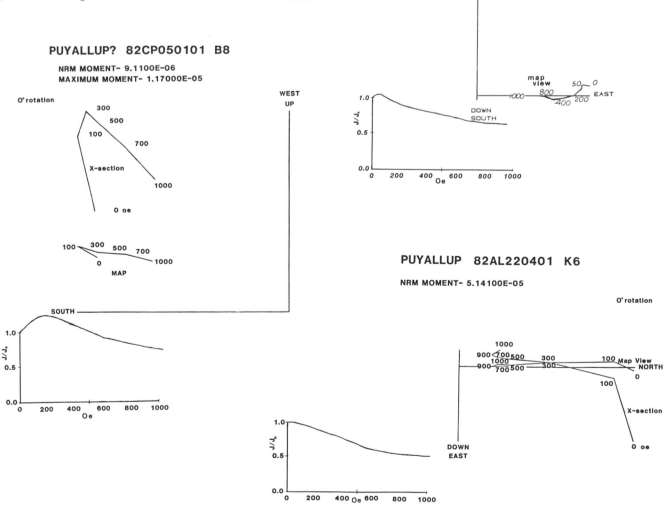

Figure 17. Orthogonal demagnetization plot and intensity curve for the Puyallup Formation.

ized by puzzling high-coercivity overprints. Three of the seven sample sites were rejected because paired specimens gave conflicting directions.

Sediment sampled from the Puyallup type locality (Fig. 16) did not demonstrate single-component directions during step a.f. demagnetization. Specimens were marked by a prominent normal overprint, which for some specimens, completely concealed the primary components. To test the possibility that percussion sampling techniques in compact silt might have caused realignment, selected sites were resampled using hand-carved extraction of specimens. Two of the four hand-carved sample sites exhibited steeper reversed (up) inclinations at or above the 500-oe level than did specimens collected by the percussion method. However, comparison demonstrated that jarring of the compact sediment during percussion sampling had no noticeable effect on declinations.

A high-coercivity normal overprint, unrelated to sampling techniques, was present in the remanent magnetism of all specimens from the Puyallup type locality. Hand-carved specimen K6 exemplifies this behavior. Its initial natural remanent magnetism (NRM) was normal with shallow inclination. Demagnetization to 100 oe removed a low-coercivity normal overprint (D = 23°, I = 70°). Throughout the remaining demagnetization, shallow, northerly down vectors were removed, resulting in the emergence of a shallow high-coercivity reversed remanence (Fig. 17). As demagnetization progressed, the emergent reversed inclination steepened, but declination remained northerly. The peak demagnetization level of 1,000 oe proved to be too weak to effectively remove the overlapping normal overprint. An attempt was made to exceed this ceiling by using a 60 Hz demagnetizer with the capacity to exceed 1,000 oe. Unfortunately, the tumbling carriage did not move randomly, and a rotational remanent magnetism was generated, which overwhelmed the remaining magnetism. Consequently, although some reversely magnetized inclinations were measured, consistent, stable, reversed components could not be isolated at the Puyallup type locality.

In order to further test the polarity of the basal Puyallup sediment, a laterally equivalent section was sampled about 200 m (660 ft) south of the type locality (Figs. 15 and 16). The reversed remanence is clearly evident in this section, although two of the three accepted sites still showed rather northerly declinations. An orthogonal plot of specimen Cl graphically displays the behavior (Fig. 17). Vector subtraction showed the removal of a normal overprint up to 100 oe. Vectors erased by the remaining step were steeply reversed (up) and easterly, but still shallower than the remaining magnetization, indicating that the characteristic component had not yet been isolated.

Sediments tentatively identified as Puyallup on the basis of lithology and provenance at the Corless and Sons gravel pit near Sumner (Fig. 2) were also sampled and measured (Figs. 15 and 16). The remanent magnetism was generally stable and clearly reversed. At the stratigraphically lowest site, silt was collected from a channel lens containing pumice clasts 7 to 10 cm in diameter. The mean magnetic direction calculated at the 700-oe

PUYALLUP? FM.

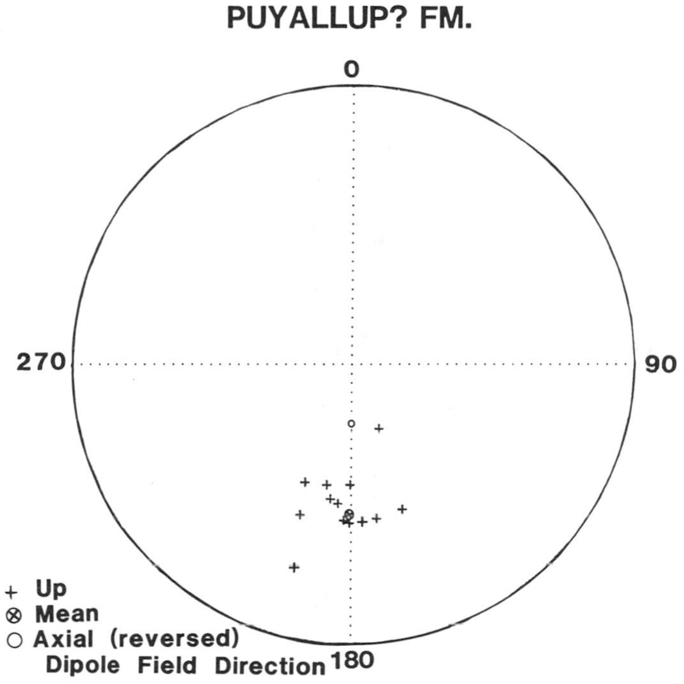

Figure 18. Stereographic projection of paleomagnetic directions used to calculate the mean formation direction for the Puyallup Formation.

demagnetization level for the three samples at the Corless and Sons gravel pit is D = 183°, I = –42° with an alpha-95 of 15° (Fig. 18).

Salmon Springs Drift

At its type locality in the Puyallup Valley near Sumner (Fig. 2), the Salmon Springs Drift unconformably overlies the Puyallup Formation and is unconformably overlain by late Wisconsin Vashon Drift (Crandell and others, 1958). The Salmon Springs Drift consists of (1) a lower glacial unit composed mostly of outwash gravel with till lenses, (2) an upper glacial unit composed mostly of outwash gravel, and (3) a thin nonglacial unit, consisting of Lake Tapps tephra (Easterbrook and others, 1981; Westgate and others, 1987), lacustrine silt, and peat, which separates the two glacial units. The Lake Tapps tephra lies directly on lower Salmon Springs outwash gravel and grades upward into the overlying 1-m-thick lacustrine silt. The silt increases in organic content upward over a thickness of several centimeters until it becomes peat. The gradational contacts of the ash-silt-peat sequence and the lack of any break in pollen spectra across the silt/peat contact indicate continuous deposition (Easterbrook, 1986; Stuiver and others, 1978).

The Salmon Springs Drift was long considered to be early Wisconsin in age on the basis of radiocarbon enrichment dates of

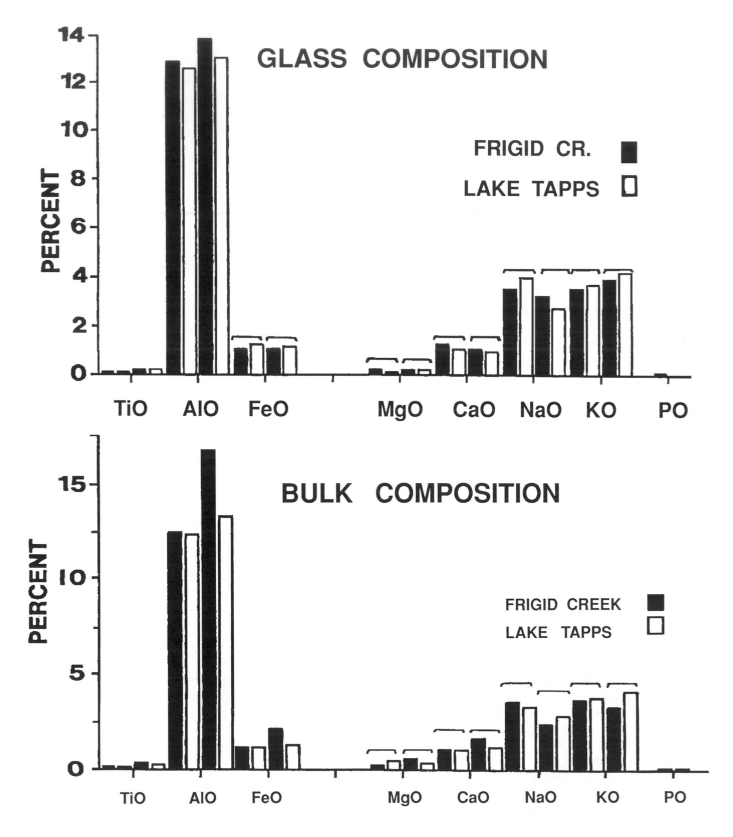

Figure 19. Geochemistry of ash at Frigid Creek and the Lake Tapps tephra type locality.

TABLE 1. FISSION-TRACK AGES OF ZIRCON AND GLASS
FROM LAKE TAPPS TEPHRA*

Locality	Sample	Material Dated	Age (Ma)	±1 Standard Deviation	Dated By
Salmon Springs	UT 88	Zircon	0.84	0.22	N. Naeser
Auburn	UT 52	Zircon	0.87	0.30	N. Naeser
Auburn	UT 52	Glass	0.66	0.04	N. Naeser
Algona	UT 462	Glass	0.65	0.08	C. van den Bogaard
Algona	UT 462[†]	Glass	1.06	0.11	J. Westgate
Frigid Creek	UT 55	Glass	0.90	0.15	N. Naeser

*See Westgate and others (1988) for additional data.

[†]Corrected for partial track fading.

50,100 + 400 years (GrN-4116c) (Armstrong and others, 1965) and 71,500 + 1,700 years (Stuiver and others, 1978) from the peat at the type locality. However, some of the silt is reversely magnetized (Easterbrook and Othberg, 1976) and zircon from the Lake Tapps tephra at the type locality has been fission-track dated at 0.84 + 0.21 m.y. (Easterbrook and others, 1981). The discrepancy between the fission-track date from the tephra and the radiocarbon dates from the peat, and the lack of any break in the depositional sequence indicate that the radiocarbon dates are invalid (Easterbrook and others, 1981). Pollen from the silt, which suggests a tundra vegetation changing progressively to a fir forest before burial by upper Salmon Springs gravel (Leopold and Crandell, 1957; Crandell and others, 1958; Crandell, 1963; Heusser, 1977; Stuiver and others, 1978), shows that the upper and lower Salmon Springs drifts are separated by a short nonglacial interval.

Paleomagnetism of the Salmon Springs Drift. Paleomagnetic profiles from the silt between the interdrift ash and peat at the type locality include both normal and reversed polarities (Easterbrook and Othberg, 1976; Easterbrook and others, 1981). Many of the remanence measurements showed high angular standard deviations indicative of low sample stability, making evaluation of the data difficult. Strong normal VRM overprinting has been found in some of the pre–Salmon Springs sediments, suggesting that the normal polarities in the Salmon Springs silt may be due to deposition in a reversed field with normal overprinting much later. The possibility exists that the remanent magnetism of the silt represents the geomagnetic field at the time of deposition, but the field was fluctuating in a transition between reversed and normal polarity. This is considered less likely because, where sample stability is good, the remanent magnetism is clearly reversed (Easterbrook, 1986). The normal polarity observed in the silt at the type locality could have been overprinted on reversely magnetized sediments by the Brunhes normal field, but the possibility of a reversed polarity overprint being imparted during a reversed event in the Brunhes normal field is considered improbable. Normal polarity was measured in samples above the reversal zone at Auburn (Easterbrook and Othberg, 1976), but those sediments were mapped as early Vashon by Mullineaux (1970).

The Lake Tapps tephra, originally defined on the basis of outcrops at the Salmon Springs Drift type locality and exposures at Auburn where the tephra is interbedded with peat in sand and gravel (Easterbrook and others, 1981), occurs at three additional localities: Frigid Creek, Peasley Canyon, and Algona. Fission-track dating and paleomagnetic analyses place the age of the lake Tapps tephra at 1.06 ± .11 Ma (Westgate and others, 1987). Its recognition at five localities makes it a valuable time-stratigraphic marker for identifying the Salmon Springs Drift in the Puget Lowland and for providing a link between sediments of the Puget lobe of the Cordilleran ice sheet and deposits of alpine glaciers from the Olympic Mountains.

The Lake Tapps tephra is a calc-alkaline dacitic to rhyolitic ash, characterized by phenocrysts of plagioclase, hornblende, hypersthene, biotite, magnetite, ilmenite, apatite, zircon, and glass (Easterbrook and others, 1981; Westgate and others, 1987). Analyses of glass and bulk samples made by J. A. Westgate are shown in Figure 19. The glass consists of (1) angular, chunky shards with few vesicles, or (2) pumiceous particles with lineated tubular vesicles. Both types of glass have rhyolitic compositions with small variation between samples or between shards, and trace and rare earth elements have similarity coefficients of 0.9 or greater. Fe-Ti oxides show small compositional variation between samples from all localities. Additional details

158

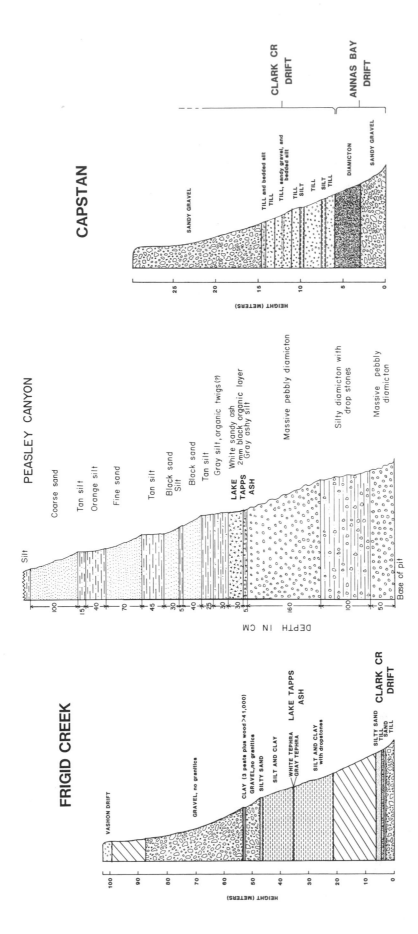

Figure 20. Stratigraphic sections at Frigid Creek, Capstan Rock, and Peasley Canyon.

concerning the geochemistry and petrography of the tephra may be found in Easterbrook and others (1981) and Westgate and others (1987).

Reversed magnetic polarity of sediments above and below the Lake Tapps tephra at all five localities indicates that it is older than the Brunhes-Matuyama boundary (0.73 Ma). Ages of 0.8 to 0.9 Ma were obtained by fission-track dating of zircon in the tephra at three localities (Table 1) (Easterbrook and others, 1981). Glass from tephra at Auburn was fission-track dated at 0.66 + 0.04 Ma, but this is a minimum age that probably reflects annealing in the glass at this locality. Plateau annealing experiments of glass gave an age of 1.0 Ma for the Lake Tapps tephra (Westgate and others, 1987).

Peasley Canyon and Algona sections. Tephra at Peasley Canyon occurs as a distinct layer about 40 cm thick (Westgate and others, 1987), interbedded in a sand/gravel section (Fig. 20) previously mapped as Salmon Springs Drift (Mullineaux, 1970). Geochemistry of the tephra confirms that it is the Lake Tapps tephra (Westgate and others, 1987). The sand/gravel unit at Peasley Canyon is considered correlative with the Salmon Springs Drift at its type locality, and till beneath the sand/gravel is probably lower Salmon Springs Drift (Easterbrook, 1986). Paleomagnetic measurements of silt in the section show a reversed magnetic polarity, but some samples have a strong VRM overprint and sample stability is not as good as at Auburn or Frigid Creek.

At Algona, (Fig. 2) a 30-cm-thick ash exposed in a silt, sand, and gravel sequence several meters thick has been identified as Lake Tapps tephra (Westgate and others, 1987). The tephra is well exposed, but vegetation covers most of the rest of the valley side. Paleomagnetic measurements of silt in the exposure show a mixture of reversed and normal polarities with poor sample stability.

Correlation of Salmon Springs Drift and Olympic Mountains Drifts. Pre-Vashon drift in the southwestern Puget Lowland was first referred to as Admiralty Drift by Bretz (1913) and Todd (1939), then "pre-Vashon" by Frisken (1965), "Salmon Springs" by Carson (1970) and Molenaar and Noble (1970), and is probably included in the "unnamed gravel" of Molenaar (1965) and the "Skokomish gravel" of Molenaar and Noble (1970).

Fission-track dating and paleomagnetic measurements now allow correlation of early Pleistocene drifts of the Puget lobe and Olympic Mountain alpine glaciers, which are complexly interfingered in the region of Hood Canal.

Clark Creek and Annas Bay Drifts. The name Clark Creek Drift is here proposed for deposits of alpine glaciers originating in the Skokomish drainage of the southeastern Olympic Mountains during the early Pleistocene. Its type locality is designated as road cuts along U.S. Highway 101 on the west side of Hood Canal 1 km south of the mouth of Clark Creek in N½,Sec.12,T22N,R4E where the drift includes several tills containing clasts dominated by basalt and graywacke derived from the Olympic Mountains, generally oxidized to red-brown hues.

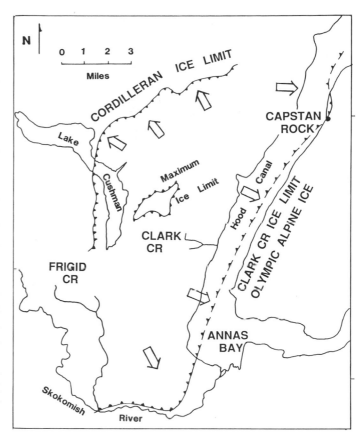

Figure 21. Map showing localities referred to in text and showing the relation between ice margins of alpine glaciers from the Olympic Mountains and the Cordilleran ice sheet.

Most Clark Creek Drift is exposed in the vicinity of the Skokomish drainage, but tills with similar characteristics also occur near the mouths of several other streams from the Olympic Mountains. Clark Creek till is also exposed at Capstan Rock on the east side of Hood Canal, 15 km east of Lake Cushman, indicating that Olympic ice advanced at least that far eastward into the Puget Lowland (Fig. 21). The "Skokomish gravel" of Molenaar and Noble (1970) and the "unnamed gravel" of Molenaar (1965) on the Kitsap Peninsula east of Hood Canal were considered by Carson (1979) to be partly Clark Creek Drift.

In places, Clark Creek Drift is interbedded with early Pleistocene sediments deposited by the Puget lobe of the Cordilleran ice sheet near Hood Canal. The name Annas Bay Drift is here proposed for early Pleistocene drift of the Puget lobe in the area near Hood Canal. Its type locality is designated as road cuts along U.S. Highway 101 south of Potlatch State Park west of Annas Bay (Fig. 21) in NW¼,Sec.35,T22N,R4W where about 2 to 5 m of compact till containing granitic clasts overlies 1 to 4 m of stratified sandy gravel, which also contains granitic clasts.

Although both the Clark Creek and Annas Bay Drifts are

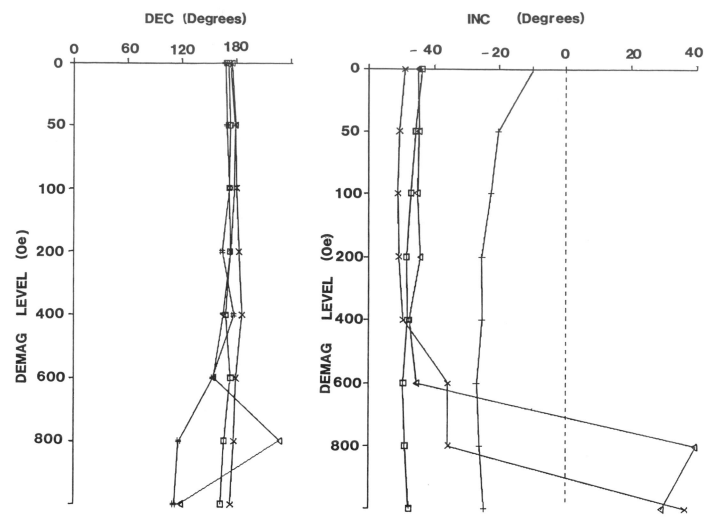

Figure 22. Changes in declination (DEC) and inclination (INC) at varying demagnetization levels for silt samples from Frigid Creek.

oxidized, most of the clasts are not rotten. The degree of oxidation of the matrix varies. Clark Creek Drift tends to be more oxidized, with stronger red-brown colors, presumably because it has more ground-up basalt than Annas Bay Drift. Complex interfingering of Olympic-derived Clark Creek Drift with Cordilleran-derived Annas Bay Drift on the north side of the Skokomish River suggests a close relationship between Olympic alpine glacier and the Puget lobe.

Sediments associated with pre-Vashon drift at two localities on opposite sides of Hood Canal were sampled for paleomagnetic analysis, and tephra at Frigid Creek (Fig. 21) was sampled for fission-track and geochemical analyses in an attempt to establish correlations with deposits in the southeastern Puget Lowland. Lacustrine silt overlying Clark Creek till at Frigid Creek on the west side of Hood Canal and silt interbedded with till at Capstan Rock (Fig. 21) on the east side of Hood Canal were studied to address the problem of the relationship between till derived from piedmont ice originating in the Olympic Mountains to the west and till from the Cordilleran ice sheet originating in British Columbia.

Lacustrine silt overlying Clark Creek Drift exposed in Frigid Creek was sampled for paleomagnetic measurement, and interbedded tephra was collected for fission-track dating and geochemical analysis. The stratigraphy at this site (Fig. 20) consists of about 24 m of gray, stratified, lacustrine silt which strikes N60 to 77E and dips 20 to 35° NW. Lake Tapps tephra occurs 11 m below the top of the silt, which is overlain by about 40 m of gravel and a few meters of Vashon till. Three peat beds within a silt/clay/peat unit about one m thick occur about 6 m above the base of the gravel sequence. Wood from the peat beds has been radiocarbon dated at >41,000 years B.P. (Carson and others, 1976).

The Lake Tapps tephra at Frigid Creek occurs in two distinct layers totaling 10 to 12 cm. The geochemistry of the two tephra beds is nearly identical, and both are believed to have been part of the same eruption (Westgate and others, 1987).

The tephra at Frigid Creek was fission-track dated at 0.9 ± 0.15 Ma (Naeser and others, 1984), corresponding with the 1.06 ± 0.11 Ma age for the Lake Tapps tephra (Westgate and others, 1987), and placing a limiting age on the Clark Creek Drift. Clark Creek Drift and the gravel unit above the lacustrine silt unit containing Lake Tapps tephra do not contain granitic clasts or other lithologies typical of Cordilleran ice from a northern provenance, indicating that their source areas were in the Olympic Mountains.

Both the Clark Creek Drift and the Annas Bay Drift are correlated with the Salmon Springs Drift on the basis of the Lake Tapps tephra (Easterbrook and others, 1985).

Paleomagnetism of sediments at Frigid Creek and Capstan Rock. Figure 22 shows measurements of remanent magnetism for six samples of lacustrine silt overlying Clark Creek Drift at Frigid Creek. The samples retain a nearly constant inclination and declination during step demagnetization to 1,000 oe. Declination and inclination of one sample began to deviate at 800 and 1,000 oe but remained reversely magnetized.

Computer-generated orthogonal projections of magnetic vectors (Zijderveld plots) were used to evaluate the stability of the samples. Figure 23 shows two representative Zijderveld plots of samples from Frigid Creek. The upper curves show progressive changes in declination with increasing peak intensity of the demagnetizing field, and the lower curves show changes in inclination. The relatively straight lines inclined toward zero for progressive levels of demagnetization indicate good sample stability. Angular standard deviation was calculated for all sample measurements and remained below 10 even for demagnetization at 800 and 1,000 oe, indicating good magnetic stability.

Figure 24a shows a plot of duplicate samples taken from a single bedding plane. All are reversely magnetized at demagnetization levels of 200 oe, but some have a weak normal overprint. Stereographic projection of paleomagnetic directions for various demagnetization levels are shown in Figure 24b.

A sea cliff section along the east side of Hood Canal at Capstan Rock was sampled for paleomagnetic measurements. Annas Bay till occurs at beach level, overlain by multiple, thin Clark Creek tills interbedded with lenses of silt (Fig. 20). Step a.f. demagnetization was carried out to 1,000 oe (Figs. 24c, 25, and 26) and gave results similar to those for Frigid Creek samples. All samples from the silt lenses interbedded with the multiple till layers were reversely magnetized. The reversed magnetic polarity of lake silts containing the Lake Tapps tephra confirms an age within the upper Matuyama Polarity Chron. Close association of Clark Creek and Annas Bay tills at Capstan Rock suggests interaction between Olympic Mountain alpine glaciers and the Puget lobe of the Cordilleran ice sheet. No Lake Tapps tephra was found at Capstan rock, but the reversed polarity indicates an age within the Matuyama Polarity Chron.

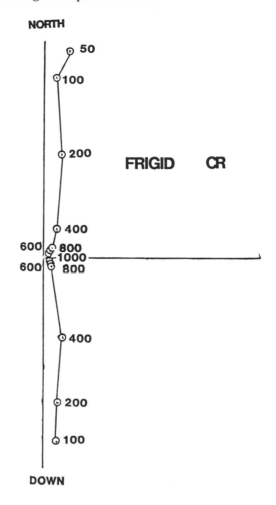

Figure 23. Orthogonal demagnetization plot for silt from Frigid Creek.

CONCLUSIONS

The Orting Drift, Alderton Formation, Stuck Drift, and Puyallup Formation are all reversely magnetized, and are believed to have been deposited during the Matuyama Reversed Polarity Chron. All are stratigraphically beneath the Salmon Springs Drift, fission-track dated at 1.0 Ma, and thus are all older than the Brunhes/Matuyama boundary. Exactly where within the Matuyama the lowermost stratigraphic units belong is yet unknown and must await additional dating of interbedded tephra.

Paleomagnetic measurements were carried out on silt at each of the five localities where Lake Tapps tephra has been found. All contain reversely magnetized silt. Paleomagnetic measurements from Auburn and Frigid Creek have clear-cut reversed magnetism with good stability, but demagnetization of samples from Salmon Springs, Peasley Canyon, and Algona showed considerably lower stability and a strong normal polarity overprint, which sometimes could not be removed by demagnetization to

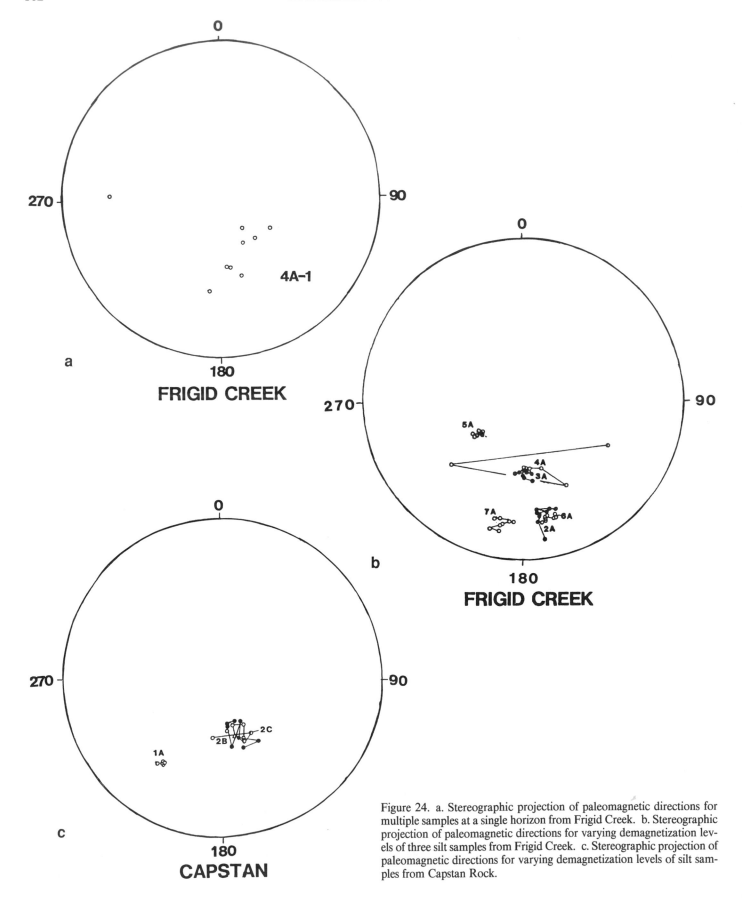

Figure 24. a. Stereographic projection of paleomagnetic directions for multiple samples at a single horizon from Frigid Creek. b. Stereographic projection of paleomagnetic directions for varying demagnetization levels of three silt samples from Frigid Creek. c. Stereographic projection of paleomagnetic directions for varying demagnetization levels of silt samples from Capstan Rock.

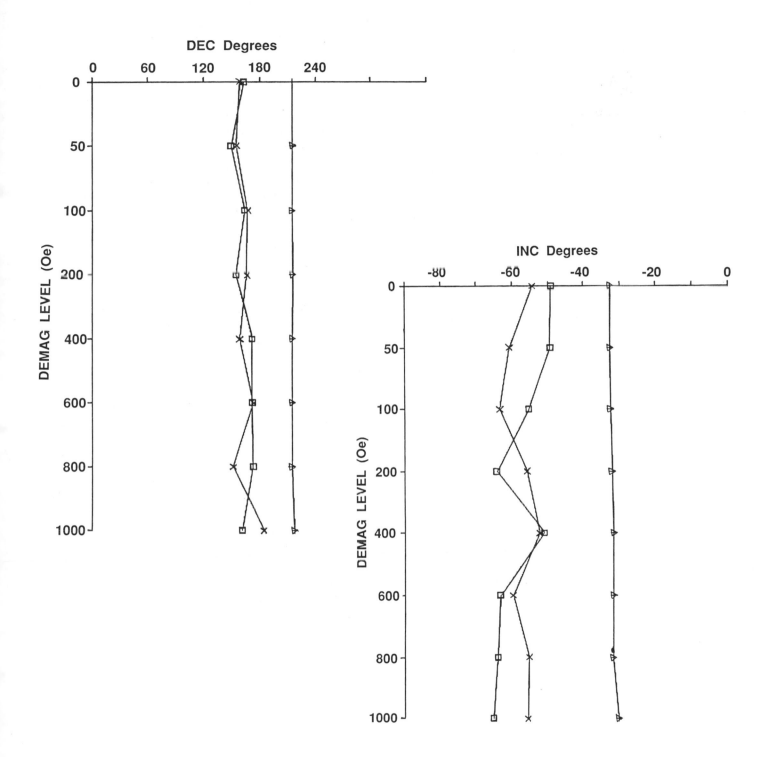

Figure 25. Changes in declination (DEC) and inclination (INC) at varying demagnetization levels for silt samples from Capstan Rock.

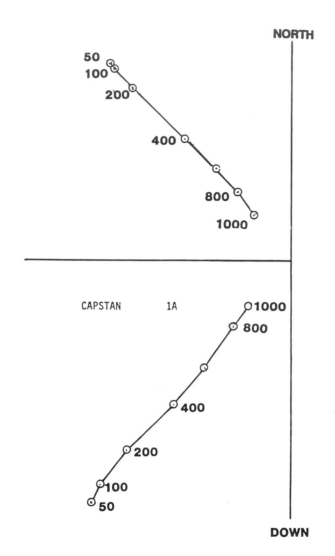

Figure 26. Orthogonal demagnetization plot for silt from Capstan Rock.

1,000 oe. The reversed magnetic polarity and association with the Lake Tapps tephra shows that the sediments at these localities were deposited during the Matuyama Polarity Chron.

Several critical issues concerning Pleistocene chronology in the Puget Lowland have been resolved by this study. (1) The antiquity of the Salmon Springs Drift demonstrated by Easterbrook and others (1981) extends to the Puyallup Formation, Stuck Drift, Alderton Formation, and Orting Drift as well. (2) The reversed paleomagnetism of these stratigraphic units is consistent with the stratigraphic positions originally mapped by Crandell and others (1958), but the tentative ages assigned by them were much too young. (3) Many correlations previously made with the Salmon Springs and older units in the Puget Lowland are invalidated. (4) None of the Salmon Springs and older units can be equivalent to the Double Bluff Drift, Whidbey Formation, or Possession Drift, which are all normally magnetized (Easterbrook and Othberg, 1976). And, (5) a chronologic gap in the Pleistocene stratigraphy of the Puget Lowland is now obvious for the interval between about 0.2 and 1.0 Ma.

The ramifications of the early Pleistocene age of the Salmon Springs Drift and older stratigraphic units are far-reaching. The term Salmon Springs Drift has been widely used in the literature for pre-Vashon drifts, but many of these previous correlations are invalid.

Clark Creek Drift, deposited by Olympic alpine glaciers in the southwestern Puget Lowland, lies beneath reversely magnetized silt containing Lake Tapps tephra fission-track dated at 1.06 ± 0.11 Ma. Clark Creek Drift complexly interfingers with Annas Bay Drift, deposited by Cordilleran ice in the southwestern Puget Lowland where the till is interbedded with reversely magnetized silt. Both the Clark Creek and Annas Bay Drifts are correlated with the Salmon Springs Drift.

Tills at several localities in the western Olympic Peninsula have been correlated with "Salmon Springs Drift" (Heusser, 1977), but these correlations were made before the early Pleistocene age of the Salmon Springs Drift was discovered. Some of these tills are clearly too young to be Salmon Springs.

REFERENCES CITED

Armstrong, J. E., Crandell, D. R., Easterbrook, D. J., and Noble, J. B., 1965, Late Pleistocene stratigraphy and chronology in southwestern British Columbia and northwestern Washington: Geological Society of America Bulletin, v. 76, p. 321–330.

Bretz, J H., 1913, Glaciation of the Puget Sound region: Washington Division of Mines and Geology Bulletin, v. 8, 244 p.

Carson, R. J., 1970, Quaternary geology of the south-central Olympic Peninsula, Washington [Ph.S. thesis]: Seattle, University of Washington, 67 p.

——, 1979, Reinterpretation of the Skokomish Gravel, southwestern Puget Lowland, Washington: Northwest Scientific Association, Program and Abstracts, p. 31.

Carson, R. J., Spence, W. H., and Birdseye, R. U., 1976, Late Pleistocene tephra layers in the western Puget Lowland, Washington: Geological Society of America Abstracts with Programs, v. 8, p. 58.

Crandell, D. R., 1963, Surficial geology and geomorphology of the Lake Tapps Quadrangle, Washington: U.S. Geological Survey Professional Paper 388–A1, 84 p.

Crandell, D. R., Mullineaux, D. R., and Waldron, H. H., 1958, Pleistocene sequence in the southeastern part of the Puget Sound Lowland, Washington: American Journal of Science, v. 256, p. 384–398.

Easterbrook, D. J., 1986, Stratigraphy and chronology of Quaternary deposits of the Puget Lowland and Olympic Mountains of Washington and the Cascade Mountains of Washington and Oregon: Quaternary Science Reviews, v. 5, p. 145–159.

Easterbrook, D. J., and Briggs, N. D., 1979, Age of the Auburn reversal and the Salmon Springs and Vashon Glaciations in Washington: Geological Society of America Abstracts with Programs, v. 11, p. 76–77.

Easterbrook, D. J., and Othberg, K., 1976, Paleomagnetism of Pleistocene sediments in the Puget Lowland, Washington, *in* Quaternary glaciations of the Northern Hemisphere: International Geological Correlation Program Project 24 Report no. 3, p. 189–207.

Easterbrook, D. J., Crandell, D. R., and Leopold, E. B., 1967, Pre-Olympia stratigraphy and chronology in the central Puget Lowland, Washington: Geological Society of America Bulletin, v. 78, p. 13–20.

Easterbrook, D. J., Briggs, N. D., Westgate, J. A., and Gorton, M., 1981, Age of the Salmon Springs Glaciation in Washington: Geology, v. 9, p. 87–93.

Easterbrook, D. J., Westgate, J. A., and Naeser, N., 1985, Pre-Wisconsin fission-track, paleomagnetic, amino-acid, and tephra chronology in the Puget Lowland and Columbia Plateau: Geological Society of America Abstracts with Programs, v. 17, p. 411.

Frisken, J. G., 1965, Pleistocene glaciation of the Brinnon area, east-central Olympic Peninsula, Washington [M.S. thesis]: Seattle, University of Washington, 75 p.

Halls, H. C., 1978, The use of converging remagnetization circles in paleomagnetism: Physics of the Earth and Planetary Interiors, v. 16, p. 1–11.

Hansen, H. P., and Mackin, J. H., 1949, A pre-Wisconsin forest succession in the Puget Lowland, Washington: American Journal of Science, v. 247, p. 833–855.

Harrison, C.G.A., 1980, Analysis of the magnetic vector in a single rock specimen: Geophysics Journal, v. 60, p. 489–492.

Henshaw, C.G.A., and Merrill, R. T., 1979, Characteristics of drying remanent magnetization of sediments: Earth and Planetary Science Letters, v. 43, p. 315–320.

Heusser, C. J., 1977, Quaternary palynology of the Pacific slope of Washington: Quaternary Research, v. 8, p. 282–306.

Leopold, D. R., and Crandell, D. R., 1957, Pre-Wisconsin interglacial pollen spectra from Washington State, USA: Geobotanisches Institut Rubel in Zurich Veroffentlichung 34, p. 76–79.

Molenaar, D., 1965, Geology and ground-water resources, *in* Water resources and geology of the Kitsap Peninsula and certain adjacent islands: Washington Division of Water Resources Water Supply Bulletin, v 18, p. 24–50.

Molenaar, D., and Noble, J. B., 1970, Geology and related ground water occurrences, southeastern Mason County, Washington: Washington Division of Water Resources Water Supply Bulletin, v. 29, p. 1–38.

Mullineaux, D. R., 1970, Geology of the Renton, Auburn, and Black Diamond Quadrangles, King County, Washington: U.S. Geological Survey Professional Paper 672, 92 p.

Naeser, N. D., Westgate, J. A., Easterbrook, D. J., and Carson, R. J., 1984, Pre-0.89 m.y. glaciation in the west-central Puget Lowland, Washington: Geological Society of America Abstracts with Programs, v. 16, p. 324.

Roland, J. L., 1984, A paleomagnetic age determination of pre-Salmon Springs drift Pleistocene deposits in the southern Puget Lowland, Washington: Geological Society of America Abstracts with Programs, v. 16, p. 330.

Stuiver, M., Heusser, C. T., and Yang, I. C., 1978, North American glacial history back to 75,000 years B.P.: Science, v. 200, p. 16–21.

Todd, M. R., 1939, The glacial geology of the Hamma Hamma Valley and its relation to the glacial history of the Puget Sound Basin [M.S. thesis]: Seattle, University of Washington, 48 p.

Westgate, J A., Easterbrook, D. J., Naeser, N. D., and Carson, R. J., 1987, Lake Tapps tephra; An early Pleistocene stratigraphic marker in the Puget Lowland, Washington: Quaternary Research, v. 28, p. 340–355.

Willis, B., 1898, Drift phenomena of Puget Sound: Geological Society of America Bulletin, v. 9, p. 111–162.

Zijderveld, J.D.A., 1967, Demagnetization of rocks; Analysis of results, *in* Collinson, D. W., Creer, K. M., and Runcorn S. K., eds., Methods in paleomagnetic methods: Newcastle upon Tyne, United Kingdom, p. 1–10.

MANUSCRIPT ACCEPTED BY THE SOCIETY FEBRUARY 16, 1988

Typeset by WESType Publishers Services, Inc., Boulder, Colorado
Printed in U.S.A. by Malloy Lithographing, Inc., Ann Arbor, Michigan